INSTRUCTOR'S RESOL

to accompany

DIFFERENTIAL EQUATIONS
GRAPHICS • MODELS • DATA

David Lomen
David Lovelock
Department of Mathematics
University of Arizona

JOHN WILEY & SONS, INC.
New York • Chichester • Weinheim • Brisbane • Singapore • Toronto

COVER ILLUSTRATION: Roy Wiemann

Copyright © 1999 by John Wiley & Sons, Inc.

Excerpts from this work may be reproduced by instructors for distribution on a not-for-profit basis for testing or instructional purposes only to students enrolled in courses for which the textbook has been adopted. *Any other reproduction or translation of this work beyond that permitted by Sections 107 or 108 of the 1976 United States Copyright Act without the permission of the copyright owner is unlawful. Requests for permission or further information should be addressed to the Permissions Department, John Wiley & Sons, Inc., 605 Third Avenue, New York, NY 10158-0012.*

ISBN 0-471-32758-1

Printed in the United States of America

10 9 8 7 6 5 4 3 2 1

Printed and bound by Odyssey Press, Inc.

Contents

0 Introduction   1

1 Ideas for Chapter 1 — "Basic Concepts"   3
    1.1 Page by Page Comments 3
    1.2 Additional Exercises 4
    1.3 Additional Materials 9

2 Ideas for Chapter 2 — "Autonomous Differential Equations"   37
    2.1 Page by Page Comments 37
    2.2 Additional Exercises 41
    2.3 Additional Materials 53

3 Ideas for Chapter 3 — "First Order Differential Equations — Qualitative and Quantitative Aspects"   89
    3.1 Page by Page Comments 89
    3.2 Additional Exercises 91
    3.3 Additional Materials 93

4 Ideas for Chapter 4 — "Models and Applications Leading to New Techniques"   99
    4.1 Page by Page Comments 99
    4.2 Additional Exercises 103
    4.3 Additional Materials 113

5 Ideas for Chapter 5 — "First Order Linear Differential Equations and Models"   157
    5.1 Page by Page Comments 157
    5.2 Additional Exercises 163
    5.3 Additional Materials 166

6 Ideas for Chapter 6 — "Interplay Between First Order Systems and Second Order Equations"   189
    6.1 Page by Page Comments 189
    6.2 Additional Exercises 192
    6.3 Additional Materials 207

7 Ideas for Chapter 7 — "Second Order Linear Differential Equations with Forcing Functions"   227
    7.1 Page by Page Comments 227
    7.2 Additional Exercises 228
    7.3 Additional Materials 229

8 Ideas for Chapter 8 — "Second Order Linear Differential Equations — Qualitative and Quantitative Aspects" — 257
- 8.1 Page by Page Comments — 257
- 8.2 Additional Exercises — 260

9 Ideas for Chapter 9 — "Linear Autonomous Systems" — 267
- 9.1 Page by Page Comments — 267
- 9.2 Additional Exercises — 268

10 Ideas for Chapter 10 — "Nonlinear Autonomous Systems" — 269
- 10.1 Page by Page Comments — 269
- 10.2 Additional Exercises — 270
- 10.3 Additional Materials — 274

11 Ideas for Chapter 11 — "Using Laplace Transforms" — 277
- 11.1 Page by Page Comments — 277
- 11.2 Additional Exercises — 278

12 Ideas for Chapter 12 — "Using Power Series" — 279
- 12.1 Page by Page Comments — 279
- 12.2 Additional Materials — 281

13 Using Matrices — 287
- 13.1 Method of Undetermined Coefficients Using Matrices — 287
- 13.2 Solutions by Variation of Parameters — 300
- 13.3 Higher Order Systems — 303

0. INTRODUCTION

This is a Resource Manual for use with the text "DIFFERENTIAL EQUATIONS Graphics • Models • Data". It is much more than an Instructor's Manual because most of it is devoted to supplying ideas exercises, and data sets beyond those in the text. Many instructors prefer to present examples in class that are different from those in the text. This manual was written in part to fulfil that preference. Some instructors like to give additional material aimed at the specific discipline of their students. This manual contains material from a wide variety of fields.

Specifically, this resource manual contains:

- Page by page comments.

- Examples that can be used by an instructor to replace specific examples in the text.

- Nonroutine exercises both computer and analytically oriented.

- Simple experiments that can be performed in or out of class. These experiments are related to the origin, the method for finding the solution, or the predictions of the differential equation being studied.

- Alternative developments that could replace those presented in the text.

- Suggestions for lengthier exercises — many worked out in detail — which can easily become projects.

- New materials that could supplement lectures or student reading. These start on a new page so they can be copied and distributed as appropriate.

We start each chapter by suggesting the number of 50 minute class periods to devote to the chapter. This is a rough estimate because every instructor will spend a different amount of time. The sections were written pursuing an idea, and were not written to be covered in exactly one class period.

After the opening suggestion, we proceed with page by page comments. These include what material is crucial for later in the text, what can safely be omitted, and what students have found especially interesting or difficult. Because the use of data sets is a significant part of our text, we have provided many additional examples which use data. These can be used as lecture material, additional exercises, projects, or for examinations. The source of this material is also noted.

The crucial task of every instructor is to help students understand the material, and very few students work very hard at understanding material that is not of interest to them. That was one reason for having material from a very wide variety of fields. If you have motivated students, you might consider replacing one exam with a project that expands on our examples.

Most data sets that are used in the text and this manual are available (either stand-alone, or preloaded into the University of Arizona's software program, Twiddle) from the World Wide Web at the web site devoted to this book, namely,
http://www.math.arizona.edu/~dsl/book.htm.
This web-site contains ideas, suggestions, links, and corrections. We can also be contacted through this site. Printing errors, and other inaccuracies, can be found at
http://www.math.arizona.edu/~dsl/perrors.htm.

The following texts are excellent sources for data sets, and references to further data sets.

- "Growth and Diffusion Phenomena : Mathematical Framework and Applications" by R. T. Banks, Springer-Verlag, 1991.
- "Models in Biology : Mathematics, Statistics and Computing" by D. Brown and P. Rothery, Wiley, 1993.
- "Problems of Relative Growth" by J. S. Huxley, Dover, 1972.
- "Expectations of Life" by H. O. Lancaster, Springer-Verlag, 1990.
- "Dynamic Energy Budgets in Biological Systems" by S. A. L. M. Kooijman, Cambridge, 1993.
- "Growth and Form" by D'Archy W. Thompson, Cambridge, 1942.
- "Statistical Abstracts of the United States 1995" U.S. Department of Commerce, Bureau of the Census.
- "Information Please Almanac 1995" Houghton Mifflin.

The suggested pronunciations of last names is based on "Asimov's Biographical Encyclopedia of Science and Technology" by I. Asimov, Doubleday, 1964.

An additional supplement, "Student Solutions Manual for DIFFERENTIAL EQUATIONS Graphics • Models • Data", with solutions to the odd-numbered exercises, is available.

We repeat what we wrote in the introduction to the text.

This book may be covered completely in a two quarter course, but contains more material than may be covered in a one semester course. The following are suggestions for a one semester course:

- Chapters 1 through 7, 8.1 through 8.4, 9, and 10.1 through 10.4.
- Chapters 1 through 7, 8.1 through 8.4, 11, and 12.

As far as selecting material from Chapters 1 through 7 is concerned, you should note the following:

- Sections 1.4, 2.6, 3.4, 3.5, 4.5, 4.6, and 5.3 are not extensively used in later sections, and so may be selectively omitted.

Some comments on the dependence of Chapters 8 through 12 follow, assuming that you have covered Chapters 1 through 7, and Sections 8.1 through 8.4, in the manner previously described.

- Sections 8.5 through 8.8 are not critical for later sections.
- Chapters 9 follows from Chapter 7. It does not use any material from Chapter 8.
- Chapter 10 follows from Section 9.4. It is not used in later chapters.
- Chapter 11 follows from Section 8.4. It does not use any material from Chapters 9 and 10. It is not used in later chapters.
- Chapter 12 follows from Section 8.4.

1. IDEAS FOR CHAPTER 1 — "BASIC CONCEPTS"

1.1 Page by Page Comments

Chapter 1 requires about 3 class meetings of 50 minutes each. It covers $y' = g(x)$ in a way that is easily generalized to $y' = g(y)$, Chapter 2, and $y' = g(x,y)$, Chapter 3. However, even though all our examples in this chapter are of the form $y' = g(x)$, we use $y' = g(x,y)$ in all definitions so that they also apply to subsequent chapters.

We always assign the software program ARE YOU READY FOR ODES? on the first day of class. It is a diagnostic tool which advises students of any weaknesses in the prerequisite material. About a week later we give a test covering prerequisite material.

Section 1.4, *Functions and Power Series Expansions*, is not extensively used in later sections, and so may be selectively omitted. If you plan to do series solutions (Chapter 12, *Using Power Series*), here is an elementary introduction, so students can get their feet wet.

Page 4. Notice that a solution is a differentiable function, so that solutions cannot include vertical tangents.

Page 5. Footnote 4 distinguishes between solutions and integrals of a differential equation. Solutions satisfy the differential equation and are differentiable in some interval $a < x < b$. Integrals merely satisfy the differential equation. In the same way we can distinguish between solution curves and integral curves.

Page 5. The Error Function is used throughout Chapter 1.

Page 5. Throughout Chapter 1, the Error Function could be replaced by the solution of $y' = 1/\cosh^2 x$ subject to $y(0) = 0$, which we call the Bombay Plague function. See the section called THE BOMBAY PLAGUE FUNCTION on page 9 of this manual.

Page 6. Exercise 1 is a good review of simple integration techniques that students will encounter in this text. Note 1(i) uses partial fractions: they are reviewed in Appendix A.2.

Pages 6 – 7. Exercises 1, 3, and 4 in Section 1.1 occur throughout this chapter in various forms. They are built upon in each section.

Page 7. Exercise 2(f). Students sometime ask how we know that $erf(x) \to 1$ as $x \to \infty$. On page 34 of this manual we show one way to do this, assuming a knowledge of double integrals.

Page 7. Exercise 3(f). Students sometime ask how we know that $S(x) \to 0.5$ as $x \to \infty$. We don't know a simple answer to this, other than to refer them to a standard application of the Residue Theorem. For example "Complex Variables and Applications" by J.W. Brown and R.V. Churchill, McGraw-Hill, 6th edition, 1996, page 215.

Page 7. There are additional exercises for Section 1.1 on page 4 of this manual.

Page 11. There are additional exercises for Section 1.2 on page 5 of this manual.

Page 11. We use the terminology **Slope Field** at this point and reserve the expression **Direction Field** for Chapter 6, and later chapters, where slope fields that have direction associated with them are discussed.

Page 14. We give the students copies of Figure 1.7, and have them draw something similar to Figure 1.8.

Page 15. We give the students copies of Figure 1.9, and have them draw something similar to Figure 1.10.

Page 16. We give the students copies of Figure 1.11, and have them draw something similar to Figure 1.12.

Page 18. We give the students copies of Figure 1.9, and have them draw something similar to Figure 1.14.

Page 18. We find some students are interested in knowing that isoclines were a great help in drawing slope fields before the advent of graphing calculators.

Page 20. We find Exercise 5 in Section 1.3 is worth assigning.

Page 20. We find Exercise 6 in Section 1.3 is worth assigning. After assigning it, we hand out copies of Figures 1.18 and 1.19, place them on top of each other, and hold them up to the light.

Page 21. Exercise 8 in Section 1.3 demonstrates that a computer drawn slope field may be misleading.

Page 22. The differential equation in Exercise 9 is equivalent to assuming that air-resistance is proportional to the velocity, because y satisfies $y'' + ky' = g$. This is a poor assumption, as we show in Chapter 4, Example 4.9 of the text.

Page 22. There are additional exercises for Section 1.3 on page 6 of this manual.

Page 22. An application of $dy/dx = -a/x$ is given on page 19 of this manual in the section called MEMORY RETENTION. Although it fits the data well it is a naive model, because it predicts that the subjects have no memory retention after a month. However, some might argue that this is true for some students!

Page 22. Other applications of $y' = g(x)$ are given on pages 27 and 29 of this manual, in sections called WATER-SKIER PROBLEM and the TAPE DECK COUNTERS. Students find these sections interesting, but not easy. However, when we have tried to include these here, we seem to spend too much time on this chapter. Using TAPE DECK COUNTERS as a group homework assignment is a possibility.

Page 22. Section 1.4, *Functions and Power Series Expansions*, is optional. If you plan to do series solutions later, here is an elementary introduction, so students can get their feet wet.

Page 24. There are additional exercises for Section 1.4 on page 6 of this manual.

1.2 Additional Exercises

Section 1.1

Page 7. Add exercise after the current Exercise 6.

7. **The Fresnel Cosine Integral.** Using (1.11) of the text, write down an integral that represents the solution of the initial value problem

$$\frac{dy}{dx} = \sqrt{\frac{2}{\pi}} \cos x^2, \qquad y(0) = 0.$$

This solution, known as the Fresnel Cosine Integral and denoted by $C(x)$, cannot be expressed in terms of familiar functions. Use the ideas from Exercise 3 to graph the solution of this differential equation for $-5 \leq x \leq 5$. What do you think happens to $C(x)$ as $x \to \infty$?

Section 1.2

Page 11. Add exercise after the current Exercise 8.

9. **The Fresnel Cosine Integral.** Use monotonicity, concavity, symmetry, singularities, and uniqueness to sketch various solution curves for the differential equation $y' = \sqrt{\frac{2}{\pi}} \cos x^2$. Then draw the graph of the particular solution that satisfies $y(0) = 0$. What do you think happens to $y(x)$ as $x \to \infty$? Compare your answer with the one you found for Exercise 7, Section 1.1.

10. **The Running Lizard:** Experiments have been performed where a lizard is encouraged to run as fast as possible, starting from rest.[1] The distance run is then measured as a function of time. The data gathered are given in Table 1.1 and shown in Figure 1.1. [Table 1.1 is in TWIDDLE. It is called LIZARD.DTA and is in the subdirectory DL-0'.]

    Table 1.1 Distance traveled by accelerating lizard as a function of time

Time (sec)	Distance (meters)
0.000	0.000
0.044	0.040
0.076	0.100
0.104	0.160
0.150	0.275
0.252	0.520
0.336	0.760
0.416	1.020
0.500	1.280
0.576	1.520
0.661	1.780
0.750	2.050

    (a) It is believed that the differential equation governing the acceleration of a lizard is
    $$\frac{d^2 x}{dt^2} = ae^{-bt} \tag{1.1}$$
    where a and b are positive constants and x is the distance traveled in time t. Explain what happens to the acceleration as t increases. Is this a reasonable model?

    (b) Integrate (1.1) once, using the fact that the lizard starts from rest at $t = 0$, to find
    $$\frac{dx}{dt} = \frac{a}{b}\left(1 - e^{-bt}\right). \tag{1.2}$$
    What happens to dx/dt as $t \to \infty$? Explain what this means physically, and show how this can be used, in conjunction with Figure 1.1 to estimate a/b.

    (c) Integrate (1.2) again, using the fact that the lizard starts from $x = 0$ at $t = 0$, to find
    $$x(t) = \frac{a}{b}\left[t - \frac{1}{b}\left(1 - e^{-bt}\right)\right]. \tag{1.3}$$

---

[1] "Effects of body size and slope on acceleration of a lizard (*Stellio Stellio*)" by R. B. Huey and P. E. Hertz, J. Exp. Biol., **110**, 1984, pages 113 – 123.

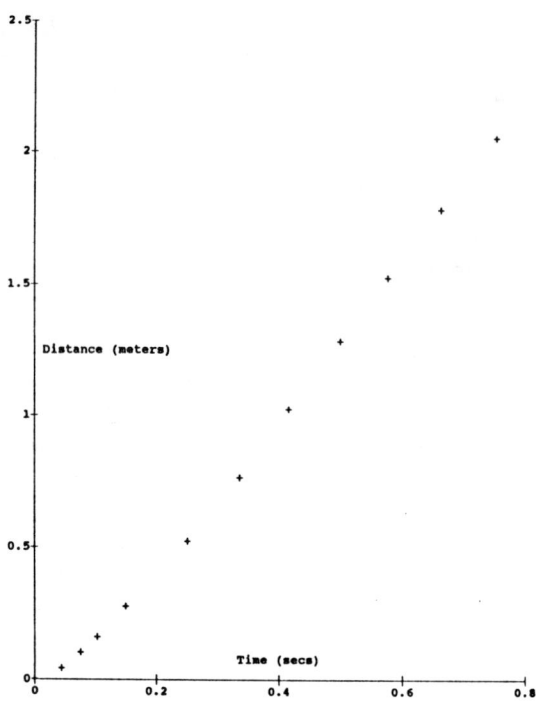

Figure 1.1 The distance run (in meters) as a function of time (in seconds) by an accelerating lizard

(d) Use one of the points from Table 1.1 — say $x(0.661) = 1.780$ — and your estimate for a/b from part (b) to estimate the value of b. Using these estimates for a and b plot (1.3) on Figure 1.1. Figure 1.2 is one possibility. How good is the differential equation (1.1) at modeling an accelerating lizard?

(e) Table 1.2 is a data set corresponding to the times achieved every 10 meters by Carl Lewis in the 100 m final of the World Championship in Rome in 1987.[2] This data set is shown in Figure 1.3. Does the model (1.1) also apply to world-class sprinters? According to this model does Carl Lewis attain his maximum speed while running the 100 m race? [Table 1.2 is in TWIDDLE. It is called LEWIS.DTA and is in the subdirectory DD-01.]

Section 1.3

Page 22. Add exercise after the current Exercise 9.

10. The Fresnel Cosine Integral. Use slope fields and isoclines for the differential equation $y' = \sqrt{\frac{2}{\pi}} \cos x^2$ to draw various solution curves. Then draw the solution curve that satisfies $y(0) = 0$. What do you think happens to $y(x)$ as $x \to \infty$? Compare your answers with those you found for Exercise 7, Section 1.1, and Exercise 9, Section 1.2.

Section 1.4

Page 24. Add exercise after the current Exercise 2.

---

[2] "Mathematical Models of Running" by W. G. Pritchard, SIAM Review, **35**, 1993, pages 359 – 379.

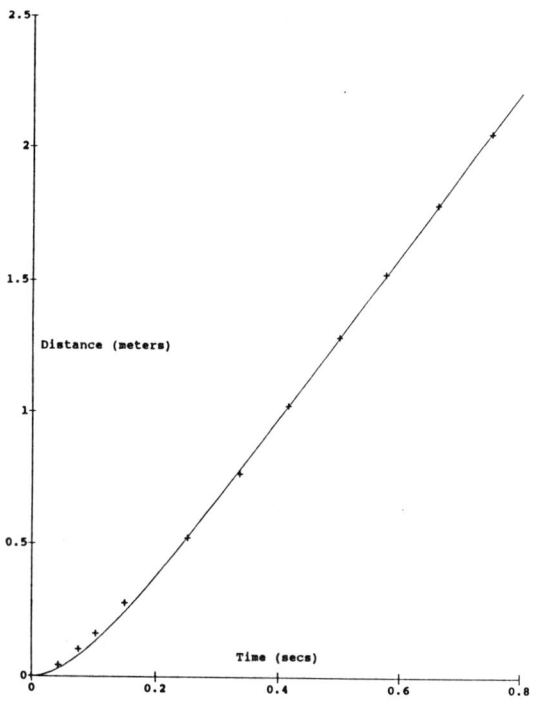

Figure 1.2   The distance run by a lizard and the function $x(t) = (a/b)[t - (1 - e^{-bt}/b)]$

Table 1.2   Distance traveled by Carl Lewis as a function of time

Time (sec)	Distance (meters)
0.00	0
1.94	10
2.96	20
3.91	30
4.78	40
5.64	50
6.50	60
7.36	70
8.22	80
9.07	90
9.93	100

ADDITIONAL EXERCISES

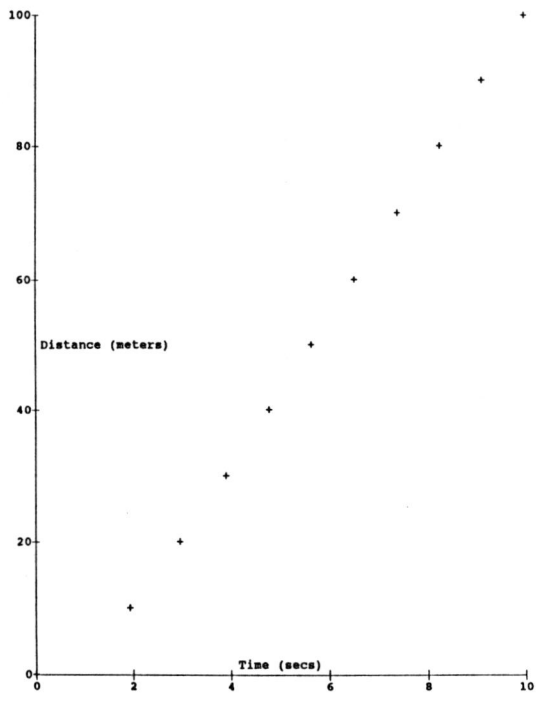

Figure 1.3 The distance run (in meters) as a function of time (in seconds) by Carl Lewis

3. **The Fresnel Cosine Integral.** Calculate the power series expansion for the Fresnel Cosine Integral $C(x)$, defined as

$$C(x) = \sqrt{\frac{2}{\pi}} \int_0^x \cos t^2 \, dt.$$

Confirm that it is consistent with the answers you found for Exercise 7, Section 1.1; Exercise 9, Section 1.2; and Exercise 10, Section 1.3.

1.3 Additional Materials

The Bombay Plague Function

You can replace the use of the Error Function throughout Chapter 1, by the solution of $y' = 1/\cosh^2 x$ subject to $y(0) = 0$, which we call the Bombay Plague function. We give the specific replacements.

Replace the text Example 1.3, page 5, by the following.

Example 1.3: The Bombay Plague Function

An important function, used to model the number of deaths in Bombay from a plague spread by rats during the period December 1905 to July 1906,[3] is the solution of the differential equation

$$\frac{dN}{dt} = \frac{a}{\cosh^2[b(t-c)]} \tag{1.4}$$

subject to the condition $N(c) = N_c$ where a, b, c, and N_c are positive constants that characterize this particular plague.[4] Here $N(t)$ is the total number of deaths at time t in weeks from December 1905. Before we go any further, use your intuition to sketch a possible graph of a function that would represent the number of deaths due to a plague.

If we use the change of variables $x = b(t-c)$, $y = b(N - N_c)/a$, we find that (1.4) can be rewritten in the simpler form[5]

$$\frac{dy}{dx} = \frac{1}{\cosh^2 x} \tag{1.5}$$

subject to the initial condition that

$$y(0) = 0. \tag{1.6}$$

This example will recur throughout this chapter.

The explicit solution of (1.5) can be written as

$$y(x) = \int \frac{1}{\cosh^2 x} \, dx + C. \tag{1.7}$$

The usual way to evaluate the constant C so (1.6) is satisfied is to substitute $x = 0$ and $y = 0$ into the solution (1.7) and solve for C. However, it is not obvious whether the integral in (1.7) can be expressed in terms of familiar functions, so this usual way to evaluate C does not work. To bypass this problem we change the form of our solution to the one given in the text (1.9) and use the fact that $x_0 = 0$ and $y_0 = 0$ to obtain

$$y(x) = \int_0^x \frac{1}{\cosh^2 t} \, dt.$$

---

[3] "A contribution to the mathematical theory of epidemics" by W. O. Kermack and A. G. McKendrick, Proc. Roy. Soc., **115A**, 1927, pages 700 – 721.

[4] Remember that $\cosh x$ is the hyperbolic cosine function which is defined by $\cosh x = (e^x + e^{-x})/2$. A summary of the properties of hyperbolic functions is in the appendix.

[5] This change of variables represents a translation of the point $t = c$, $N = N_c$ to the origin of the xy system, and then a rescaling of the vertical and horizontal axes. Translations and scaling do not change the general shape of a curve (increasing, decreasing, maxima, minima, concavity, points of inflection), merely where they are located.

We call the integral form of this solution the Bombay Plague Function, and denote it by $bpf(x)$, that is

$$bpf(x) = \int_0^x \frac{1}{\cosh^2 t} \, dt. \tag{1.8}$$

We might ask how we can determine the graph of this function from its form in (1.8). One way would be to construct a table of values of $(x, bpf(x))$ by using a numerical method of approximating the integral for specific choices of x (see Exercise 1). However, numerical techniques require considerable computation to plot enough points to be confident of the shape of the graph (see the text Exercises 3, 4, and 5). For that reason, in the next sections we develop methods for obtaining the graph of the solution of our differential equation directly from the differential equation. □

Exercises

1. The purpose of this exercise is to graph the Bombay Plague Function defined by (1.8), namely

$$bpf(x) = \int_0^x \frac{1}{\cosh^2 t} \, dt,$$

   by constructing a table of values of $(x, bpf(x))$.

   (a) What is the value of $bpf(0)$?

   (b) What is the relationship between $bpf(x)$ and $bpf(-x)$?

   (c) Use a computer/calculator package that performs numerical integration to obtain approximate values (say to 3 decimal places) for $bpf(x)$ at $x = 2, 4, 6, 8$, and 10. Use this information to plot $bpf(x)$ in the interval $[-10, 10]$. How confident are you that the graph you have is fairly accurate?

   (d) Now repeat part (c) for $x = 1, 3, 5, 7$, and 9. Did this change the accuracy of your previous graph for $bpf(x)$?

   (e) Now repeat part (c) for $x = 0.5, 1.5, 2.5$, and 3.5. Did this change the accuracy of your previous graph for $bpf(x)$?

   (f) What do you think happens to $bpf(x)$ as $x \to \infty$?

Replace the text Example 1.5, page 10, by the following.
Example 1.5 : The Bombay Plague Function
Using the techniques of calculus, sketch the family of solutions of the differential equation

$$\frac{dy}{dx} = \frac{1}{\cosh^2 x}. \tag{1.9}$$

Before proceeding further, we rewrite (1.9) in terms of e^x, as follows:

$$\frac{dy}{dx} = \frac{1}{\cosh^2 x} = \frac{1}{\left[\frac{1}{2}(e^x + e^{-x})\right]^2} = \frac{4}{(e^x + e^{-x})^2}. \tag{1.10}$$

- *Monotonicity*. The derivative of y is always positive, so all solutions are increasing.

- *Concavity*. If we differentiate (1.10) with respect to x we find

$$\frac{d^2y}{dx^2} = \frac{d}{dx}\left[4(e^x + e^{-x})^{-2}\right] = 4(-2)(e^x + e^{-x})^{-3}\frac{d}{dx}(e^x + e^{-x}) = -\frac{8}{(e^x + e^{-x})^3}(e^x - e^{-x}).$$

From this we see that $y'' > 0$ when $-(e^x - e^{-x}) > 0$. Now $-(e^x - e^{-x}) > 0$ can be rewritten as $e^x < e^{-x}$ or $e^{2x} < 1$, which is only satisfied if $x < 0$. Thus, $y'' > 0$ when $x < 0$. In a similar way we find that $y'' < 0$ when $x > 0$. Thus, all solutions are concave up when $x < 0$ and concave down when $x > 0$. Because $y''(0) = 0$, this means that all points of inflection lie on the vertical line $x = 0$.

- *Symmetry.* If we replace x with $-x$ on both sides of (1.10), the right-hand side is unchanged but the left-hand side changes sign. So the family of solutions is not symmetric about the y-axis. However, if we simultaneously replace x with $-x$ and replace y with $-y$, then we obtain (1.10) back again. So the family of solutions of (1.10) is unchanged under simultaneous interchange of x with $-x$ and y with $-y$. This means that the family of solutions is symmetric about the origin.[6]

- *Singularities.* There are no obvious points where the derivative fails to exist.

- *Uniqueness.* From our arguments at the end of the text Example 1.4, we see that solutions cannot intersect.

Based on this information, we can sketch by hand[7] the family of solutions of (1.10), which is shown in **Figure 1.4**. This family of solutions contains the particular solution curve that passes through the point P with coordinates $(0,0)$ — namely, $bpf(x)$. □

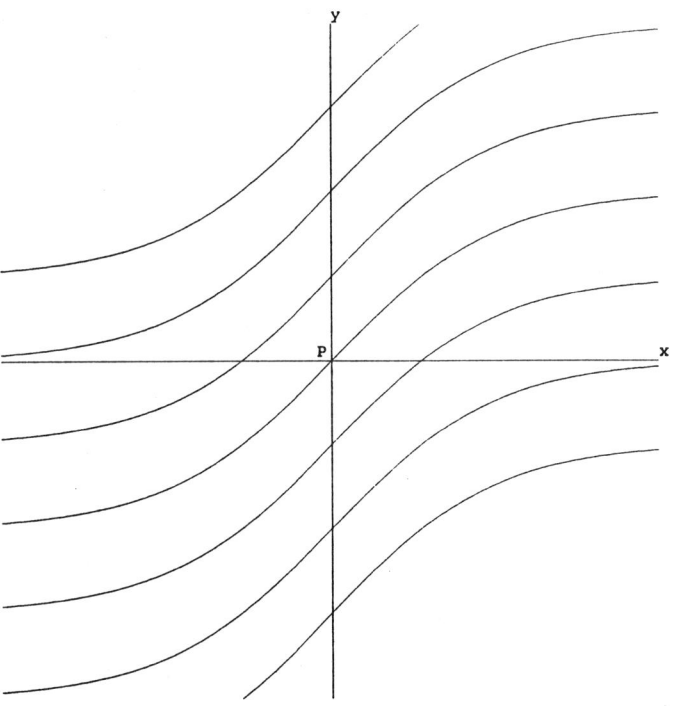

Figure 1.4 Some hand-drawn solution curves of $y' = 1/\cosh^2 x$

---

[6] A graph is symmetric about the origin if it is unchanged when rotated $180°$ about the origin.
[7] Throughout the text we make reference to hand-drawn solutions. Of course, they were drawn by machine.

ADDITIONAL MATERIALS

Replace the text Example 1.9, page 16, by the following.

Example 1.9 : The Bombay Plague Function

We already have an example of this need in the case of the differential equation that generates the Bombay Plague Function,

$$\frac{dy}{dx} = \frac{1}{\cosh^2 x}. \tag{1.11}$$

Because we want the solution of this equation, which starts at the point $(0,0)$, we construct the slope field that includes this point, as shown in Figure 1.5.

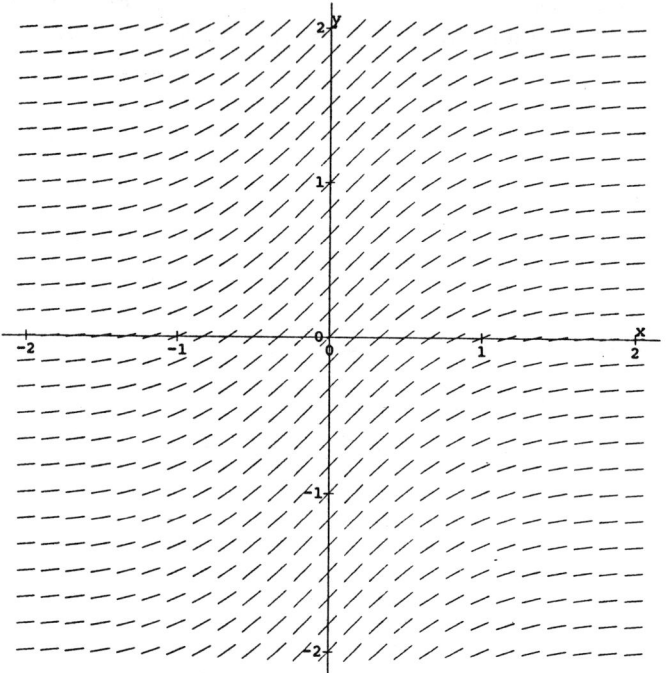

Figure 1.5   Slope field for $y' = 1/\cosh^2 x$

As expected, the slope field indicates that the solution curve that passes through $(0,0)$ is increasing, concave up when $x < 0$ and concave down when $x > 0$. Also notice that the slope field appears to be symmetric about the origin. Figure 1.6 shows a hand-drawn solution curve of (1.11) that passes through the origin, the graph of $y = bpf(x)$.

We could also use a numerical integration technique to obtain values for $bpf(x)$ at different values of x from its definition, namely,

$$bpf(x) = \int_0^x \frac{1}{\cosh^2 t}\, dt.$$

For example, we used Simpson's rule[8] to create Table 1.3. (Here we set the number of subintervals to 16 and rounded the answers to three decimal places.) Figure 1.7 shows the slope field, these numerical values, and a hand-drawn solution curve. Notice the agreement between this solution curve and these numerical values. □

Replace the text Example 1.11, page 19, by the following.

---

[8] See a calculus text to remind yourself of Simpson's rule.

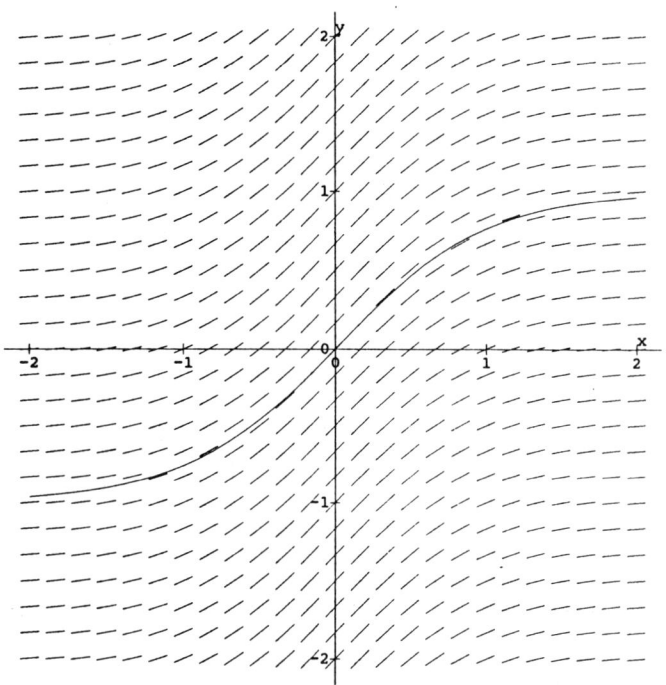

Figure 1.6 Hand-drawn graph of the solution of $y' = 1/\cosh^2 x$, $y(0) = 0$

Table 1.3 Simpson's rule for $y' = 1/\cosh^2 x$, $y(0) = 0$

x	$y(x)$
0.0	0.000
0.5	0.462
1.0	0.762
1.5	0.905
2.0	0.964

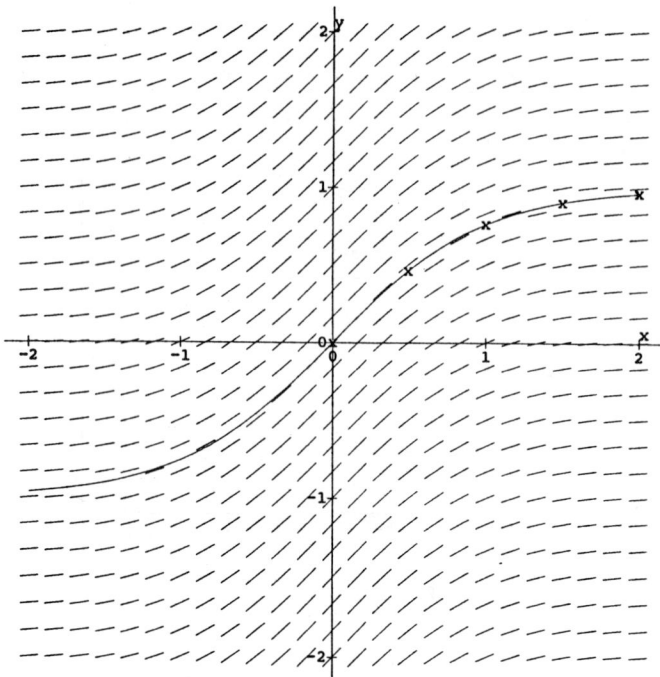

Figure 1.7 Numerical values and hand-drawn graph of the solution of $y' = 1/\cosh^2 x$, $y(0) = 0$

Example 1.11 : The Bombay Plague Function

Now we return to the differential equation giving rise to the Bombay Plague Function,

$$\frac{dy}{dx} = \frac{1}{\cosh^2 x}.$$

The isocline corresponding to slope m is given by

$$\frac{1}{\cosh^2 x} = m, \tag{1.12}$$

Notice that this means that there are no isoclines for slope $m \leq 0$. (Why?). To find more information about the isoclines, we need to solve (1.12) for x as a function of m. To do this, we use $\cosh x = \frac{1}{2}\left(e^x + e^{-x}\right)$ to rewrite (1.12) as

$$\frac{4}{\left(e^x + e^{-x}\right)^2} = m,$$

or

$$\frac{4}{e^{2x} + 2 + e^{-2x}} = m.$$

We multiply both sides of this equation by $e^{2x} + 2 + e^{-2x}$ and bring all terms to the left-hand side to find

$$me^{2x} + 2m - 4 + me^{-2x} = 0.$$

Multiplying by e^{2x} leads to the equation

$$m\left(e^{2x}\right)^2 + (2m - 4)e^{2x} + m = 0.$$

This is a quadratic equation in e^{2x} with solutions

$$e^{2x} = \frac{-(2m-4) \pm \sqrt{(2m-4)^2 - 4m^2}}{2m} = \frac{-(2m-4) \pm \sqrt{-16m + 16}}{2m},$$

or

$$e^{2x} = \frac{-(2m-4) \pm 4\sqrt{1-m}}{2m} = -1 + \frac{2}{m} \pm \frac{2}{m}\sqrt{1-m}. \qquad (1.13)$$

Notice that this means that there are no isoclines for slope $m > 1$. (Why?) If we solve (1.13) for x we obtain

$$x = \frac{1}{2}\ln\left[-1 + \frac{2}{m} \pm \frac{2}{m}\sqrt{1-m}\right]$$

as the equation of the isocline corresponding to slope m. The isoclines for slope 0.1, 0.3, and 0.7 are shown in **Figure 1.8**. Notice that in this case each isocline consists of two vertical lines. □

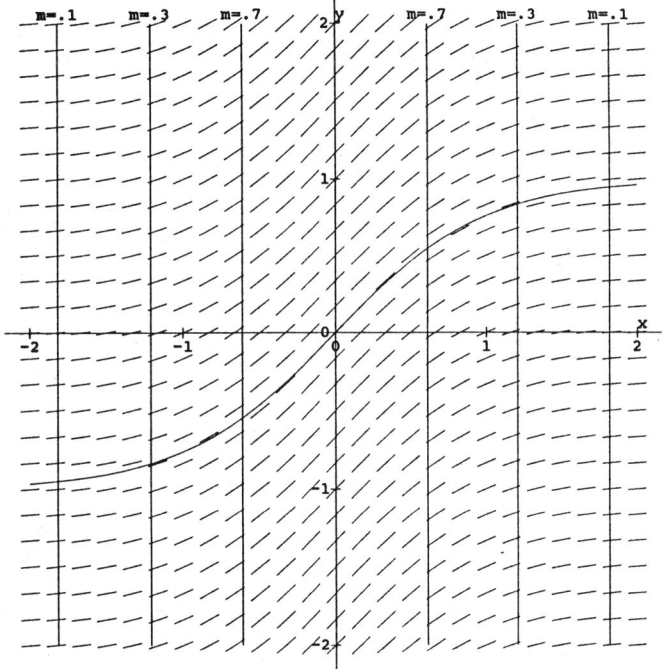

Figure 1.8 Isoclines ($m = 0.1, 0.3, 0.7$) and slope field for $y' = 1/\cosh^2 x$

At the end of this you should point out how much information we have obtained about the Bombay Plague function, all from the differential equation. In fact it is possible to find an explicit form for this function, because the integral can be evaluated by rewriting the integrand as follows:

$$y(x) = \int \frac{1}{\cosh^2 x}\,dx = \int \frac{4}{(e^x + e^{-x})^2}\,dx = \int \frac{4}{e^{-2x}(e^{2x}+1)^2}\,dx = \int \frac{4e^{2x}}{(e^{2x}+1)^2}\,dx.$$

Now we make the substitution $u = e^{2x} + 1$ to find

$$y(x) = 2\int \frac{du}{u^2} = -\frac{2}{u} + C = \frac{-2}{e^{2x}+1} + C.$$

If we impose the initial condition $y(0) = 0$, we have $C = 1$, so

$$y(x) = \frac{-2}{e^{2x}+1} + 1 = \frac{-2 + e^{2x}+1}{e^{2x}+1} = \frac{e^{2x}-1}{e^{2x}+1}.$$

If we multiply the numerator and denominator by e^{-x} we see that

$$y(x) = \frac{e^x - e^{-x}}{e^x + e^{-x}} = \tanh x.$$

Thus, in terms of the original variables — N and t — the number of deaths due to the Bombay Plague is given by

$$N(t) = N_c + \frac{a}{b} \tanh\left[b\left(t - c\right)\right].$$

(You should check this.) If we assume that a long time ago nobody had the plague, so that $N \to 0$ as $t \to -\infty$, we find $N_c = a/b$, because $\lim_{t \to -\infty} \tanh\left[b\left(t-c\right)\right] = -1$ for $b > 0$. Thus, we have

$$N(t) = \frac{a}{b} + \frac{a}{b} \tanh\left[b\left(t - c\right)\right].$$

The actual data for the total number of deaths due to the Bombay Plague is given in Table 1.4 and shown in Figure 1.9.[9] [Table 1.4 is in TWIDDLE. It is called BOMBAY.DTA and is in the subdirectory DD-01.] The general shape of this data and the solution of the differential equation agree well. We want to estimate the values of the constants a, b, and c to see how well they agree.

Table 1.4 The Bombay Plague data

Week	Deaths	Week	Deaths	Week	Deaths
1	4	12	900	23	8129
2	14	13	1290	24	8480
3	29	14	1738	25	8690
4	47	15	2379	26	8803
5	68	16	3150	27	8868
6	99	17	3851	28	8920
7	150	18	4547	29	8971
8	203	19	5414	30	9010
9	300	20	6339	31	9043
10	425	21	7140		
11	608	22	7720		

We know that point of inflection occurs at $x = 0$, which corresponds to $t = c$. At the point of inflection, the number of deaths is thus $N(c) = a/b$. We also know that $N'(c) = a$, and that $\lim_{t \to \infty} N(t) = 2a/b$, because $\lim_{t \to \infty} \tanh\left[b\left(t-c\right)\right] = 1$ for $b > 0$. Thus, if we can estimate $N(t)$ as $t \to \infty$, then we have an estimate for $2a/b$. By estimating where N takes on half the value of $2a/b$, that is, where $N = a/b$, we can estimate c. Finally, by estimating the slope of the data at c, namely, $N'(c)$, we can estimate a. In this way we can estimate the constants a, b, and c. From Figure 1.9 we estimate that $N(t) \to 9100$ as $t \to \infty$. This gives us the estimate $2a/b = 9100$, so the point of inflection occurs $N_c = a/b = 4550$. We estimate that this occurs when $t = 17.9$, so that $c = 17.9$. The slope at $t = 17.9$ appears to be about 865 so $a = 865$. From this information we estimate $b = 0.19$. Thus, our function that models the Bombay Plague is

$$N(t) = 4550 + 4550 \tanh\left[0.19\left(t - 17.9\right)\right]. \tag{1.14}$$

Figure 1.10 shows the data set from Figure 1.9 and the function (1.14). The agreement is good.

---

[9] "A contribution to the mathematical theory of epidemics" by W. O. Kermack and A. G. McKendrick, Proc. Roy. Soc., **115A**, 1927, pages 700 – 721. Data extracted from the figure on page 714. It is not uncommon for data in the literature to be presented graphically without a table of numbers. An easy way to estimate the numerical data from a graph is to first photocopy a piece of graph paper onto a transparency and then place the transparency on top off the graph containing the data. (You may need to enlarge the graph by magnifying a photocopy of it.) That is how we estimated a number of the data sets in this book.

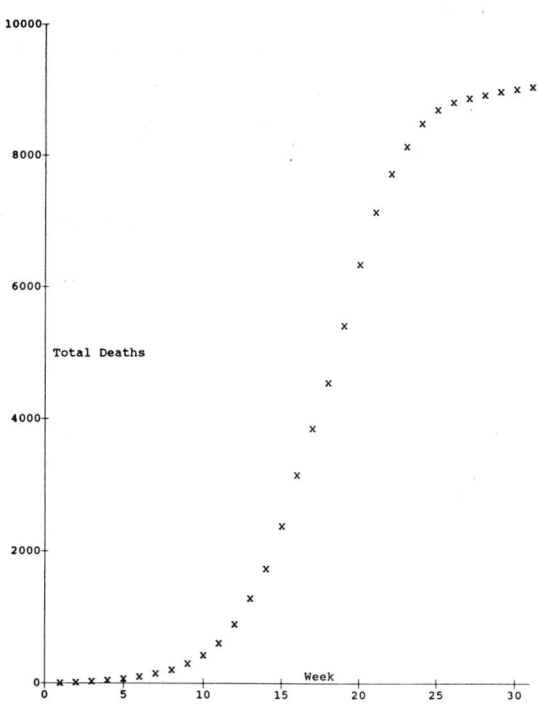

Figure 1.9 The number of deaths due to the Bombay Plague

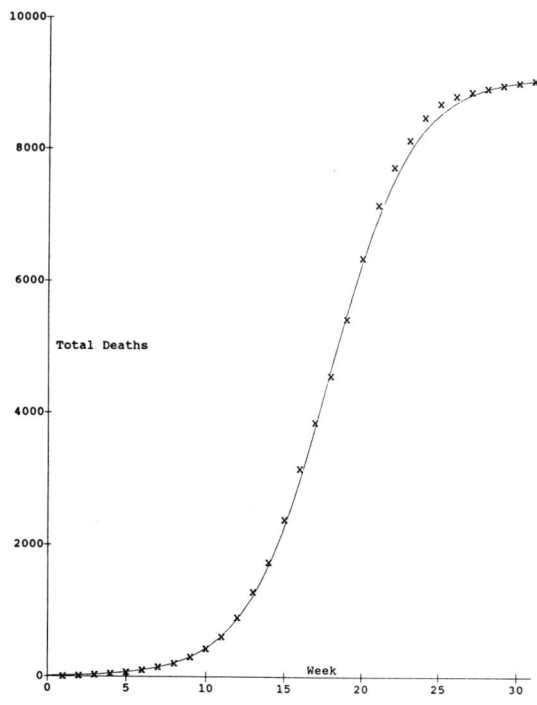

Figure 1.10 The number of deaths due to the Bombay Plague and the function $4550 + 4550 \tanh\left[0.19\left(t - 17.9\right)\right]$

ADDITIONAL MATERIALS

Exercises

1. The weekly deaths during the Great Plague of London from August 1665 to October 1665 are given in Table 1.5 and sketched in Figure 1.11.[10] [Table 1.5 is in TWIDDLE. It is called PLAGUE.DTA and is in the subdirectory DD-01.] Can these data be modeled by (1.4)?

Table 1.5   The Great Plague of 1665

Week	Total Deaths
0	3880
1	8117
2	14219
3	21207
4	27751
5	34916
6	40449
7	45378
8	49705

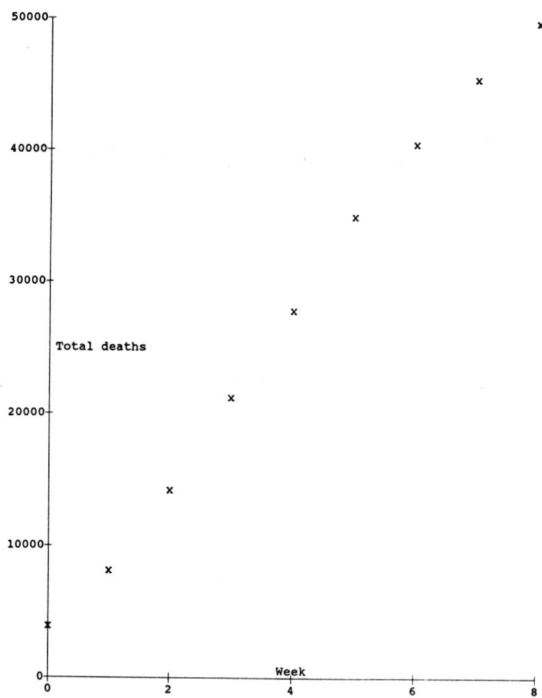

Figure 1.11   The number of deaths due to the Great Plague

---

[10] "A Handbook of Small Data Sets" by D. J. Hand, F. Daly, A. D. Lunn, K. J. McConway and E. Ostrowski, Chapman and Hall, 1994, page 144.

Memory Retention

Experiments have been done in which the average percentage memory retention is measured against time.[11] The subjects memorized a list of items, and then the percentage of these items that the subjects remembered was measured at various time intervals. The results of this experiment are tabulated in Table 2.27 and shown graphically in Figure 1.12. [Table 2.27 is in TWIDDLE. It is called MEMORY.DTA and is in the subdirectory DD-01.]

Table 1.6 Percentage memory retention at various times

Time (minutes)	Memory retention (percentage)
1	84
5	71
15	61
30	56
60	54
120	47
240	45
480	38
720	36
1440	26
2880	20
5760	16
10080	8

It has been suggested, possibly naively, that the percentage memory retention, $y(x)$, where x is time measured in minutes, is modeled by the differential equation

$$\frac{dy}{dx} = -\frac{a}{x}$$

where a is a positive constant and $x \geq 1$.

The solution of this equation is

$$y(x) = -a \ln x + C,$$

which suggests that if we plot y versus $\ln x$ we should see a line with slope $-a$ and intercept C, from which we can estimate a and C. Figure 1.13 shows this and a line with slope -7.9 and intercept 84.6 which seems to give a reasonable fit. This allows us to estimate $a = 7.9$ and $C = 84.6$. The data set and the function

$$y(x) = -7.9 \ln x + 84.6$$

are shown in Figure 1.14, where the fit is also reasonable.

However, this model does have some problems. We might expect that at time $x = 0$ the subjects would remember 100% of the items. Furthermore, according to this model there is a time when $y(x) = 0$ — namely, when $7.9 \ln x = 84.6$. Solving for x we find $x \approx 44,750$ min ≈ 31 days. Thus, this model predicts that the subjects retained nothing after 31 days.

A more reasonable model might be

$$\frac{dy}{dx} = -be^{-ax}$$

where a and b are positive constants. Unfortunately the data set cannot be modeled reasonably by the solution of this equation — namely, $y = be^{-ax}/a + C$.

---

[11] "Probability with statistical applications" by F. Mosteller, R. E. K. Rourke, and G. B. Thomas, Addison-Wesley, 1970, Table 11-1, page 383, as reported in "A Handbook of Small Data Sets" by D. J. Hand, F. Daly, A. D. Lunn, K. J. McConway and E. Ostrowski, Chapman and Hall, 1994, page 128.

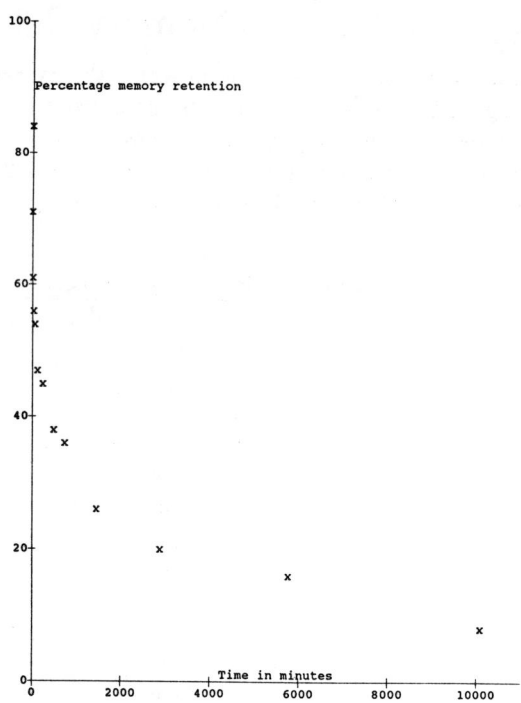

Figure 1.12 Percentage memory retention at various times

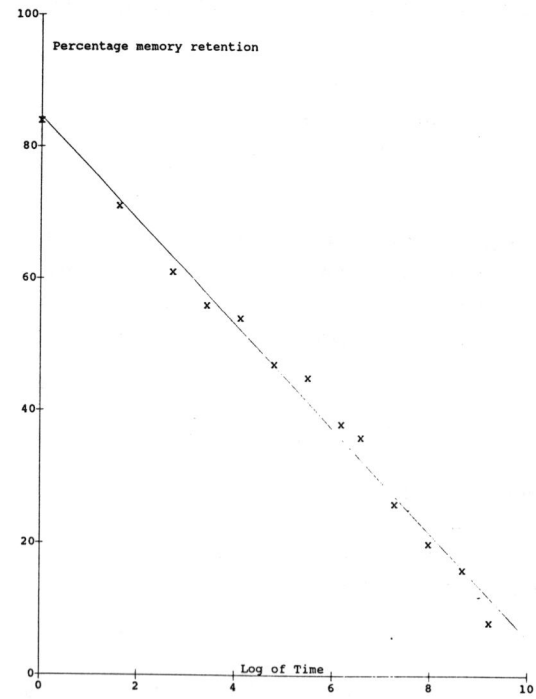

Figure 1.13 Percentage memory retention versus log(time) and the line with slope −7.9 and intercept 84.6

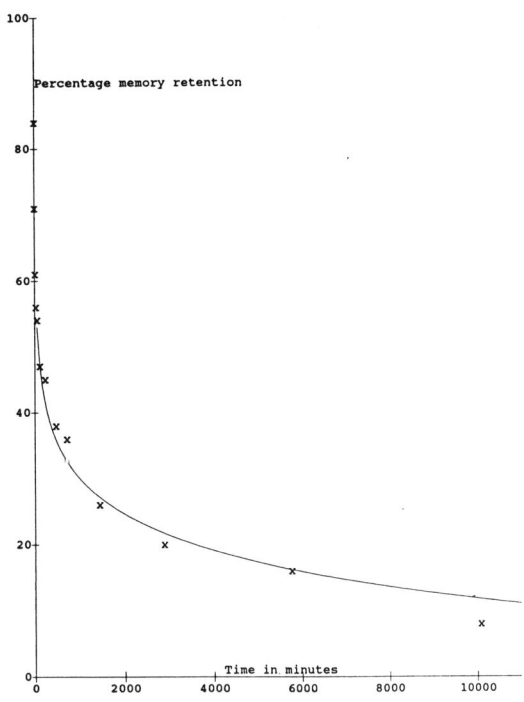

Figure 1.14 Percentage memory retention at various times and the function $y(x) = -7.9 \ln x + 84.6$

Exercises

1. **Motion of an Accelerating Porsche.** Table 1.7 is part of a driving test report for a Porsche 944 Turbo.[12] The table shows the velocity $v = dy/dx$ in mph of an accelerating Porsche as a function of the time x in seconds measured from when the Porsche started from rest at $y = 0$. These data are plotted in Figure 1.15. [Table 1.7 is in TWIDDLE. It is called PORSCHE.DTA and is in the subdirectory DD-01.] By plotting $\ln v$ versus $\ln x$ obtain a relationship between v and x — that is, $v(x)$. Integrate $dy/dx = v(x)$ to find $y(x)$. [Caution: Check that your units are consistent before integrating.] According to the driving test report, the Porsche covered 100 ft in 3.0 secs, 500 feet in 8.1 secs, and 1320 feet (1/4 mile) in 14.6 secs. The velocity at the 1/4 mile mark was 97.5 mph. How well does your model fit these data?

Table 1.7 The velocity of a Porsche 944 Turbo as a function of time

Time (sec)	Velocity (mph)
2.1	30
4.7	50
6.0	60
8.1	70
10.1	80
15.3	100

---

[12] "Road and Track Special Series: Porsche", CBS Magazines, 1987, page 59 as reported in "The motion of an accelerating automobile" by M. G. Calkin, The American Journal of Physics, **58**, 1990, pages 573 through 575, Table I.

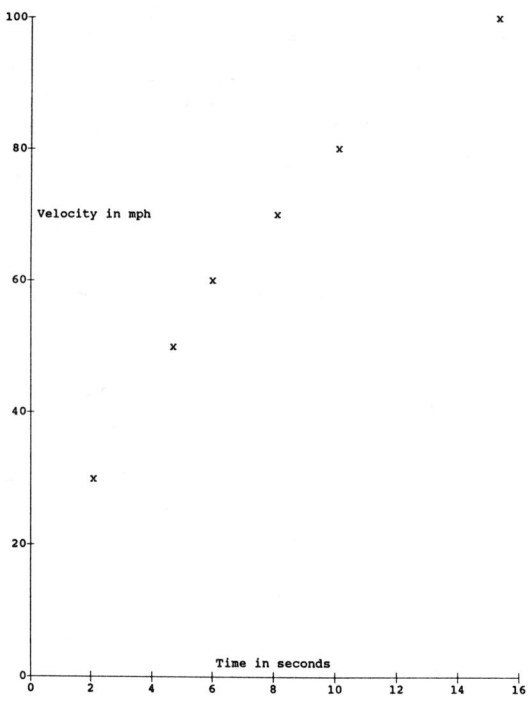

Figure 1.15 The velocity of a Porsche 944 Turbo as a function of time

2. **Horizontal Infiltration of Water in Soil.**[13] Consider a tube of soil held horizontally which is going to be wetted at one end. If $y(x)$ is the position of the wetting front at time x then a model for the movement of the wetting front is

$$\frac{d^2y}{dx^2} = ax^b$$

where a and b are constants which vary with the sample of soil, and $y = dy/dx = 0$ when $x = 0$. The data for such an experiment is given in Table 1.8 and shown in Figure 1.16. [Table 1.8 is in TWIDDLE. It is called WET.DTA and is in the subdirectory DD-01.] What values for a and b make this a reasonable model? [Hint: Find $y(x)$ and then plot $\ln y$ versus $\ln x$.]

Table 1.8 The movement of the wetting front as a function of time

Time (minutes)	Distance (cm)
1	3.7
2	4.5
4	6.0
8	8.0
16	10.6
32	13.8
64	18.6
128	24.3
256	32.0

---

[13] "An example of motion in a course of physics for agriculture" by I. A. Guerrini, The Physics Teacher, February 1984, pages 102 through 103.

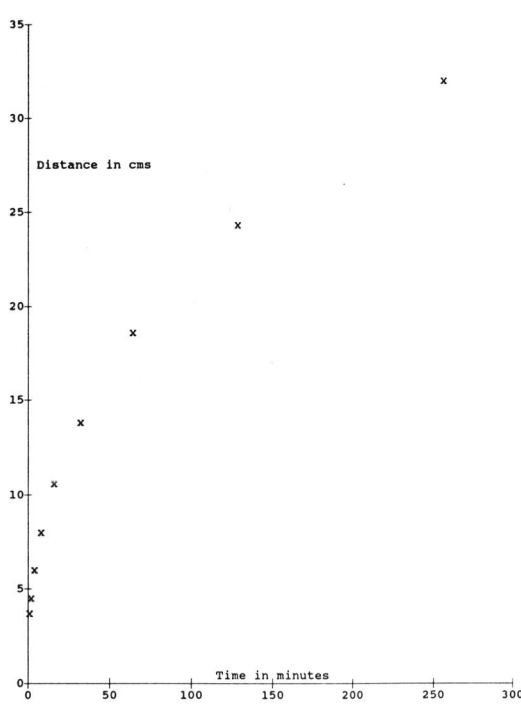

Figure 1.16 The movement of the wetting front as a function of time

3. **Determining the Shape of a Vessel.**[14] Imagine we were to construct a vessel using the following technique. We start with a 40 cm by 40 cm slab of plastic 5 cm thick. We use a band saw to remove waste leaving a shape which is symmetric about the x-axis. See Figure 1.17 for an example. Now we take two pieces of 40 cm by 40 cm plastic sheeting and glue them onto the two 40 cm by 40 cm sides of the slab creating a vessel inside this object when the x-axis is vertical. Finally we paint the entire exterior of the object with opaque paint, except for two vertical strips 1 cm wide along the x-axis. This allows us to see the height of any liquid in the vessel, but hides its shape. We are going to add water to this vessel, measuring its volume as a function of height and then use these data to determine the shape of the flask.

   (a) Add water in 100 cc units to the vessel, measuring the height each time. Construct a table which shows the volume as a function of height. A typical data set is given in Table 1.9 and shown in Figure 1.18. [Table 1.9 is in TWIDDLE. It is called SHAPE.DTA and is in the subdirectory DD-01.]

   (b) Show that if the shape of the upper half of the curve cut by the band saw is $S(x)$ then the volume $V(x)$ contained in the flask when the water level is at height x is

   $$V(x) = 2 \int_0^x S(h)\,dh.$$

   The units of x are centimeters and those of V are cubic centimeters.

   (c) Explain why

   $$S(x) = \frac{1}{2}\frac{dV}{dx}.$$

   (d) Use the above information to estimate the shape of the vessel for Table 1.9.

[14] "Indirect determination of the shape of a vessel" by J. M. Dundon, The Physics Teacher, May 1980, pages 376 through 377.

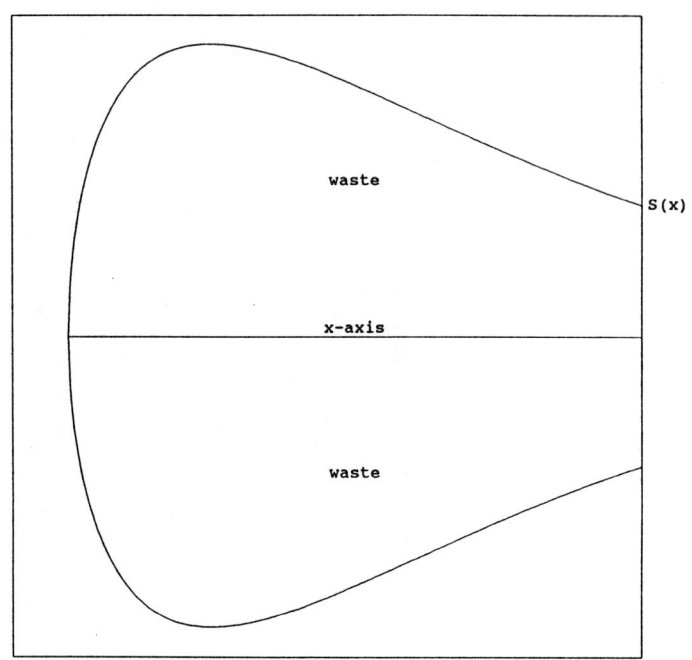

Figure 1.17 Constructing the vessel

Table 1.9 The volume of the vessel as a function of height

Height (cm)	Volume (cc)	Height (cm)	Volume (cc)
0.00	0	12.55	675
2.55	100	13.20	700
4.25	200	13.85	725
5.90	300	14.55	750
7.55	400	15.40	775
9.20	500	16.35	800
11.05	600	17.45	825
11.55	625	18.65	850
12.05	650	20.00	875

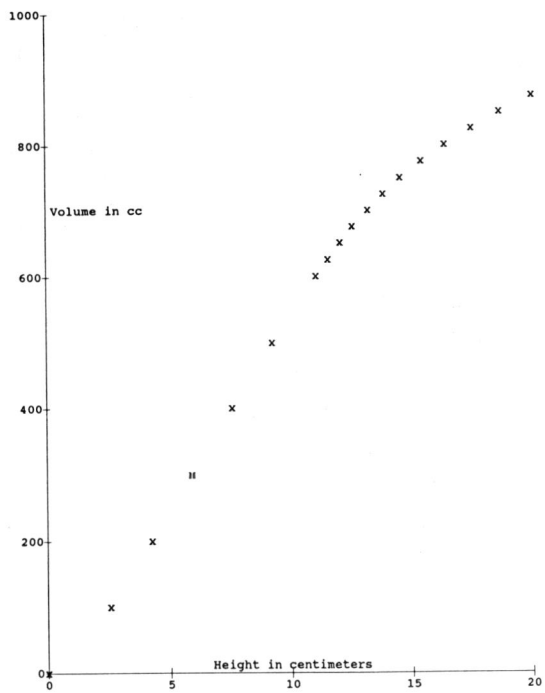

Figure 1.18 The volume of the vessel as a function of height

4. **Simple Experiment.**[15] Take a piece of string 8 feet long and lie it on the ground so the string is straight. Measuring from one end of the string — the "bottom" — attach 5 marbles to it with sticky tape at 0, 6, 24, 54, and 96 inches from the bottom. Stand on a chair and hold the marble at the "top" of the string in front of you so the string hangs vertically over a metal plate. The bottom marble should be just above the plate. Now release the top marble and listen to the clicks as the marbles strike each other. Explain what is happening, assuming that the marbles are falling according to Newton's Law of motion

$$\frac{d^2x}{dt^2} = -g.$$

5. **Quicksort.** In computer programming a common procedure is the sorting of a list of names, or numbers, etc. Many sorting algorithms have been developed to reduce the time the computer spends in the sorting process. One such algorithm is Quicksort.[16] The time y it takes to sort a list of x elements is approximated by the solution of the differential equation

$$\frac{dy}{dx} = a + b \ln x \qquad (1.15)$$

where a and b are constants depending on the computer, the language the program is written in, and the type of list being sorted (names, numbers, etc.) Typical times (in milliseconds) for a list of various lengths are shown in Table 1.10. These are average times generated using a PASCAL program running on a University of Arizona VAX computer.[Table 1.10 is in TWIDDLE. It is called QCKSORT.DTA and is in the subdirectory DD-01.]

(a) Solve the differential equation (1.15) to find

$$y(x) = (a - b) x + bx \ln x + c \qquad (1.16)$$

---

[15] "String and sticky tape experiments" by R. D. Edge, The Physics Teacher, April 1978, page 233.

[16] For a discussion of Quicksort, see for example "Algorithms" by R. Sedgewick. Addison Wesley, 1983, Chapter 9.

ADDITIONAL MATERIALS

Table 1.10 The time for sorting lists of different lengths

Length	Time
50	280
100	660
150	1080
200	1530
250	1990

where c is a constant.

(b) Explain why the initial condition of $\lim_{x \to 0} y(x) = 0$ is reasonable and why this implies that $c = 0$ in (1.16).

(c) By using the two data points $x = 100$, $y = 660$ and $x = 200$, $y = 1530$, estimate a and b in (1.16).

(d) With the estimates made for a and b in part (c), plot (1.16) and the data set in Table 1.10. Is the differential equation (1.15) a reasonable model for timing Quicksort?

The Water-Skier Problem

Consider the situation in which a boat and water-skier are alongside a dock, connected by a tightly stretched rope. At time $t = 0$, the boat moves with a constant velocity v away from the dock in a direction perpendicular to the dock. We construct a coordinate system so that initially the skier is at $(a, 0)$ and the boat at $(0, 0)$. Later, at time t, the boat is at $(0, vt)$, and the skier at $(x(t), y(t))$. Thus, the rope has length a, and the boat is traveling up the y-axis with velocity v (Figure 1.19).

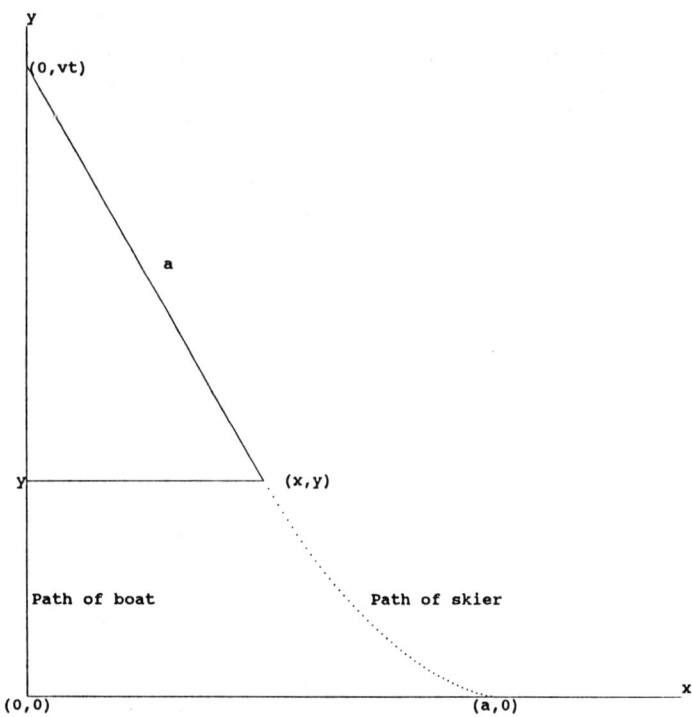

Figure 1.19 The water-skier problem

We want to determine what curve the skier follows. There are two key assumptions that we make to tackle this problem: (i) the rope is always tangent to the skier's trajectory; (ii) the rope has a constant length. We can see from Figure 1.19 that the requirement that the rope is tangent to the trajectory leads to

$$\frac{dy}{dx} = -\frac{vt - y}{x}, \tag{1.17}$$

whereas requiring the rope to have constant length gives

$$(vt - y)^2 + x^2 = a^2. \tag{1.18}$$

Eliminating $(vt - y)$ between (1.17) and (1.18) gives the equation

$$\frac{dy}{dx} = -\frac{\sqrt{a^2 - x^2}}{x}. \tag{1.19}$$

Because the problem requires that x always be positive, from (1.19) we see that the slope of the y versus x trajectory will always be negative, so y will be a decreasing function of x. We also note that the only place the trajectory can have a horizontal tangent is $x = a$.

If we differentiate (1.19) we obtain

$$\frac{d^2y}{dx^2} = \frac{a^2}{x^2\sqrt{a^2-x^2}},$$

which is never negative, so the trajectory will always be concave up. (Does this agree with your intuition about what will happen to the skier?)

If we look at the differential equation (1.19), we see that the isoclines are given by

$$-\sqrt{a^2-x^2}/x = m,$$

which can be rewritten as

$$x = \frac{a}{\sqrt{1+m^2}}. \quad (1.20)$$

Because a and m are constants, this is a vertical line. Note from (1.20) that the slopes are steeper at points on the curve that are closer to $x = 0$ (that is, the y-axis).

Figure 1.20 shows the slope field as well as the hand-drawn solution curve for the case $a = 1$. Notice that we have obtained all the information about the solution of (1.19) from its slope field and an analysis of the differential equation by means of isoclines and concavity, without obtaining the explicit solution.

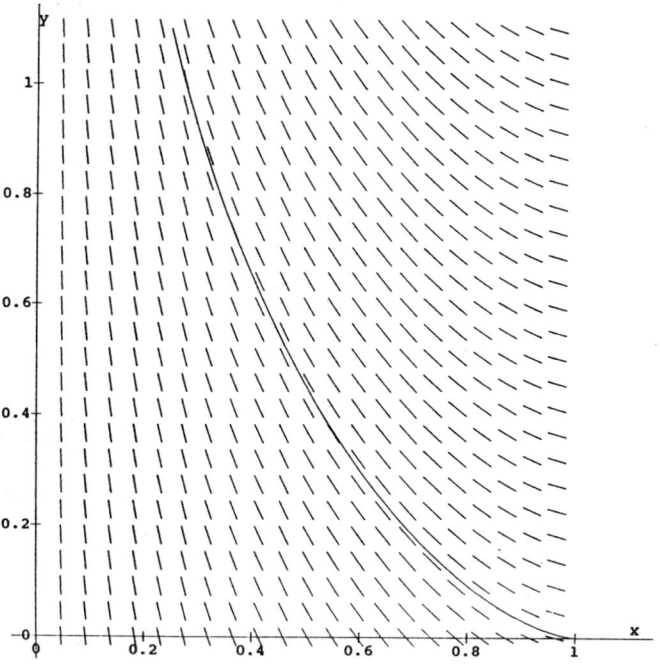

Figure 1.20 Slope field and hand-drawn solution curve for $dy/dx = -(1-x^2)^{1/2}/x$, $y(1) = 0$

To find an explicit form of the trajectory we must integrate (1.19), which requires some skill. To find an antiderivative of $-\sqrt{a^2-x^2}/x$, we use the substitution $u = \sqrt{a^2-x^2}$, so $u^2 = a^2 - x^2$. We then integrate $u^2/(a^2-u^2) = -1 + a^2/(a^2-u^2)$. The result is

$$y = -\sqrt{a^2-x^2} + \frac{a}{2}\ln\left(\frac{a+\sqrt{a^2-x^2}}{a-\sqrt{a^2-x^2}}\right) + C.$$

Because $y = 0$ when $x = a$, then $C = 0$. (This curve is called a tractrix, from the Latin "trahere", to pull.) □

Tape Deck Counters

Conduct the following experiment using either a VCR or a tape deck equipped with a digital tape-counter. Put a tape in the unit, set the tape-counter to zero, and, as you start the tape playing, note the time. Then record the time every 5 to 10 minutes along with the number on the tape-counter. A data set obtained in this way is shown in Table 1.11, which shows the number of revolutions n, the number on the tape-counter, as a function of time t. [Table 1.11 is in TWIDDLE. It is called TAPE.DTA and is in the subdirectory DD-01.] This is presented graphically in Figure 1.21. Can we model this situation? To do so requires finding a relationship between the number of revolutions n and the time t. The number of revolutions is related to the angular velocity of either the take-up reel or the feeder reel because the tape passes with constant velocity past the read-write head.

Table 1.11 Tapecounter versus time

Time (secs)	Counter
0	0
360	525
720	985
1080	1395
1380	1709
1680	2003
1980	2278
2280	2540
2520	2741
2820	2982
3120	3214
3420	3436

Let's concentrate on the take-up reel as sketched from above in Figure 1.22. Imagine that the reel starts running at time $t = 0$ and that at any time t, the radius of the tape on the reel is $R(t)$. Thus, $R(0)$ is the radius of the take-up reel when empty. The area of the tape on the reel (as seen from above) at time t is that of the washer-shaped region in Figure 1.22, namely, $\pi \left[R^2(t) - R^2(0) \right]$. If c is the thickness of the tape, the amount of tape on the reel at time t is $\pi \left[R^2(t) - R^2(0) \right] / c$. On the other hand, if v is the constant velocity of the tape past the read-write head, then after the tape has been running for time t the amount of tape on the take-up reel will be vt. From these two expressions for the amount of tape on the reel at time t, we have

$$vt = \frac{\pi}{c} \left[R^2(t) - R^2(0) \right],$$

from which

$$R(t) = R(0)\sqrt{bt + 1}, \qquad (1.21)$$

where b is the constant

$$b = \frac{cv}{\pi R^2(0)}.$$

If $A(t)$ is the angle (in radians) of the take-up reel at time t measured from an initial angle of zero at time $t = 0$, we have

$$A(0) = 0. \qquad (1.22)$$

Now let's see how A depends on t by looking at the motion of the reel. Consider what happens as the time increases by a small amount from t to $t + h$. The change in the angle will be $A(t + h) - A(t)$. When $A(t + h) - A(t)$ is multiplied by an appropriate radius $\mathcal{R}$, this is the amount of tape added to the reel in this time interval—namely, vh —so we will have

$$[A(t + h) - A(t)] \mathcal{R} = vh,$$

ADDITIONAL MATERIALS

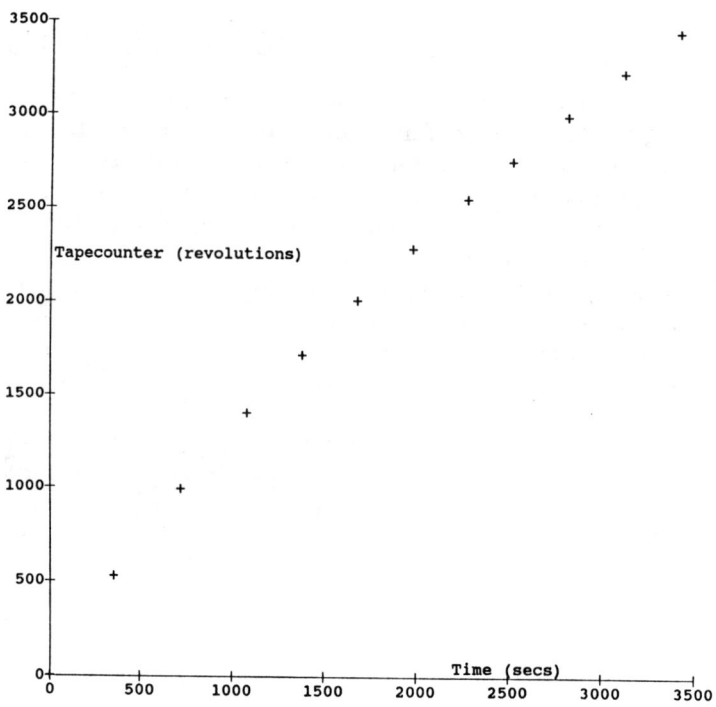

Figure 1.21 Tapecounter versus time

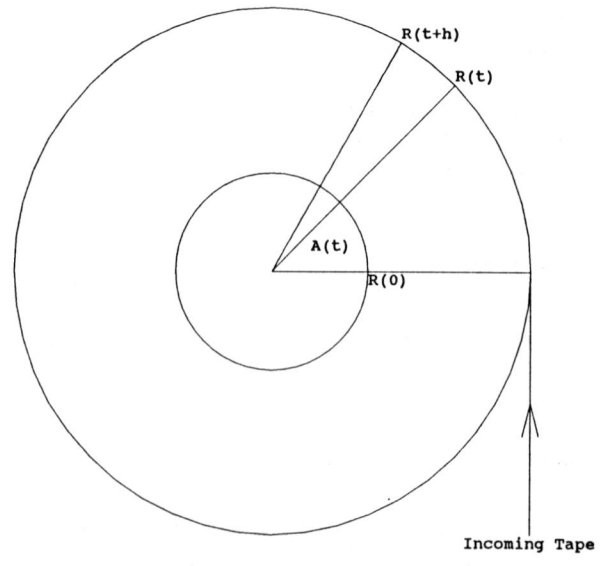

Figure 1.22 The take-up reel (seen from above)

which can be written as
$$\frac{A(t+h) - A(t)}{h}\mathcal{R} = v. \qquad (1.23)$$

There are a variety of functions we could use for the radius $\mathcal{R}$, such as

the radius at the beginning of the time interval $R(t)$,
the radius at the end of the time interval $R(t+h)$,
the mean of the radius at the beginning and end $\frac{1}{2}\left[R(t+h) + R(t)\right]$, $\qquad (1.24)$
the average value of the radius from t to $t+h$, namely, $\frac{1}{h}\left(\int_t^{t+h} R(u)du\right)$.

However, from (1.23) it is obvious that we are interested in what happens to $\mathcal{R}$ as $h \to 0$. It is easy to see that all of the possibilities for $\mathcal{R}$ suggested in (1.24) have the common property that
$$\lim_{h \to 0} \mathcal{R} = R(t). \qquad (1.25)$$

From (1.23) and (1.25) we thus find
$$\frac{dA}{dt} = \frac{v}{R(t)}. \qquad (1.26)$$

From (1.21) and (1.26) we have the differential equation
$$\frac{dA}{dt} = \frac{v}{R(0)\sqrt{bt+1}}, \qquad (1.27)$$
subject to (1.22). The solution of (1.27) subject to (1.22) is
$$A(t) = \frac{2v}{bR(0)}\left(\sqrt{bt+1} - 1\right),$$
which relates the angle $A(t)$ to the time t.

If we assume that the number of revolutions (the number on the tape-counter) is proportional to the angle $A(t)$ —that is, $n(t) = kA(t)$ —we find
$$n(t) = \frac{2vk}{bR(0)}\left(\sqrt{bt+1} - 1\right). \qquad (1.28)$$

If we introduce the constant a by
$$a = \frac{2vk}{bR(0)},$$
(1.28) can be rewritten in the final form
$$n(t) = a\left(\sqrt{bt+1} - 1\right), \qquad (1.29)$$
where a and b are constants depending on the tape and the machine used. This is the relation for which we have been looking.

The question now is whether this equation fits the experimental data in Table 1.11. One way to test this is to rewrite (1.29) by solving it for t, to find
$$t = \frac{1}{ba^2}n^2 + \frac{2}{ba}n.$$
This equation implies that t is a quadratic function of n. If we rewrite it (for $n \neq 0$) as
$$\frac{t}{n} = \frac{1}{ba^2}n + \frac{2}{ba} \qquad (1.30)$$
and plot t/n along the vertical axis and n along the horizontal axis, the result will be a straight line with slope $1/ba^2$ and vertical intercept $2/(ba)$. We can then see whether the data set (Table 1.11) is consistent with (1.29). Figure 1.23 shows the data set plotted in this manner along with a straight line with slope 0.000107 and t/n-intercept 0.626179, which, from (1.30), suggests that $a = 2926$ and $b = 0.00109$. Figure 1.24 shows the original data set (Table 1.11) plotted with the function (1.29), where $a = 2926$ and $b = 0.00109$. □

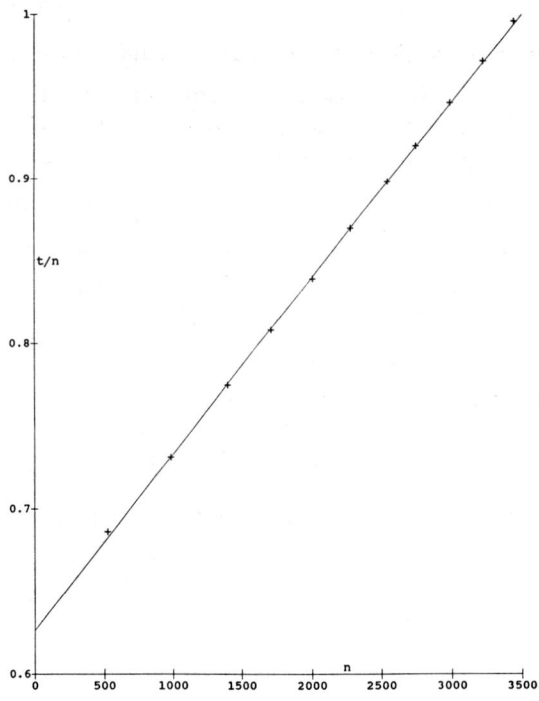

Figure 1.23 t/n versus n

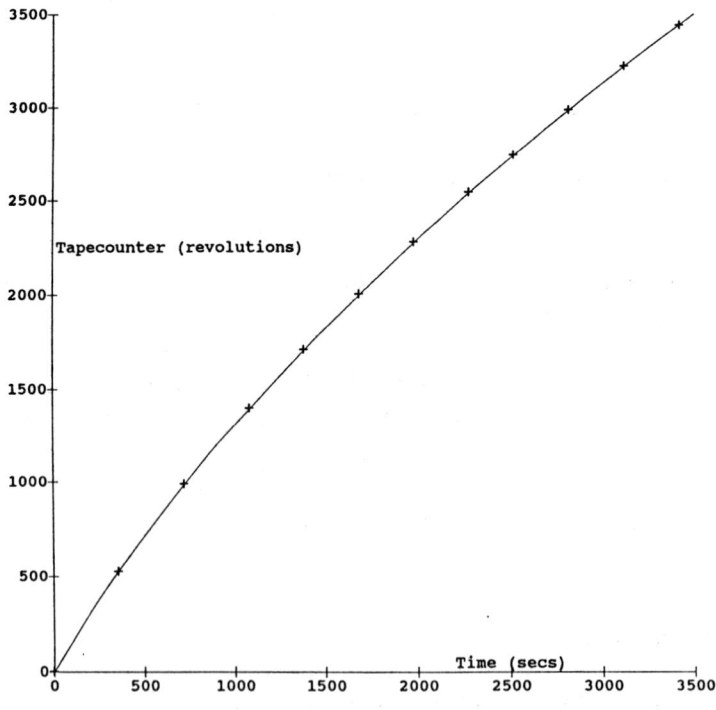

Figure 1.24 Tapecounter versus time and $n(t) = a[(bt + 1)^{1/2} - 1]$

IDEAS FOR CHAPTER 1 — "BASIC CONCEPTS"

Exercises

1. The experiment described in Example 1.3 used a 120-minute tape. I have used the tape to record some programs, and the present tape-counter reading is 4000. What is the longest program I can still record, if initially the tape-counter was set to 0?

2. The analysis of the tape deck described in Example 1.3 assumed that the digital tape-counter measured the rotation of the take-up reel. However, some tape decks are designed so that the tape-counter measures the rotation of the feeder reel. How does this change the analysis and graphs presented in Example 1.3?

3. Perform the experiment described in Example 1.3. Use either a VCR or a tape deck equipped with a digital tape-counter. Put a tape in the unit, set the tape-counter to zero, and, as you start the tape playing, note the time. Then record the time every 5 to 10 minutes along with the number on the tape-counter. Compare the data set obtained in this way with (1.29).

4. A student[17] conducted the Tape Deck experiment and found the data in Table 1.12, shown graphically in Figure 1.25. Does this data set fit the model described previously? [Table 1.12 is in TWIDDLE. It is called DENECKE.DTA and is in the subdirectory DD-01.]

Table 1.12 Tapecounter versus time

Time (secs)	Counter
0	0
600	113
1200	238
1800	383
2100	465
2400	556
2700	662

---

[17]Paul Denecke, University of Arizona.

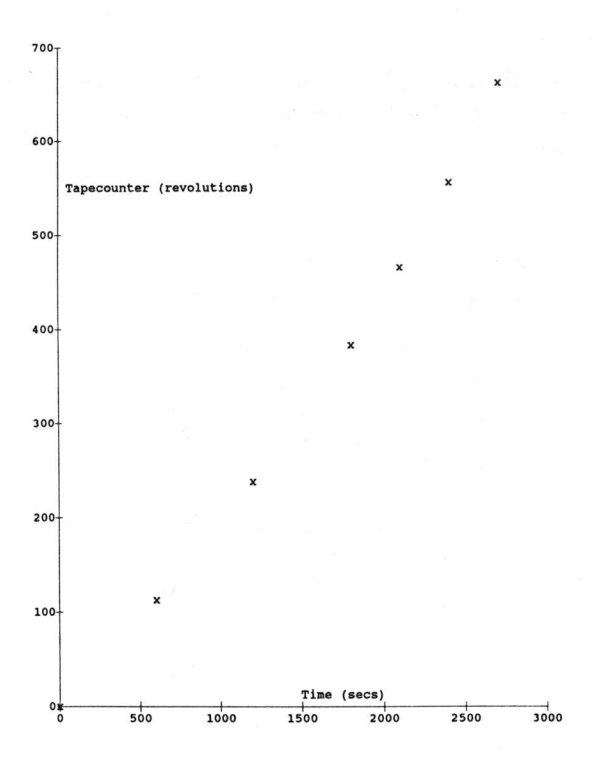

Figure 1.25 Tapecounter versus time

Showing that $erf(x) \to 1$ as $x \to \infty$

Students sometime ask how we know that $erf(x) \to 1$ as $x \to \infty$. Here is one way to show this, assuming a knowledge of double integrals.[18]

Define
$$I(R) = \int_0^R e^{-x^2} dx,$$

so that

$$I^2(R) = \left(\int_0^R e^{-x^2} dx\right)\left(\int_0^R e^{-x^2} dx\right) = \left(\int_0^R e^{-x^2} dx\right)\left(\int_0^R e^{-y^2} dy\right),$$

or

$$I^2(R) = \int_0^R \int_0^R e^{-(x^2+y^2)} dx dy = \iint_S e^{-(x^2+y^2)} dx dy,$$

where S is the square of side length R in the first quadrant, with sides along the axes, and one corner at the origin. See Figure 1.26.

Now, in the first quadrant, construct two quarter circles, S_1 and S_2, both with center the origin, the first having radius R, and the second radius $\sqrt{2}R$. See Figure 1.26.

---

[18] A sloppier version of this argument is frequently given in textbooks on multivariable calculus. This tighter argument was shown to one of the authors (Lovelock) by David Burgess in 1962, when they were faculty members in the Department of Mathematics at the University of Bristol, England.

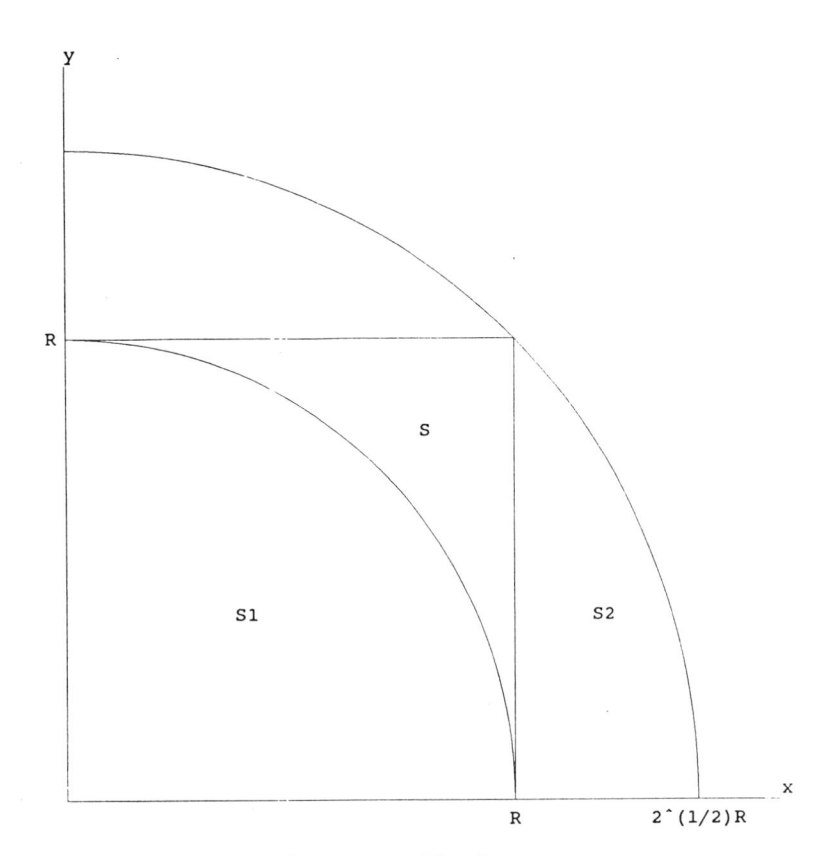

Figure 1.26 The three areas

Because the area inside S_1 is less than the area inside S, and the area inside S is less than the area inside S_2, and because $e^{-(x^2+y^2)} > 0$, we have

$$\iint_{S_1} e^{-(x^2+y^2)}\,dxdy < \iint_{S} e^{-(x^2+y^2)}\,dxdy = I^2(R) < \iint_{S_2} e^{-(x^2+y^2)}\,dxdy.$$

But

$$\iint_{S_1} e^{-(x^2+y^2)}\,dxdy = \int_0^{\pi/2} \int_0^R re^{-r^2}\,drd\theta = \frac{\pi}{4}\left(1 - e^{-R^2}\right),$$

and

$$\iint_{S_2} e^{-(x^2+y^2)}\,dxdy = \int_0^{\pi/2} \int_0^{\sqrt{2}R} re^{-r^2}\,drd\theta = \frac{\pi}{4}\left(1 - e^{-2R^2}\right),$$

so

$$\frac{\pi}{4}\left(1 - e^{-R^2}\right) < I^2(R) < \frac{\pi}{4}\left(1 - e^{-2R^2}\right).$$

Thus, $\lim_{R\to\infty} I^2(R) = \pi/4$, so $\lim_{x\to\infty} I(R) = \sqrt{\pi}/2$, that is,

$$\int_0^\infty e^{-x^2}\,dx = \frac{\sqrt{\pi}}{2}.$$

ADDITIONAL MATERIALS

2. IDEAS FOR CHAPTER 2 — "AUTONOMOUS DIFFERENTIAL EQUATIONS"

2.1 Page by Page Comments

Chapter 2 requires 4 or 5 class meetings of 50 minutes each.

Section 2.6, *Bifurcation Diagrams*, is not used in later sections, and so may be selectively omitted.

Page 27. Autonomous means 'independent of time'.

Page 27. The chain rule is used several times in this chapter. Be sure the students know how to use it.

Page 32. The steps between (2.12) and (2.13) are used repeatedly later in the text, without dwelling on them. This is the place to dwell on them.

Page 32. We give the students copies of Figures 2.4 and 2.5, and have them draw something similar to Figures 2.6 and 2.7.

Page 33. The way in which the equilibrium solution (2.14) can be absorbed into (2.13) producing (2.15) is used repeatedly later in the text, without dwelling on it. This is the place to dwell on it.

Page 34. Equilibrium solutions play a major role throughout this text.

Page 35. It is possible to demonstrate the difference between a stable and an unstable solution as follows. Construct a pendulum with a magnet as the mass, with its south side down. Place two other magnets below the mass. One should be south side up and directly below the mass when at rest. The other should be north side up. There are two stationary positions for the mass. The first is when the mass is put over north-side-up magnet. This is a stable solution and it is easy to have the mass remain in this position. Any slight disturbance from this solution remains close to this solution. The second is when the mass is placed on the line of the two magnets near the south-side-up magnet on the other side from the north-side-up magnet. This is an unstable solution and it is very difficult to have the mass remain in this position. Any slight disturbance from this solution moves away this solution.

Page 35. Exercises 1(c), (e), and (f), in Section 2.1, are worth assigning, because we do them in depth in Section 2.4, and the results are surprising.

Page 35. Exercise 5 in Section 2.1 is worth assigning. After assigning it, we hand out copies of Figures 2.11 and 2.12, place them on top of each other, and hold them up to the light.

Page 36. Exercise 10. Many students find it difficult to think about two parameters. If you assign Exercise 10 you may find it worthwhile to do part (a) in class, leaving parts (b) and (c) for homework.

Page 37. Example 2.3, the Population of Botswana, will be used in Chapter 5 to motivate linear differential equations. Be careful if you leave it out. You could replace it by THE POPULATION OF THE WORLD (1950–1990) on page 53 of this manual.

Page 37. Table 2.1 is in TWIDDLE adjusted so $x = 0$ corresponds to the year 1970. It is called BOTSWANA.DTA and is in the subdirectory DD-02.

Page 38. There are many ways to estimate the parameters in an equation when trying to fit data. Throughout this text we usually use the simplest technique, but recognize that this may not always be as accurate as more sophisticated techniques. See AN INTRODUCTION TO CURVE/DATA FITTING on page 56 of this manual.

Page 38. If we use nonlinear regression to find the parameters C and k that best fit the Population of Botswana (starting from the ones we found, namely, $C = 0.755$ and $k = 0.0355$, and using the program Kalkulator) we find $C = 0.7551$ and $k = 0.03547$. The graph with these revised values of C and k does not differ appreciably from Figure 2.15 of the text.

Page 39. Table 2.2 is in TWIDDLE. It is called DRUGS.DTA and is in the subdirectory DD-02.

Page 39. If we use nonlinear regression to find the parameters C and k that best fit the theophylline data set (starting from the ones we found, namely, $C = 12$ and $k = -0.167$, and using the program Kalkulator) we find $C = 11.8869$ and $k = -0.17045$. The graph with these revised values of C and k does not differ appreciably from Figure 2.18 of the text.

Page 39. Example 2.4, Administering Drugs, could be replaced by THE HALF-LIFE OF ^{64}CU on page 66 of this manual. This is Exercise 8 on page 41 of the text.

Page 40. Exercises 1 and 2 in Section 2.2 introduce students to terminology that they might find in other courses.

Page 40. Many students find Exercises 6 and 7 in Section 2.2 interesting.

Page 41. Table 2.3 is in TWIDDLE. It is called HOUSTON.DTA and is in the subdirectory DD-02.

Page 41. Table 2.4 is in TWIDDLE. It is called HALFLIFE.DTA and is in the subdirectory DD-02.

Page 41. Exercise 8 in Section 2.2 is interesting. Most half-lives are either milliseconds or decades, making them difficult to measure. This one is measurable. It is worked out in detail in THE HALF-LIFE OF ^{64}CU on page 66 of this manual.

Page 41. Many students find Exercise 9 in Section 2.2 interesting, but difficult.

Page 42. We recommend that you assign Exercise 11 in Section 2.2.

Page 42. Table 2.5 is in TWIDDLE. It is called DUCKWEED.DTA and is in the subdirectory DD-02.

Page 42. There are additional exercises for Section 2.2 on page 41 of this manual.

Page 42. The logistic equation occurs repeatedly later in the text. Do not omit it.

Page 42. Example 2.5, Growth of Sunflower Plants, could be replaced by THE BOMBAY PLAGUE on page 69 of this manual. This is Exercise 17 on page 52.

Page 42. Table 2.6 is in TWIDDLE. It is called FLOWER.DTA and is in the subdirectory DD-02.

Page 43. The use of the word "logistic" in this context is due to P.F. Verhulst. It was believed that this was the true law of population growth and thereby indicated the limits above which the population would not grow.[1]

Page 46. Some students need reminding about partial fractions. See Appendix A.2.

Page 48. The manner in which we estimated the parameters in the exponential case (see text page 38) and the logistic case (see text page 48) are similar. This is a general property of differential equations of the type

$$\frac{dy}{dx} = ag(y),$$

because the general (nonequilibrium) solution of this equation is

$$\int \frac{1}{g(y)} \, dy = ax + C.$$

By plotting the left-hand side against x we should see a straight line with slope a and intercept C.

Page 48. If we use nonlinear regression to find the parameters a, b and C that best fit the Population of Botswana (starting from the ones we found, namely, $ab = -0.087$, $bC = 12.94$ and $C = 0.0498$, and using the program Kalkulator) we find $ab = -0.0877$, $bC = 12.9184$ and $C = 0.04949$. The graph with these revised values of a, b and C does not differ appreciably from Figure 2.29 of the text.

Page 49. Students enjoy Exercise 1 in Section 2.3.

Page 49. Exercise 4 in Section 2.3 shows that the solution of the logistic equation is symmetric about the point of inflection.

Page 49. Exercise 7 in Section 2.3 is a good example to see whether students remember to check for equilibrium solutions.

Page 50. Table 2.7 is in TWIDDLE. It is called PLANT.DTA and is in the subdirectory DD-02.

---

[1] "On Growth and Form" by D. Thompson, Dover, 1992, page 145.

Page 51. Table 2.8 is in TWIDDLE. It is called BACTERIA.DTA and is in the subdirectory DD-02.

Page 51. Table 2.9 is in TWIDDLE. It is called DOVES.DTA and is in the subdirectory DD-02.

Page 52. Table 2.10 is in TWIDDLE. It is called BOMBAY.DTA and is in the subdirectory DD-02.

Page 52. Table 2.11 is in TWIDDLE. It is called BEANS.DTA and is in the subdirectory DD-02.

Page 52. Exercises 18 and 21 in Section 2.3 are worth assigning.

Page 53. Exercises 19 and 20 in Section 2.3 can be considerably enhanced, perhaps during lectures. If covered, then many subsequent calculations can be simplified or checked for consistency. See this manual, page 72, PARAMETERS IN ODEs — RESCALING and page 74, PARAMETERS IN ODEs — UNITS ANALYSIS.

Page 53. There are additional exercises for Section 2.3 on page 42 of this manual.

Page 53. Section 2.4 is the cornerstone of the book. Don't leave it out.

Page 53. You may have assigned Example 2.6 as homework (Exercise 1(c), Section 2.1).

Pages 53 – 55. It needs emphasizing that it is almost impossible to tell from the slope field whether a solution has a vertical asymptote. One way to demonstrate this is to plot the slope fields for $y' = y$ and $y' = y^2$ on the same figure in the window $0 < x < 10, 0 < y < 10$. It is almost impossible to distinguish them apart, and yet the first has no vertical asymptotes and the second has an infinite number.

Page 56. The comments following the theorem should be stressed.

Page 57. You may have assigned Example 2.7 as homework (Exercise 1(e), Section 2.1).

Page 60. You may have assigned Example 2.9 as homework (Exercise 1(f), Section 2.1).

Page 64. We always assign Exercise 3 in Section 2.4.

Page 64. Exercises 5, 8, 9, and 11 are very rich.

Page 65. Table 2.12 is in TWIDDLE. It is called WORLDPOP.DTA and is in the subdirectory DD-02.

Page 65. Exercise 10, the Doomsday Model is worked out in detail under THE DOOMSDAY MODEL on page 75 of this manual.

Page 65. There are additional exercises for Section 2.4 on page 49 of this manual.

Page 65. Section 2.5 is used heavily in the text.

Page 73. Students enjoy Exercise 3 in Section 2.5.

Page 73. Exercise 6. You might point out that it is not necessary to use the explicit solution in part (b) to answer the question regarding the choice of c.

Page 74. Table 2.14 is in TWIDDLE. It is called MILE.DTA and is in the subdirectory DD-02.

Page 75. There are additional exercises for Section 2.5 on page 49 of this manual.

Page 75. Section 2.6, *Bifurcation Diagrams*, is optional. It is not used in the rest of the book.

2.2 Additional Exercises

Section 2.2

Page 42. Add exercises after the current Exercise 11.

12. **The Population Density of Boston.**[2] Table 2.1 and Figure 2.1 show the average population density (in people per square mile) of Boston in 1940 as a function of the distance from the center of the city (in miles). How well does the exponential model fit this data set? [Table 2.1 is in TWIDDLE. It is called BOSTON.DTA and is in the subdirectory DD-02.]

Table 2.1 The population density of Boston in 1940 as a function of the distance from the city center

Distance (miles)	Population (people/mi^2)	Distance (miles)	Population (people/mi^2)
0.5	26300	8.5	3200
1.5	25100	9.5	2300
2.5	19900	10.5	1700
3.5	15500	11.5	1200
4.5	11500	12.5	900
5.5	9800	13.5	700
6.5	5200	14.5	600
7.5	4600	15.5	500

13. **Feeling Exponential Growth — A Simple Experiment.**[3] There is an interesting way to get students to "feel" exponential growth. Divide the class into groups of 4 to 6 students. Give each group a large plastic bucket, a stopwatch, an eyedropper, and a collection of containers and graduated cylinders. The instructions are: Introduce one drop of water into the bucket and then cause the quantity of water in the bucket to double every 10 seconds. Measure the time it takes to fill the bucket.

---

[2] "Urban Population Densities" by C. Clark, Journal of the Royal Statistical Society, 64, 1951, pages 490 through 496, as reported in "Growth and Diffusion Phenomena" by R.T. Banks, Springer-Verlag, 1991, page 386.

[3] "Understanding exponential growth: As simple as a drop in the bucket" by F. Goldberg and J. Shumann, The Physics Teacher, October 1984, pages 444 through 445.

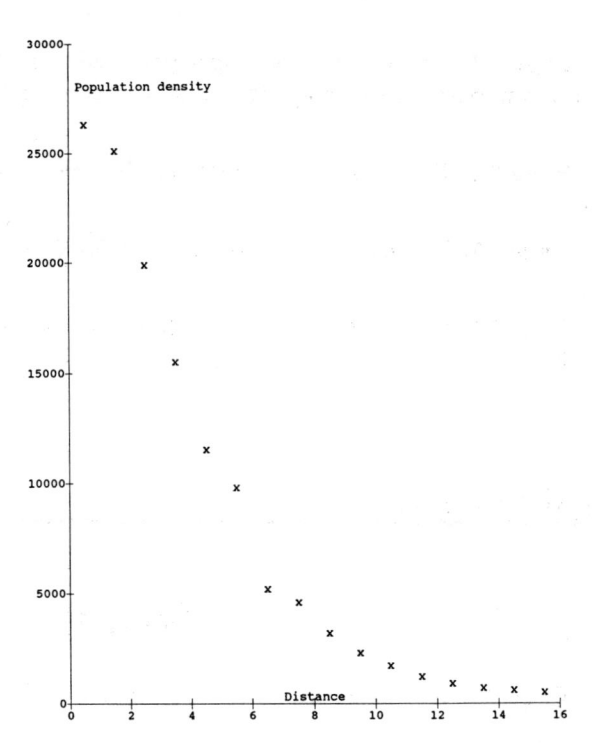

Figure 2.1 The population density of Boston in 1940 as a function of the distance from the city center

14. **Gypsy Moths.**[4] Here are some data on an outbreak of the gypsy moth, which devastated forests in Massachusetts in the US. Rather than count the number of moths, the number of acres defoliated by the moths was counted. Table 2.2 shows the acres defoliated as a function of the year. How well does the exponential model fit this data set? [Table 2.2 is in TWIDDLE. It is called MOTHS.DTA and is in the subdirectory DD-02.]

Table 2.2 The acres defoliated by gypsy moths as a function of the year

Year	Acres
1978	63,042
1979	226,260
1980	907,075
1981	2,826,095

Section 2.3

Page 53. Add exercises after the current Exercise 21.

22. Describe the behavior of the solutions of $y' = a + by$ if

   (a) $a < 0$ and $b < 0$, $b = 0$, $b > 0$.

   (b) $a = 0$ and $b < 0$, $b = 0$, $b > 0$.

   (c) $a > 0$ and $b < 0$, $b = 0$, $b > 0$.

---

[4]This data was supplied by Chuck Schwalbe, U.S. Dept of Agriculture, to the Exploring Data website at http://curriculum.qed.qld.gov.au/kla/eda/.

23. **Chemical Kinetics.** See page 78 of this manual, CHEMICAL KINETICS.

24. **Lupine Growth.**[5] Table 2.3 shows the growth of a Lupine from day 4 to day 21. This data set is plotted in Figure 2.2. How well does the logistic model fit this data set? [Table 2.3 is in TWIDDLE. It is called LUPINE.DTA and is in the subdirectory DD-02.]

Table 2.3   Growth of Lupine

Days	Height (mm)	Days	Height (mm)
4	10.5	13	120.1
5	16.3	14	132.3
6	23.3	15	140.6
7	32.5	16	149.7
8	42.2	17	155.6
9	58.7	18	158.1
10	77.9	19	160.6
11	93.7	20	161.4
12	107.4	21	161.6

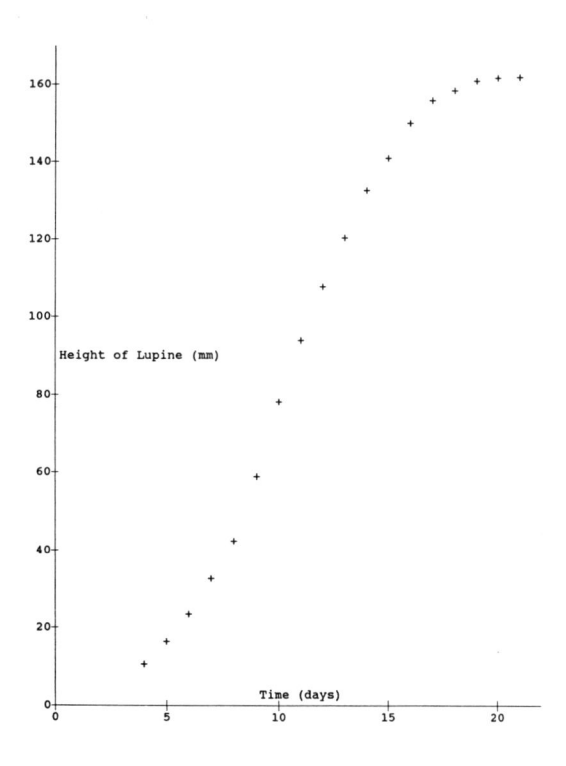

Figure 2.2   Growth of Lupine

25. **Married Australian Men.**[6] The age at which 41,000 Australian men, born in the same year, were married is shown in Table 2.4. Figure 2.3 shows these data where the vertical axis shows the percentage of the 41,000 men who were married, while the horizontal axis shows the age, measured from 22 years, at which they were married. Thus, the point (4, 55) tells us that 55% of these men were married by the age of 26.

---
[5] "Growth and Form" by D'Arcy W. Thompson, Cambridge, 1942, page 160.
[6] "What is a Growth Cycle?" by S.A. Coutis, Growth, 1, 1937, pages 155 through 174.

It is claimed that the Gompertz equation models this situation very well. Investigate this claim. [Table 2.4 is in TWIDDLE. It is called MARRIED.DTA and is in the subdirectory DD-02.]

Table 2.4 Percentage of married men as a function of age at marriage

Age	Percentage
16	0.01
17	0.11
18	0.67
19	1.85
20	4.34
21	10.57
22	17.73
23	26.20
24	36.01
25	45.37
26	54.76
27	63.26
28	70.42
29	76.44
30	80.99
31	84.19
32	87.66

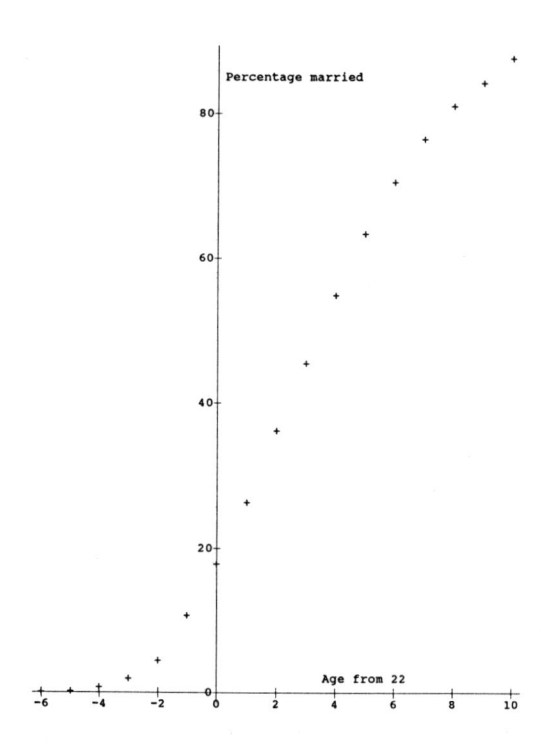

Figure 2.3 The percentage of married male Australians as a function of marrying age

26. **Water Bugs**. Experiments with water bugs suggest a model of population growth of a species that is limited by the food supply. If $P(t)$ represents the population of water bugs at time t, the basic equation says the rate of change of $P(t)$ is propor-

tional to the difference between the available food, $A(t)$, and the food necessary for subsistence, $N(t)$, as well as being proportional to the current population — that is, $\frac{dP}{dt} = k(A(t) - N(t))P(t)$. It is assumed that the available food is a constant, $A(t) = a$, and that the food necessary for subsistence, $N(t)$, is equal to the sum of two terms. The first term gives the need for food as proportional to the existing population (bP), whereas the second term accounts for the need of more food during periods of rapid growth ($r\,dP/dt$). Under these assumptions the differential equation becomes $\frac{dP}{dt} = k\left(a - bP - r\frac{dP}{dt}\right)P$, where k, a, b, and r are positive constants. (If $r = 0$ we have the standard logistic equation.) We may rearrange this equation as

$$\frac{dP}{dt} = k\frac{aP - bP^2}{1 + krP}.$$

What can you tell about the behavior of the solution subject to an initial condition $P(0) = P_0$? (The slope field might be of some help here.) Does this behavior depend on the value of P_0? Please explain fully.

27. **AIDS in the USA.**[7] Table 2.5 shows the total number of cases of AIDS in the USA in 4 month intervals from 1982. This data set is plotted in Figure 2.4. How well does the logistic model fit this data set? [Table 2.5 is in TWIDDLE. It is called AIDS.DTA and is in the subdirectory DD-02.]

Table 2.5 Total number of cases of AIDS in the USA from 1982

Quarter since 1982	Number	Quarter since 1982	Number
0	374	13	12260
1	559	14	14785
2	759	15	17736
3	1052	16	20896
4	1426	17	24715
5	1980	18	29036
6	2693	19	33899
7	3456	20	39091
8	4313	21	45246
9	5460	22	52062
10	6829	23	59553
11	8392	24	67279
12	10118	25	75762

28. **Improved Pastures in Uruguay.**[8] In 1961 fertilized grass-legume pastures were introduced into Uruguay. Table 2.6 shows the total number of ranchers who adopted this new technology from 1965 to 1976. This data set is plotted in Figure 2.5. How well does the logistic model fit this data set? [Table 2.6 is in TWIDDLE. It is called URUGUAY.DTA and is in the subdirectory DD-02.]

29. **The Logistic Equation.** Show that the solution of the logistic equation $y' = ay(b-y)$ can be expressed in terms of $\tanh(x)$ by proving the identity

$$\frac{bC}{e^{-abx} + C} = \frac{b}{2}\left[\tanh\left(\frac{1}{2}abx + \frac{1}{2}\ln C\right) + 1\right].$$

---

[7] "Backcalculation of flexible linear models of the Human Immunodeficiency Virus infection" by P. S. Rosenburg and M. H. Gail, Applied Statistics, **40**, 1991, pages 269 through 282, Table 1, as reported in "Growth and Diffusion Phenomena" by R. T. Banks, Springer-Verlag, 1991, page 372.

[8] "Predicting the Diffusion of Improved Pastures in Uruguay" by L. S. Jarvis, American Journal of Agricultural Economics, **63**, 1981, pages 495 through 502.

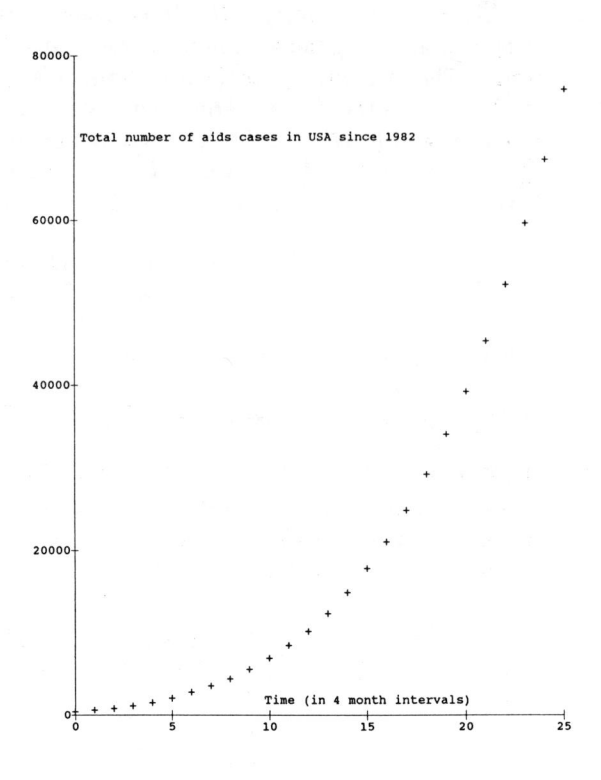

Figure 2.4 The total number of cases of AIDS in the USA from 1982

Table 2.6 Total number of ranchers in Uruguay adopting new technology

Year since 1965	Total Adopting
0	944
1	1455
2	2060
3	3247
4	5284
5	7999
6	9554
7	11465
8	13767
9	14678
10	14998
11	15473

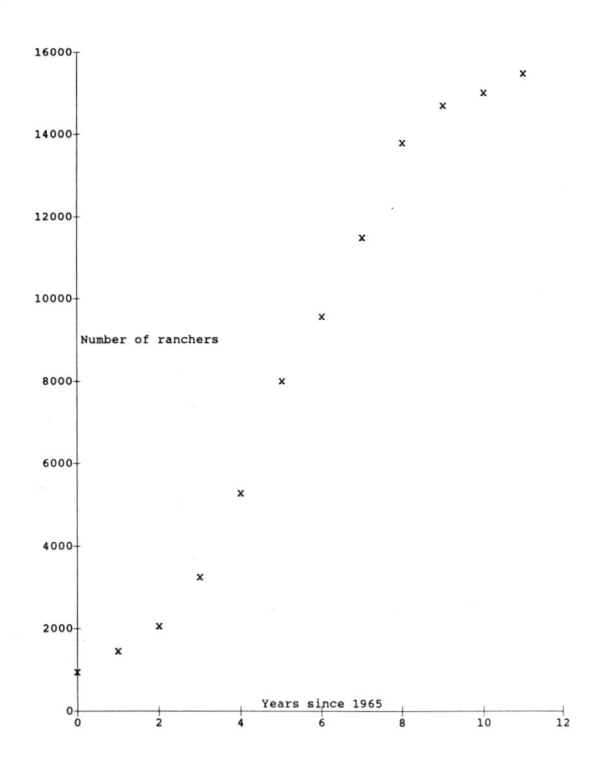

Figure 2.5 The total number of ranchers in Uruguay adopting new technology from 1965

30. A differential equation is symmetric about the point (a, b) if rotating the slope field 180 degrees about this point gives the original slope field. Show that if an autonomous equation is symmetric about the point (a, b) it is also symmetric about $(a + c, b)$ for every c.

31. **Rats.**[9] In this experiment 144 seven-month-old rats were observed over a period of 30 months. Table 2.7 shows the cumulative number that died every month. How well does the logistic model fit this data set? [Table 2.7 is in TWIDDLE. It is called RATS.DTA and is in the subdirectory DD-02.]

Table 2.7 The cumulative number of rats that died each month

Month	Number	Month	Number	Month	Number
1	1	11	30	21	114
2	1	12	36	22	115
3	3	13	47	23	122
4	5	14	54	24	125
5	6	15	64	25	133
6	7	16	74	26	136
7	8	17	89	27	139
8	13	18	94	28	141
9	22	19	96	29	141
10	26	20	108	30	142

32. **Guinea-Pig.**[10] Table 2.8 shows the pre-natal growth of a Guinea-Pig. The length in inches is given as a function of the time in hours. How well does the logistic model

---

[9] "Growth and Diffusion Phenomena" by R.B. Banks, Springer-Verlag, 1994, page 222.

[10] Data obtained by R.L. Draper, as reported in "On Growth and Form" by D'Arcy Thomas, Dover, 1992, page 115.

ADDITIONAL EXERCISES

fit this data set? [Table 2.8 is in TWIDDLE. It is called GUINEA.DTA and is in the subdirectory DD-02.]

Table 2.8   The length (inches) of a pre-natal guinea pig as a function of the time (hours)

Hour	Length
6	0.3
16	1.7
42	12.6
54	15.4
65	16.1
77	16.7
88	17.1

33. **Maize.**[11] Table 2.9 shows the weight of an ear of maize in grams as a function of time in days. How well does the logistic model fit this data set? [Table 2.9 is in TWIDDLE. It is called MAIZE.DTA and is in the subdirectory DD-02.]

Table 2.9   The weight of an ear of maize in grams as a function of time in days

Time	Weight
6	1
18	4
30	9
39	17
46	26
53	42
60	62
74	71
93	74

34. **Infant.**[12] Table 2.10 shows the mean length in mms from month to month of an unborn infant. How well does the logistic model fit this data set? [Table 2.10 is in TWIDDLE. It is called PRENATAL.DTA and is in the subdirectory DD-02.]

Table 2.10   The mean length in mms from month to month of an unborn child

Month	Length
1	7.5
2	40.0
3	84.0
4	162.0
5	275.0
6	352.0
7	402.0
8	443.0
9	472.0
10	500.0

35. **Bean Root.**[13] Table 2.11 shows the length of the root of a bean in mms as a function

---

[11]Data obtained by G. Backman, as reported in "On Growth and Form" by D'Arcy Thomas, Dover, 1992, page 115.

[12]Data obtained by W. His, as reported in "On Growth and Form" by D'Arcy Thomas, Dover, 1992, page 111.

[13]Data obtained by J. Sachs, as reported in "On Growth and Form" by D'Arcy Thomas, Dover, 1992, page 115.

of time in days. How well does the logistic model fit this data set? [Table 2.11 is in TWIDDLE. It is called BEANROOT.DTA and is in the subdirectory DD-02.]

Table 2.11 The length of the root of a bean in mms as a function of time in days

Day	Length
0	1.0
1	2.8
2	6.5
3	24.0
4	40.5
5	57.5
6	72.0
7	79.0
8	79.0

Section 2.4

Page 65. Add exercise after the current Exercise 11.

12. This exercise deals with the two differential equations

$$y' = p(x)y - 1 \tag{2.1}$$

$$y' = [p(x)y - 1]y \tag{2.2}$$

where $p(x)$ is a given continuous function of x.

(a) Find all equilibrium solutions (if any) of (2.1) and sketch them on Figure 2.6.

(b) Find all equilibrium solutions (if any) of (2.2) and sketch them on Figure 2.6.

(c) The already printed graph in Figure 2.6 shows a solution of one of the differential equations (2.1) or (2.2). Which one? Explain how you came to your conclusion. Do not attempt to find $p(x)$. [Hint: The results from parts (a) and (b) may be useful.]

Section 2.5

Page 75. Add exercises after the current Exercise 13.

14. **The Generalized Logistic Equation.** See page 82 of this manual, THE GENERALIZED LOGISTIC EQUATION.

15. **The Michaelis-Menten Rate-Law.**[14] The Michaelis-Menten rate-law is a standard way of modeling enzyme-catalyzed reactions. We use it here to model the hydrolyzing of Sucrose by the catalytic action of the enzyme Sucrase. With this rate-law the differential equation for the consumption of Sucrose is

$$\frac{dy}{dx} = -\frac{ay}{y+b}$$

where $y(x)$ is the concentration of Sucrose (in mmol/l) at time x, and a and b are positive constants.

(a) Use a phase line analysis to predict the behavior of y as $x \to \infty$. Is this reasonable?

---

[14] We would like to thank Joel Statkevicus, University of Arizona, for this information.

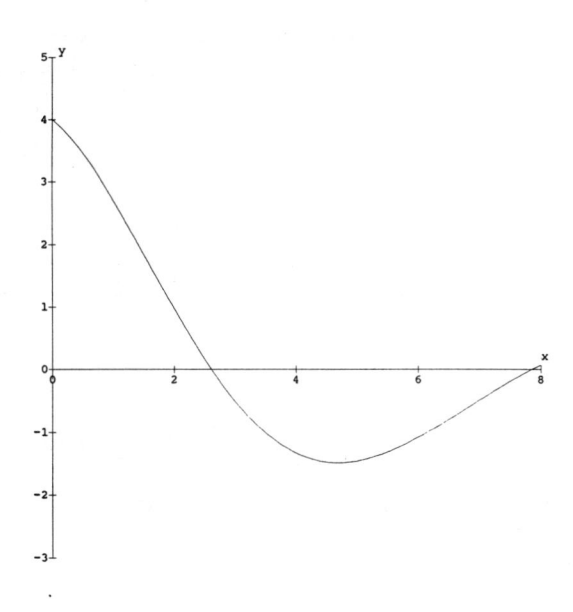

Figure 2.6 The mystery function

(b) Plot the slope field for this differential equation for $y \geq 0$. What does it suggest about the monotonicity and concavity of solutions? Confirm your observations directly from the differential equation, by using calculus.

(c) Solve the differential equation to show that

$$y + b \ln y = -ax + c,$$

where c is a constant. Is this solution consistent with your prediction from part (a)?

(d) Why is it easier to plot x as a function of y than to plot y as a function of x?

(e) Table 2.12 shows the concentration (in mmol/l) of Sucrose over an 11 hour period.[15] This data set is plotted in Figure 2.7. Use the initial condition $y(0) = 1$ to evaluate the constant c in part (c). Use the values of y at $x = 6$ and $x = 11$ to estimate the values of the constants a and b in part (c). With these values of a, b, and c, how well does the Michaelis-Menten model fit this data set? [Table 2.12 is in TWIDDLE. It is called SUCROSE.DTA and is in the subdirectory DD-02.]

16. **The Gompertz Equation.** See page 84 of this manual, THE GOMPERTZ EQUATION.

17. **The Sterile Male Technique of Insect Control.**[16] The sterile male technique for the suppression of insect pest populations has been used to control the screwworm fly. A specific number of sterile male flies, s, were released each week, and then the number of egg masses were counted periodically. It was found that if s was small the number

---

[15] "Physical Chemistry" by P.W. Atkins, W.H. Freeman, 1982, 2nd edition.

[16] "Mathematical Models of the Sterile Male Technique of Insect Control" by W. G. Costello and H. M. Taylor in "Mathematical Analysis of Decision Problems in Ecology" edited by A. Charnes and W. R. Lynn, Springer-Verlag, 1975, pages 318 through 359.

Table 2.12 Concentration of Sucrose as a function of time

Time (hours)	Concentration (mmol/l)	Time (hours)	Concentration (mmol/l)
0	1.0000	6	0.1600
1	0.8400	7	0.0900
2	0.6800	8	0.0400
3	0.5300	9	0.0180
4	0.3800	10	0.0060
5	0.2700	11	0.0025

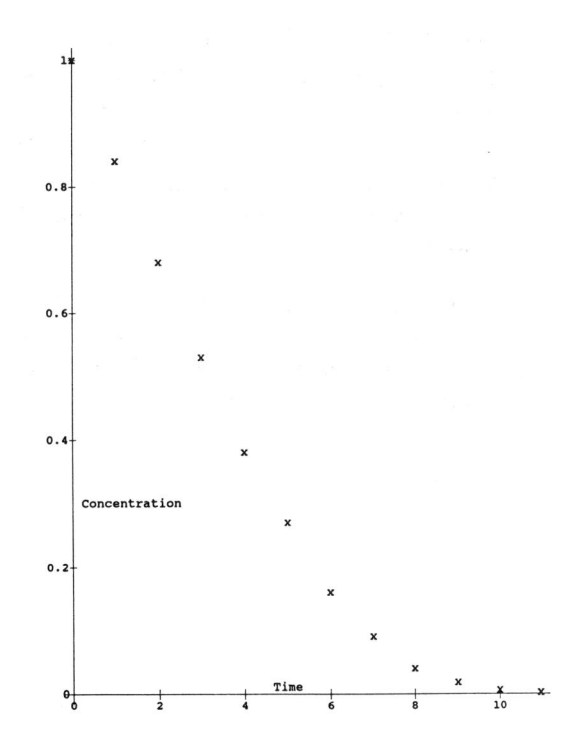

Figure 2.7 Concentration of Sucrose as a function of time

ADDITIONAL EXERCISES

of egg masses grew indicating an increase in the population of fertile flies despite the presence of sterilized flies. However, if s was large the number of eggs found decreased, and in fact was zero after 9 weeks, indicating the population of flies had died out. The differential equation

$$\frac{dy}{dx} = y\left(b\frac{y}{y+s} - d\right) \qquad (2.3)$$

has been suggested as a model for this situation. Here $y(x)$ is the number of fertile male flies at time x. The constants b and d are the natural birth and death rates of male flies where $b > d$. The constant s is the number of sterile male flies that are released.

(a) Explain what happens to the fertile male fly population if $s = 0$.

(b) Based on your answer to part (a) and the observation that $y+s$ is the total number of male flies, so that $y/(y+s)$ is the proportion of male flies capable of producing offspring, motivate the differential equation (2.3).

(c) By doing a phase line analysis, sketch typical solutions of (2.3).

(d) For what values of s will $y(x) \to \infty$ as $x \to \infty$.

(e) For what values of s will $y(x) \to 0$ as $x \to \infty$.

(f) Is this consistent with the experimental observations mentioned previously?

18. Many classical growth laws are modeled by $y' = ay - h(y)$, where $h(y)$ is a nonnegative increasing function that is either concave up or concave down.[17]

    (a) Show that this differential equation can have at most two equilibrium points.

    (b) Give specific formulas for $h(y)$ such that this differential equation has (i) no equilibrium points, (ii) one equilibrium point, (iii) two equilibrium points. For parts (ii) and (iii) determine their stability.

    (c) What condition on $h(y)$ is necessary for Theorem 2.1 to apply to this differential equation? For this condition, show that any solution can have at most one inflection point.

19. The logistic equation has an unstable equilibrium at $y = 0$ and stable equilibrium at $y = b$, where $b > 0$. Its only points of inflection lie along $y = b/2$. Write down another autonomous differential equation with these properties. Confirm your answer by using slope fields and numerical solutions.

---

[17] This exercise is based on "A General Solution for the Complete Richards Function" by R.I. Fletcher, Mathematical Biosciences, Volume 27, 1975, pages 349–360.

2.3 Additional Materials

The Population of the World (1950–1990)

The population of the world[18] from 1950 to 1990 is shown in Table 2.13 and Figure 2.8. [Table 2.13 is in TWIDDLE. It is called WORLDPOP.DTA and is in the subdirectory DD-02.] It is claimed that this data set is consistent with the model where the population growth per unit population is equal to a constant. Let y be the population in millions at time x in years. Because the population growth per unit population is the quantity $(1/y)dy/dx$, we have the differential equation studied before, namely,

$$\frac{1}{y}\frac{dy}{dx} = k,$$

where k is a positive constant (the growth rate). We know the solution of this differential equation is

$$y(x) = Ce^{kx}, \tag{2.4}$$

where C is a constant and x is the year.

Table 2.13 Population of the World

Year	Population (billions)
1950	2.565
1960	3.050
1970	3.721
1980	4.476
1990	5.320

How do we estimate values for C and k from the data set? If the data set is approximated by a line then we could fit the data set by eye and estimate the slope and intercept. Unfortunately these data are not linear. However, if we take the logarithm of (2.4) we find

$$\ln y(x) = \ln C + kx.$$

Consequently, if we plot $\ln y(x)$ versus x, we would have a straight line[19] with slope k and y-intercept $\ln C$, from which we can estimate both C and k. This is done in Figure 2.9, from which we estimate $\ln C \approx -34.99$ (so $C \approx e^{-34.99} \approx 6.368 \times 10^{-16}$) and $k \approx 0.018426$.

Figure 2.10 shows the population data as well as the function

$$y(x) = e^{-34.99}e^{0.018426x}, \tag{2.5}$$

which seem to be in good agreement.

Let's see where this model leads. It predicts that the population of the world is doubling about every 37.6 years (see Exercise 1 on page 55) and that by the year 2743 the population density will be 1 person per square foot. 37.6 years later it will be 2 persons per square foot. Although this model seems reasonable for 1950–1990, it is unlikely to be valid forever. □

---

[18] "The Universal Almanac 1990" edited by J.W. Wright, Andrews and McMeel, Kansas City, 1989.

[19] Throughout the text we will be fitting straight lines through data sets. This can be done by eye, or by using the least squares method described in the Appendix. We will consistently use the latter without further reference to it.

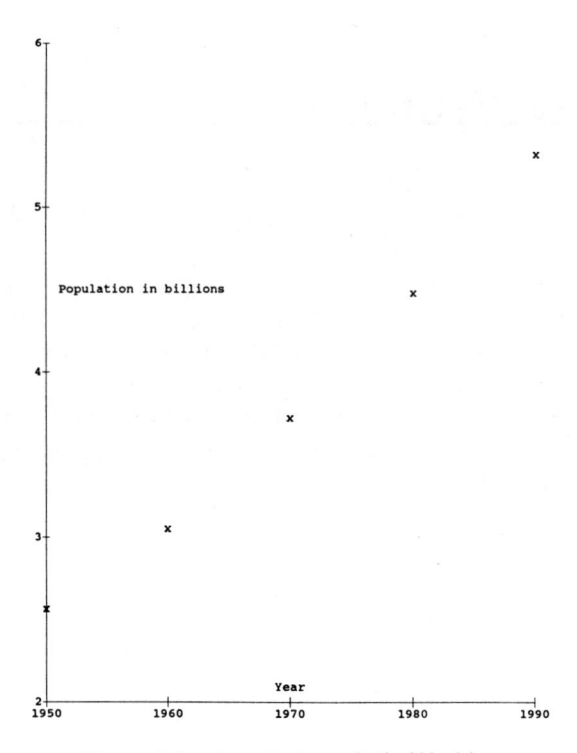

Figure 2.8 Population of the World

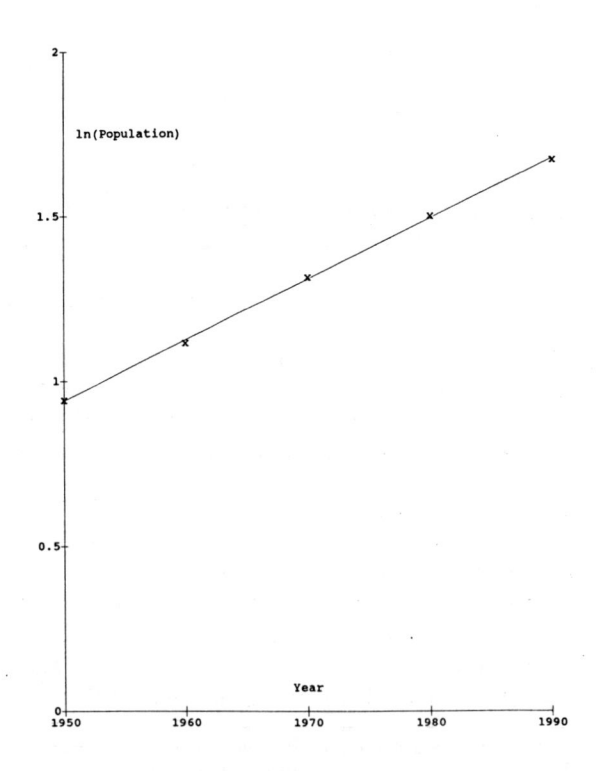

Figure 2.9 Logarithm of population of world versus time

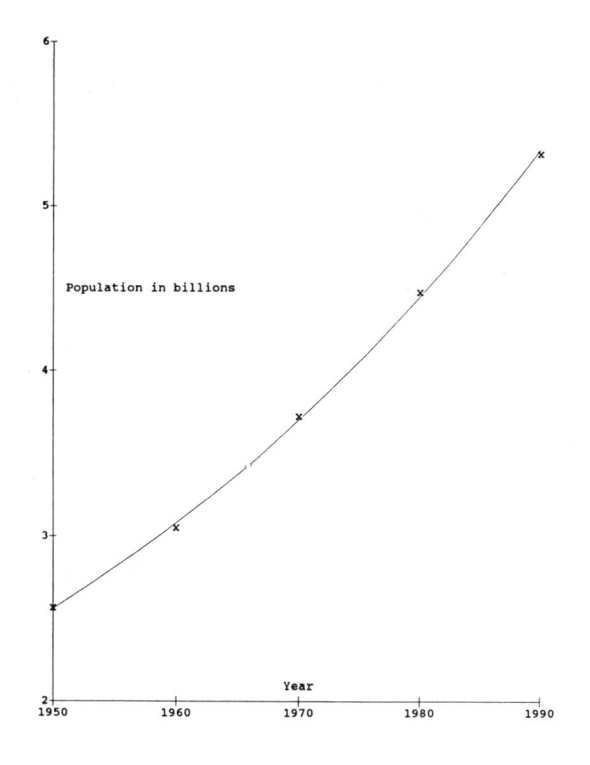

Figure 2.10   Actual and theoretical population of the world

Exercises

1. Show that according to (2.5) the world population is doubling every 37.6 years. The earth is approximately a sphere with radius 4,000 miles. Show that based on the model characterized by (2.5), the time when the population density of the world will be one person per square foot will occur about the year 2743.

An Introduction to Curve/Data Fitting

A spring was suspended vertically from one end. A mass, x, was attached to the other end and the length of the spring, y, was measured. This was repeated for different masses giving rise to Table 2.14, which is also shown in Figure 2.11.[20] It is believed that the length of the spring is related to the mass by Hooke's Law, namely $y = a + bx$, where a and b are parameters. What are the values of a and b that give the best fit?

Table 2.14 The length of a spring as a function of the mass

Mass (kg)	Length (m)	Mass (kg)	Length (m)
0.000	0.100	0.030	0.190
0.005	0.115	0.035	0.204
0.010	0.130	0.040	0.217
0.015	0.145	0.045	0.232
0.020	0.160	0.050	0.246
0.025	0.175		

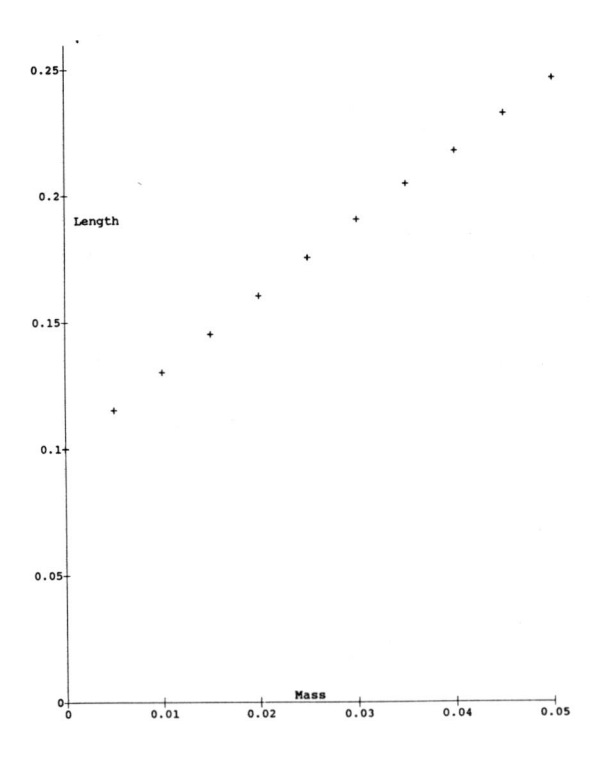

Figure 2.11 The length of a spring as a function of the mass

There are various ways we could do this.

1. First, we recognize that there are two constants to be determined, namely a and b, so we could select two representative data points, say the first $(0.00, 0.100)$ and the last

---

[20]This experiment was conducted by David Harman and Mark Zerella while students at the University of Arizona.

(0.05, 0.246), set up two equations in two unknown,

$$a + 0.00b = 0.100$$
$$a + 0.05b = 0.246,$$

solve for a and b, $a = 0.100$, $b = 2.92$, giving $y = 0.100 + 2.92x$. One problem with this approach is that it ignores all data points except two. Thus, the other data points could have any values, and we would still find the same values for a and b. A second problem is that the two points we have selected may not be representative of the data set.

2. Second, we could plot the data set and fit the set by eye using a ruler. In this way we would draw a straight line which best approximates the data, and then measure the slope and y-intercept of the line, which will give us a and b. One problem with this approach is that any two persons will generally fit the data with two different lines.

3. The third way is to select the line $f(x) = a + bx$ in such a way that it minimizes the error between the line and the data points. However, there is more than one "error" that can be used, two of them being the

$$\text{Average error} = \frac{1}{n}\sum_{i=1}^{n}|f(x_i) - y_i|,$$

and the

$$\text{Root mean square error} = \sqrt{\frac{1}{n}\sum_{i=1}^{n}[f(x_i) - y_i]^2}.$$

The error of choice is the root mean square error, and the line we obtain by choosing a and b that minimizes this error is called the "line of best fit" or "the least-squares line". Minimizing $\sqrt{\frac{1}{n}\sum_{i=1}^{n}[f(x_i) - y_i]^2}$ is equivalent to minimizing $\sum_{i=1}^{n}[f(x_i) - y_i]^2$, which is done in Appendix A.6. There it is shown that the least-squares line passes through the point $(\bar{x}, \bar{y})$, where $\bar{x}$ is the average of the x data points, and $\bar{y}$ is the average of the y data points. The slope of this line is the weighted average of the slopes of the lines from $(\bar{x}, \bar{y})$ to each of the data points. Many calculators and programs perform this calculation. In the case of Table 2.14, we find $a = 0.1010$ and $b = 2.9218$, in good agreement with our first estimate. Figure 2.12 shows the length y of a spring as a function of the mass x, together with the line of best-fit $y = 0.1010 + 2.9218x$.

This is a typical example of curve fitting, or data fitting. We have some experimental results and an equation that contains parameters, and we want to select the parameters in the best possible way. Thus, we are given n data points (x_1, y_1), (x_2, y_2), (x_3, y_3), $\cdots$, (x_n, y_n), and a function of x containing parameters $a, b, \cdots$, namely, $y(x) = f(x, a, b, \cdots)$. We want to select $a, b, \cdots$, to give the best-fit to the data set.

However, most data sets, and the functions that model them, are not linear, so what do we do in this case? For example Table 2.15 and Figure 2.13 show the decay of the ^{64}Cu isotope as a function of time in hours.[21] It is not linear and the standard model for this is exponential decay, namely, $y(x) = ae^{-bx}$ How well does the exponential model fit this data set?

We could mimic the linear case, as follows.

1. First, we recognize that there are two constants to be determined, namely a and b, so we could select two representative data points, say the first $(0, 16744)$ and the last

---

[21] "Data analysis in the undergraduate nuclear laboratory" by B. Curry, D. Riggins, and P. B. Spiegel, *American Journal of Physics* 63, 1995, pages 71–76. Table 1, page 73.

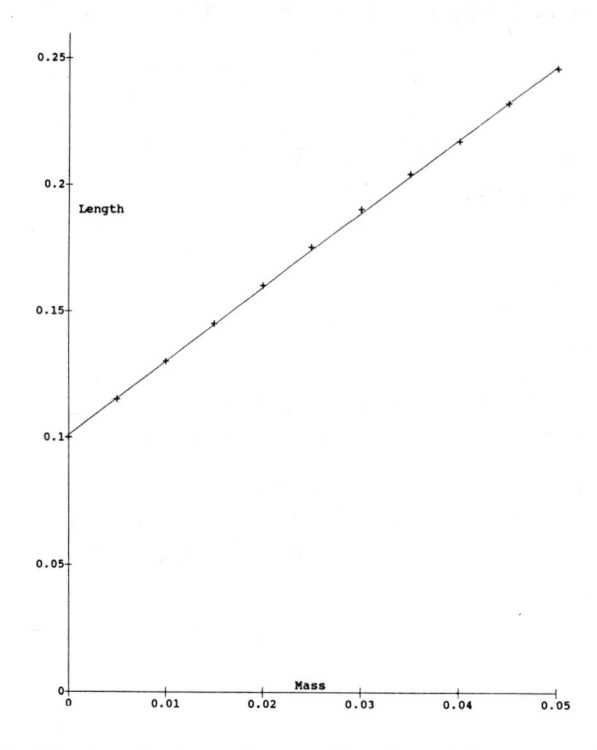

Figure 2.12 The length of a spring as a function of the mass and the line of best-fit

Table 2.15 Data for ^{64}Cu decay

Time	Counts	Time	Counts	Time	Counts
0	16744	7	11277	30	3184
1	15596	8	10949	31	3177
2	15120	9	10174	32	2910
3	14325	25	4317	33	2766
4	13723	26	4181	34	2598
5	12788	27	3784		
6	12141	29	3454		

$(34, 2598)$, set up two equations in two unknown,

$$ae^{-0b} = 16744$$
$$ae^{-34b} = 2598,$$

solve for a and b, $a = 16744$, $b = 0.0548$, giving $y = 16744e^{-0.0548x}$. The half-life is $-\ln(2)/b \approx 12.65$ hours.

2. Second we could try to fit the data set by eye, by experimenting with different values for a and b. This method is not recommended!

3. Third we could try to find a and b in $f(x) = ae^{-bx}$ that minimizes $\sum_{i=1}^{n} [f(x_i) - y_i]^2$. This is called **nonlinear regression**, and can be done in two different ways. First, we could try to perform a calculation similar to that in Appendix A.6. Doing so leads to a nonlinear system of algebraic equations for a and b, which would have to be solved by a numerical technique.[22] Second, we could use a numerical method

---
[22] See "Numerical Methods" by J.H. Mathews, Prentice-Hall, New Jersey, 1987, page 224.

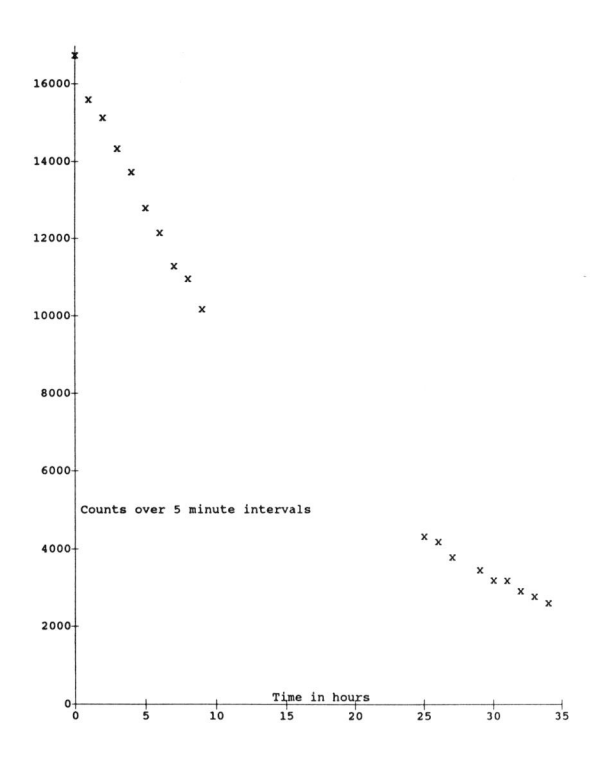

Figure 2.13 The decay of the ^{64}Cu isotope as a function of time in hours

to minimize $\sum_{i=1}^{n}\left[f\left(x_{i}\right)-y_{i}\right]^{2}$ as a function of a and b.[23] A computer program[24] is the method of choice here, giving $a = 16759$ and $b = 0.0544$. The half-life is $-\ln(2)/b \approx 12.74$ hours. In both cases, a good initial guess for a and b is needed.

4. A fourth method — called **data linearization** — is to try to use a transformation from (x, y) to (X, Y) so that the equation $y(x) = ae^{-bx}$ is transformed into the linear equation $Y = A + BX$, where A and B are parameters determined in terms of a and b. Then we could find the line of best-fit, which will determine A and B, and in turn, a and b. In the present case, by taking logs, we can rewrite $y(x) = ae^{-bx}$ in the form $\ln y = \ln a - bx$, or $Y = A + BX$, where $X = x$, $Y = \ln y$, $A = \ln a$, and $B = -b$. Thus, instead of plotting y against x, we should plot $\ln y$ against x. If our data is modeled by $y(x) = ae^{-bx}$, we should then see an approximate straight-line. (See Figure 2.14.) In this case, the line of best-fit will determine A and B, from which we can determine a and b from $a = e^A$ and $b = -B$. Doing this gives $a = 16783$ and $b = 0.0546$. The half-life is $-\ln(2)/b \approx 12.70$ hours. Figure 2.15 shows the decay of the ^{64}Cu isotope as a function of time in hours, together with the line of best-fit $y = 16783 \exp(-0.0546x)$. We can compare these half-life estimates to the value given in the *CRC Handbook of Chemistry and Physics* for the half-life of the ^{64}Cu isotope — 12.701 hours.

Table 2.16 shows how certain equations $y = f(x)$ can be recast into the linear equation $Y = A + BX$ by a change of variables and/or parameters.[25] Note that there is often more than one way to do this.

---

[23] See "Numerical Recipes in C" by W.H. Press, S.A. Teukolsky, W.T. Vetterling, and B.P. Flannery, Cambridge University Press, 2nd edition, 1992, page 681.

[24] Two excellent, easy to use, Windows programs that do this type of fit, (and much more) are *Kalkulator*, written by Andrzej Wrotniak (available for evaluation at the website http://www.freefloght.com/wrotniak/kalkul.html) and *Curve Expert*, written by Daniel Hyams (available for evaluation at the website http://www.ebicom.net/~dhyams/cvxpt.html).

[25] Adapted from "Numerical Methods" by J.H. Mathews, Prentice-Hall, New Jersey, 1987.

ADDITIONAL MATERIALS

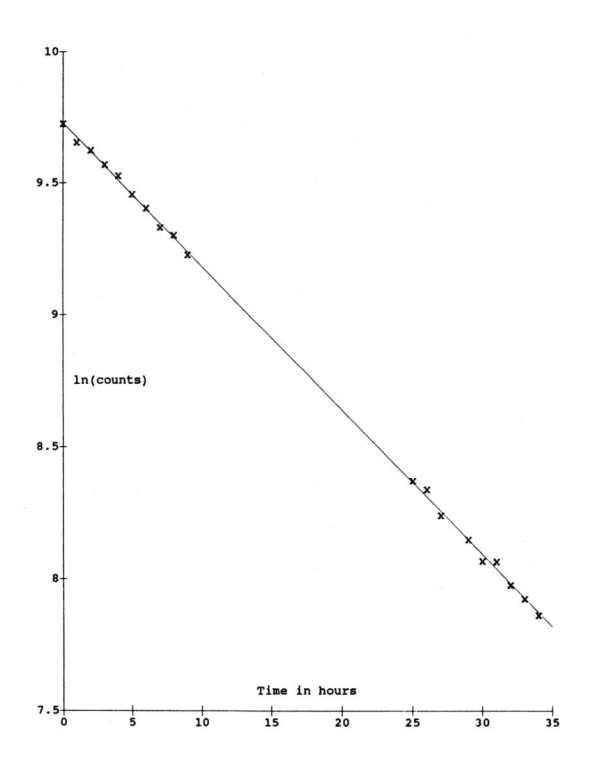

Figure 2.14   The ln of the decay as a function of time

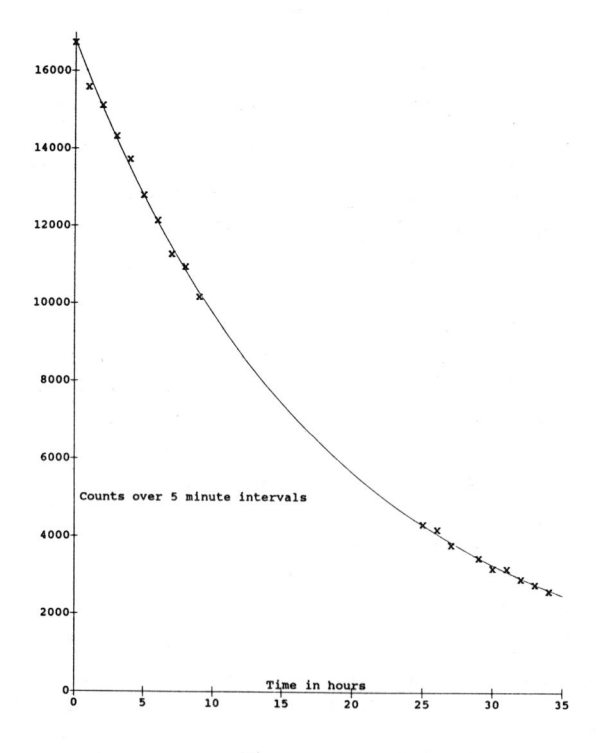

Figure 2.15   The decay of the ^{64}Cu isotope as a function of time in hours

IDEAS FOR CHAPTER 2 — "AUTONOMOUS DIFFERENTIAL EQUATIONS"

Table 2.16 Data linearization transformations

$y = f(x)$	$Y = A + BX$	X	Y	a	b
$y = ae^{bx}$	$\ln y = \ln a + bx$	x	$\ln y$	e^A	B
$y = ax^b$	$\ln y = \ln a + b \ln x$	$\ln x$	$\ln y$	e^A	B
$y = axe^{bx}$	$\ln\left(\frac{y}{x}\right) = \ln a + bx$	x	$\ln\left(\frac{y}{x}\right)$	e^A	B
$y = ax + bx \ln x$	$\frac{y}{x} = a + b \ln x$	$\ln x$	$\frac{y}{x}$	A	B
$y = \frac{a}{x+b}$	$y = \frac{a}{b} - \frac{1}{b}(xy)$	xy	y	$-\frac{A}{B}$	$-\frac{1}{B}$
$y = \frac{ax}{b+x}$	$\frac{x}{y} = \frac{b}{a} + \frac{1}{a}x$	x	$\frac{x}{y}$	$\frac{1}{B}$	$\frac{A}{B}$
$y = \frac{ax}{b+x}$	$\frac{1}{y} = \frac{b}{a}\left(\frac{1}{x}\right) + \frac{1}{a}$	$\frac{1}{x}$	$\frac{1}{y}$	$\frac{1}{A}$	$\frac{B}{A}$
$y = \frac{1}{(a+bx)^2}$	$\frac{1}{\sqrt{y}} = a + bx$	x	$\frac{1}{\sqrt{y}}$	A	B
$y = \frac{1}{1+a\exp(bx)}$	$\ln\left(\frac{1}{y} - 1\right) = \ln a + bx$	x	$\ln\left(\frac{1}{y} - 1\right)$	e^A	B
$y = a + bf(x)$	$y = a + bf(x)$	$f(x)$	y	A	B
$y = \frac{1}{a+bf(x)}$	$\frac{1}{y} = a + bf(x)$	$f(x)$	$\frac{1}{y}$	A	B

Additional Comments

Methods for estimating the parameters of $c + ae^{bx}$ are outlined in Exercise 4 of Section 4.3 of the text.

Methods for estimating the parameters of ax^2 are outlined in Example 4.10 of Section 4.4 of the text.

Methods for estimating the parameters of $a_1 e^{b_1 x} + a_2 e^{b_2 x}$ are outlined in Example 9.19 of Section 9.6 of the text.

Exercises

1. Each of the equations $y = f(x)$ in Table 2.16 contains no more than two parameters a and b. Explain why, in general, it is not possible to recast an equation $y = f(x)$ containing more than two parameters, such as $y = a + be^{cx}$, into the linear equation $Y = A + BX$ by a change of variables and/or parameters.

2. If you were trying to estimate a, b, and c, in $y = a + be^{cx}$ from a given data set which of a, b, and c would you try to estimate first, before using Table 2.16?

3. Show that the equation $y = a + b \ln(cx)$, which contains three parameters a, b, and c, can be recast into the linear equation $Y = A + BX$ by the change of variables $Y = y$, $X = \ln x$, $A = a + b \ln c$, $B = b$. Explain why this does not violate the principle you put forward in Exercise 1.

4. Table 2.17 shows the air temperature T in degrees Celsius as a function of the height h in meters above the earth's surface.[26] How well does the model $T(h) = a + bh$ fit this data set? Complete the following rule of thumb: "For every 1000 meters rise in height, the air temperature ...." [Table 2.17 is in TWIDDLE. It is called AIRTEMP.DTA and is in the subdirectory DD-02.]

Table 2.17 Air temperature as a function of height

Height (m)	0	1000	2000	3000	4000	5000	6000	7000	8000	9000	10,000
Temp (C)	15.0	8.5	2.0	−4.5	−11	−17.5	−24.0	−30.5	−37.0	−43.5	−50.0

5. Insulation in a house is frequently measured using R-Values. Table 2.18 shows the R-Value as a function of the thickness of the insulation material in inches, in this case Nature Guard insulation.[27] Find the line that best fits this table and use this

---

[26] "The Simple Science of Flight" by H. Tennekes, MIT Press, Cambridge, 1996.
[27] From the pamphlet "Do It Yourself — Nature Guard Insulation" published by Louisiana-Pacific, available from Home Depot.

to construct a rule that translates R, the R-Value of the house, to the thickness t of the insulation. If the R-Value is 35 how thick is the insulation? If the insulation is 8 inches thick, what is the R-Value, rounded to the nearest integer? What does your rule suggest when the R-Value is zero? What would you expect it to suggest? [Table 2.18 is in TWIDDLE. It is called RVALUE.DTA and is in the subdirectory DD-02.]

Table 2.18 Insulation thickness as a function of R-Value

R-Value	11	13	19	24	30	32	38	40
Thickness	3.5	4.1	6.0	7.6	9.5	10.1	12.0	12.6

6. Table 2.19 shows the weight (in pounds) of 4 people on the Earth and the Moon.[28] Find the line that best fits these points and use this to construct a rule to translate the weight E of an object on the Earth to its weight M on the Moon. What does this rule say about the weight of an object on the Moon that weighs 0 pounds on the Earth? Is that reasonable? Suggest why this happened and what could be done in future to avoid this problem.

Table 2.19 Weight on Earth and Moon

Weight on Earth	100	125	150	175
Weight on Moon	16.1	20.1	24.1	28.1

7. A group of five people worked continuously for 4 hours grading papers. Every 20 minutes they were each asked to estimate "the satisfyingness or tolerability or zest or interest in the work" on a scale of 0 to 10, where 0 meant extreme distaste and 10 extreme pleasure. Table 2.20 shows the average satisfaction S of the five as a function of the 20-minute time interval t.[29] Find the straight line that best fits these data. What does the slope of this line represent? Estimate the t-intercept. What time does this correspond to and what does this imply? If you had the choice of having your paper graded near the beginning or the end of the 4 hour grading period, which would you choose? [Table 2.20 is in TWIDDLE. It is called SATIS.DTA and is in the subdirectory DD-02.]

Table 2.20 The average satisfaction S with grading as a function of the time interval t

t	1	2	3	4	5	6	7	8	9	10	11	12
S	6.0	5.7	5.2	5.0	4.6	4.3	3.6	3.3	2.8	2.4	2.5	2.2

8. Suppose you are given an instrument that looks like a thermometer. You are asked to find out what the instrument does and you place it in liquids at various known temperatures in degrees F and record the readings on the unknown instrument. Table 2.21 shows the results of those experiments.[30] What is the instrument?

9. Suppose you are given an instrument that looks like it might measure weight. You are asked to find out what the instrument does. With nothing on the instrument you notice that it reads 46.0. You place various known weights in pounds on the instrument and record the readings on the unknown instrument. Table 2.22 shows the results of those experiments.[31] What is the instrument?

---

[28] Adapted from "Space Mathematics" by B. Kastner, Dale Seymour, Palo Alto, 1987.

[29] "The Curve of Work and the Curve of Satisfyingness" by E.L. Thorndike, Journal of Applied Psychology, 1917, 1, 265 – 267, as reported in "Quantitative Methods in Psychology" by D. Lewis, McGraw-Hill, New York, 1960.

[30] Adapted from "Elementary Theoretical Psychology" by J.G. Greeno, Addison-Wesley, Massachusetts, 1968.

[31] Adapted from "Elementary Theoretical Psychology" by J.G. Greeno, Addison-Wesley, Massachusetts, 1968.

Table 2.21 Readings from an unknown instrument

Temperature (°F)	30	50	70	90	110	130
Instrument reading	−1.1	10.0	21.1	32.2	43.3	54.4

Table 2.22 Readings from an unknown instrument

Weight in pounds	0.0	0.5	1.0	1.5	2.0	2.5
Instrument reading	46.0	272.8	499.6	726.4	953.2	1180

10. A subject was given some objects and asked to judge their weights where a 98 gram object was assigned the value 100. Table 2.23 shows the result of such an experiment where w is the physical weight of the object in grams, and J is the judged weight of the object.[32] This is an example of a "judgment of magnitude" experiment, which is modeled by the power law $J = aw^b$, where a and b are constants. Estimate a and b. What does the w-intercept suggest about the value implicitly assigned to an object with no weight? Someone observes that a doubling of the physical weight results in the judged weight being multiplied by 2.6. Is that an accurate observation? [Table 2.23 is in TWIDDLE. It is called JUDGE1.DTA and is in the subdirectory DD-02.]

Table 2.23 The judged weight as a function of the physical weight

Physical weight, w	19	33	46	63	74	98	136	212
Judged weight, J	10	25	35	60	90	100	150	280

11. A subject was shown a 400 cm² square and advised that its area was assigned the number 20 units as area. The subject was then given various numbers and asked to draw a squares with those numbers as the area when compared to the 20 units area. Table 2.24 shows the result of such an experiment where p is the physical area of the object in cm², and J is the judged area of the object.[33] This is another example of a "judgment of magnitude" experiment, which is modeled by the power law $J = ap^b$, where a and b are constants. Estimate a and b. What does the p-intercept suggest about the value implicitly assigned to an object with no area? Someone observes that a doubling of the physical area results in the judged area being multiplied by 1.6. Is that an accurate observation? In what way do the two power functions from this and the previous problem differ? [Table 2.24 is in TWIDDLE. It is called JUDGE2.DTA and is in the subdirectory DD-02.]

12. Very high energy particles are found in the radiation belts of Earth, Jupiter, and Saturn. During *Pioneer 10*'s flyby of Jupiter in 1973, the data in Table 2.25 was obtained.[34] It shows the number of particles per unit time, n, at different energies E in Jupiter's radiation belt and is called a spectrum. If the spectrum obeys a power law $n = kE^b$, where k and b are constants, the spectrum is called a power spectrum. Is this a power spectrum? [Table 2.25 is in TWIDDLE. It is called SPECTRUM.DTA and is in the subdirectory DD-02.]

13. Table 2.26 shows the results of an experiment involving human decision making in which the time R in seconds it took to respond was measured when faced with N choices.[35] These data are sometimes modeled by the equation $R = a + b \ln N$, where a and b are constants. Estimate a and b. What does a measure? Comment on replacing this model with the model $R = a + b \ln(cN)$, where c is a constant.

---

[32] "Elementary Theoretical Psychology" by J.G. Greeno, Addison-Wesley, Massachusetts, 1968.
[33] "Elementary Theoretical Psychology" by J.G. Greeno, Addison-Wesley, Massachusetts, 1968.
[34] "Space Mathematics" by B. Kastner, Dale Seymour, Palo Alto, 1987.
[35] Adapted from "Space Mathematics" by B. Kastner, Dale Seymour, Palo Alto, 1987.

Table 2.24 The judged area as a function of the physical area

Physical area, p	16	52	130	400	900	3300
Judged area, J	2.5	5.0	10.0	20.0	40.0	80.0

Table 2.25 The number of particles per unit time, n, at different energies E

Energy E	0.16	0.30	0.60	1.00	1.60	4.50	10.00	20.00
Number n	1,000,000	150,000	13,000	6800	1000	20	1	0.1

14. Experiments have been performed in which the average percentage memory retention is measured against time.[36] The subjects memorized a list of items, and then the percentage of these items that the subjects remembered was measured at various time intervals. The results of this experiment are tabulated in Table 2.27. It has been suggested, possibly naively, that the percentage memory retention, $P(t)$, where t is time measured in minutes, is modeled by the equation $P(t) = a + b \ln t$, where a and b are constants. Plot P versus $\ln t$ and from this estimate a and b. Is there a time when $P(t) = 0$? What does this mean? Is there a time when $P(t) = 100$? What does this mean? Is this a reasonable model? [Table 2.27 is in TWIDDLE. It is called MEMORY.DTA and is in the subdirectory DD-02.]

15. To study how recognition memory decreases with time, the following experiment was conducted. The subject read a list of 20 words slowly and aloud and then at different time intervals was shown a list of 40 words, containing the 20 words that had been read. The percentage P of words recognized was recorded as a function of t the time elapsed in minutes. Table 2.28 shows the averages for 5 different subjects.[37] This is modeled by $P = a + b \ln t$. Estimate a and b. When does this model predict the subjects recognize no words? All words? [Table 2.28 is in TWIDDLE. It is called RECOGN.DTA and is in the subdirectory DD-02.]

16. A light, flashing regularly, consists of cycles, each cycle having a dark phase and a light phase. The frequency of this light is measured in cycles per second. As the frequency is increased the eye initially perceives a series of flashes of light, then a course flicker, a fine flicker, and ultimately steady light. The frequency at which flickering disappears is called the *fusion frequency*.[38] Table 2.29 shows the results of an experiment[39] in which the fusion frequency F was measured as a function of the light intensity I. It is modeled by $F = a + b \ln I$. Estimate a and b. The units of I are described as arbitrary. If the units of I were changed, which of the constants a and b would be affected, and in what way?

17. Shotgun bores are sometimes measured in gauges, leading to expressions like "a 12 gauge shotgun." Table 2.30 shows the relation between the gauge, G, and diameter, b, of a shotgun's bore in mms.[40] It is claimed that this data set can be modeled by the equation $b = aG^n$, where a and n are constants. Comment on this claim and estimate a and n.

---

[36] "Probability with statistical applications" by F. Mosteller, R. E. K. Rourke, and G. B. Thomas, Addison-Wesley, 1970, Table 11-1, page 383, as reported in "A Handbook of Small Data Sets" by D. J. Hand, F. Daly, A. D. Lunn, K. J. McConway and E. Ostrowski, Chapman and Hall, 1994, page 128.

[37] Adapted from "Quantitative Methods in Psychology" by D. Lewis, McGraw-Hill, New York, 1960.

[38] "Experimental Psychology" by R.S. Woodworth, Holt and Company, New York, 1948.

[39] "Quantitative Methods in Psychology" by D. Lewis, McGraw-Hill, New York, 1960.

[40] Data from "The World Almanac and Book of Facts: 1996" edited by R. Famighetti, Funk and Wagnalls, New Jersey, 1995.

Table 2.26 Time R in seconds it took to respond when faced with N choice

Number of choices N	3	4	5	6	7	8	9
Response time R	0.37	0.42	0.48	0.52	0.56	0.58	0.59

Table 2.27 Percentage memory retention at various times

Time (minutes)	1	5	15	30	120	1440	2880	5760
Memory retention (percentage)	84	71	61	56	47	26	20	16

Table 2.28 The percentage P of words recognized as a function of the time elapsed in minutes

t	5	15	30	60	120	240	480	720	1440	2880	5760	10080
P	73.0	61.7	58.3	55.7	50.3	46.7	40.3	38.3	29.0	24.0	18.7	10.3

18. In 1619, Kepler published his third law, which relates D, the distance of a planet from the sun, to P, the period of the planet — the time it takes for the planet to orbit the sun, a "year". Kepler conjectured that $P = kD^b$, where k and b are constants he determined empirically from experimental data. Table 2.31 represents the up-to-date observational data.[41] Does this model fit the data? What are your estimates for k and b, what is the final form of Kepler's equation? [Table 2.13 is in TWIDDLE. It is called PLANETS.DTA and is in the subdirectory DD-02.]

19. A tube of soil is held horizontally and wetted at one end. If x is the position of the wetting front at time t then a model for the movement of the wetting front is $x = at^b$, where a and b are constants which vary with the sample of soil. The data for such an experiment is given in Table 2.32.[42] What values for a and b make this a reasonable model? [Table 2.32 is in TWIDDLE. It is called WET.DTA and is in the subdirectory DD-02.]

20. Table 2.33 shows the results of an experiment from biochemical enzyme kinetics involving glucose-6-phosphatase.[43] The rate of phosphate release, R, is recorded as a function of the concentration, C. These data are sometimes modeled by the Michaelis-Menten equation $R = aC/(b + C)$, where a and b are constants. Estimate a and b. What does a measure? [Table 2.33 is in TWIDDLE. It is called MENTEN.DTA and is in the subdirectory DD-02.]

---

[41] "The Macmillan Dictionary of Measurement" by M. Darton and J. Clark, Macmillan, New York, 1994.
[42] Adapted from "An example of motion in a course of physics for agriculture" by I. A. Guerrini, The Physics Teacher, February 1984, pages 102 through 103.
[43] Adapted from "C Curve Fitting and Modeling" by J.G. Reich, McGraw-Hill, New York, 1992.

Table 2.29 The fusion frequency F as a function of the light intensity I

I	0.8	1.9	4.4	10.0	21.4	48.4	92.5	218.7	437.3	980.0
F	8.0	12.1	15.2	18.5	21.7	25.3	28.3	31.9	35.2	38.2

ADDITIONAL MATERIALS

Table 2.30 The gauge of a shotgun as a function of the diameter in mm.

Diameter	23.34	19.67	18.52	17.60	16.81
Gauge	6	10	12	14	16

Table 2.31 The period and distance of a planet from the sun

Planet	Distance (kms $\times 10^6$)	Period (days)
Mercury	57.9	88
Venus	108.2	225
Earth	149.6	365
Mars	227.9	687
Jupiter	778.3	4329
Saturn	1427.0	10753
Uranus	2870.0	30660
Neptune	4497.0	60150
Pluto	5900.0	90670

The Half-Life of ^{64}Cu

Table 2.34 show the decay of the ^{64}Cu isotope as a function of time in hours.[44] Figure 2.16 shows these data plotted against time.

Figure 2.17 shows the logarithm of the counts plotted against time with a straight line fit. This suggests that the counts $y(x)$ at time x might decay exponentially and so is governed by the differential equation

$$\frac{dy}{dx} = -ky,$$

with solution

$$y(x) = Ce^{-kx}.$$

Figure 2.17 allows us to estimate $\ln C$ because it is the y-intercept. Thus we have $\ln C \approx 9.728$, so $C \approx 16781$. Similarly $-k$ is the slope so $k \approx 0.05457$. This gives

$$y(x) = 16781 e^{-0.05457x}. \tag{2.6}$$

Figure 2.18 shows these data plotted against time with the exponential function (2.6). □

This allows us to estimate the half-life of ^{64}Cu at about 12.702 hours. The value given in the "CRC Handbook of Chemistry and Physics" for the half-life of the ^{64}Cu isotope is 12.701 hours.

---

[44] "Data analysis in the undergraduate nuclear laboratory" by B. Curry, D. Riggins, and P. B. Spiegel, American Journal of Physics, 63, 1995, pages 71 through 76. Table 1, page 73.

Table 2.32 The movement of the wetting front as a function of time

Time (minutes)	0	1	2	4	8	16	32	64	128	256
Distance (cm)	0	3.7	4.5	6.0	8.0	10.6	13.8	18.6	24.3	32.0

Table 2.33 Rate of phosphate release as a function of the concentration

Concentration C	0.12	0.20	0.51	0.84	1.25	1.98	3.4	4.8
Rate R	13.8	21.0	39.5	50.6	58.6	66.3	72.7	75.5

Table 2.34 Data for ^{64}Cu decay

Time	Counts	Time	Counts	Time	Counts
0	16744	7	11277	30	3184
1	15596	8	10949	31	3177
2	15120	9	10174	32	2910
3	14325	25	4317	33	2766
4	13723	26	4181	34	2598
5	12788	27	3784		
6	12141	29	3454		

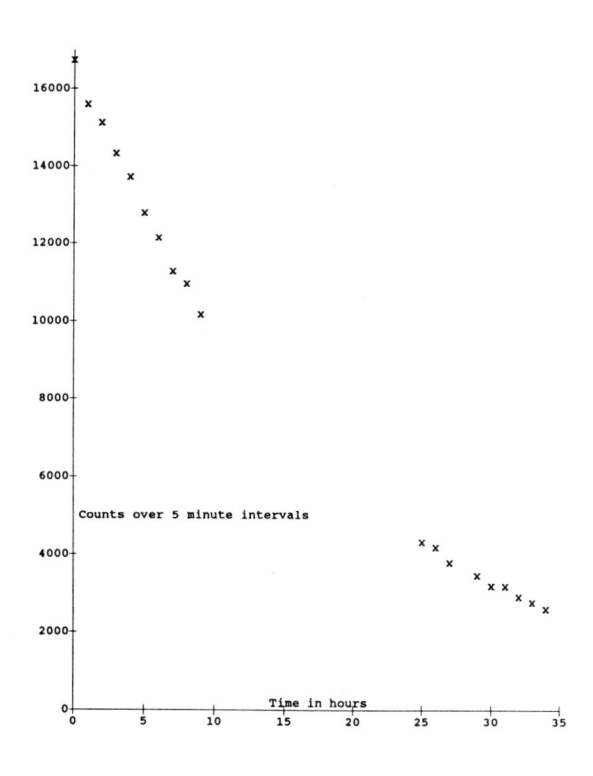

Figure 2.16 The decay of the ^{64}Cu isotope as a function of time in hours

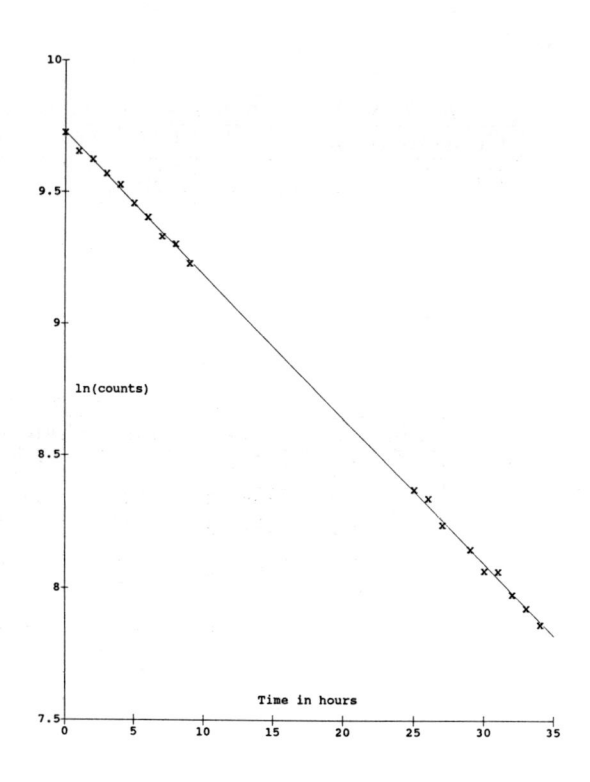

Figure 2.17 Logarithm of counts of the decay of ^{64}Cu isotope versus time with straight line fit

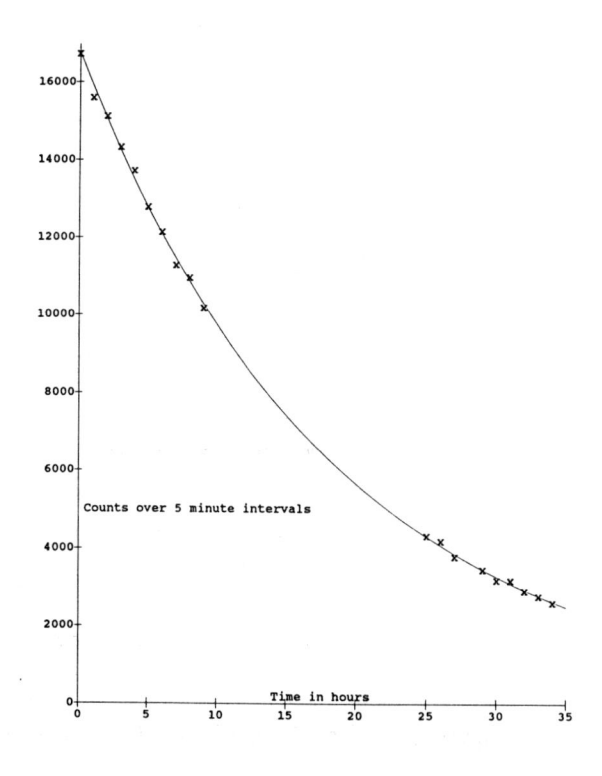

Figure 2.18 The decay of the ^{64}Cu isotope versus time with exponential fit

The Bombay Plague

Table 2.35 shows the number of deaths in Bombay from a plague spread by rats during the period December 1905 to July 1906.[45] We have plotted this data set in Figure 2.19.

Table 2.35 The Bombay Plague data

Week	Deaths	Week	Deaths	Week	Deaths
1	4	12	900	23	8129
2	14	13	1290	24	8480
3	29	14	1738	25	8690
4	47	15	2379	26	8803
5	68	16	3150	27	8868
6	99	17	3851	28	8920
7	150	18	4547	29	8971
8	203	19	5414	30	9010
9	300	20	6339	31	9043
10	425	21	7140		
11	608	22	7720		

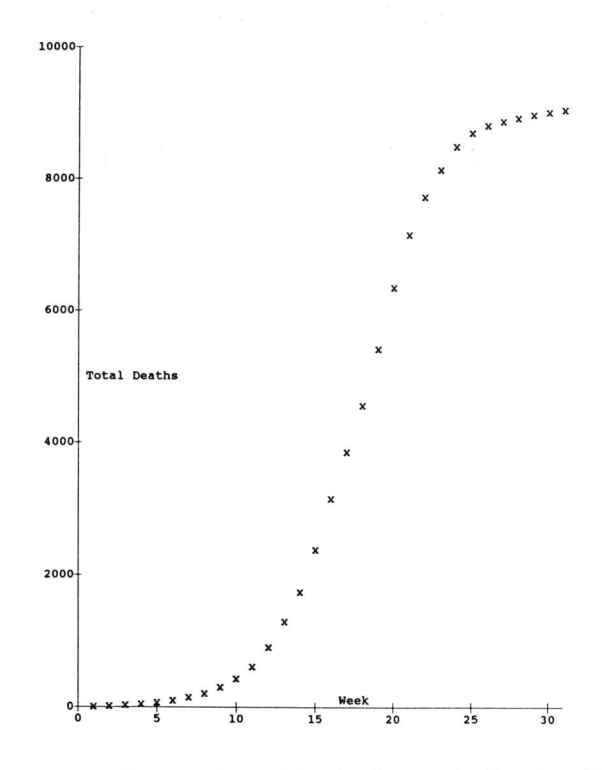

Figure 2.19 The number of deaths due to the Bombay Plague

Let us see if the explicit solution of the logistic equation

$$\frac{dy}{dx} = ay(b - y),$$

---

[45] "A contribution to the mathematical theory of epidemics" by W. O. Kermack and A. G. McKendrick, Proc. Roy. Soc., **115A**, 1927, pages 700 – 721.

namely,
$$y(x) = \frac{bC}{e^{-abx} + C} \tag{2.7}$$

where C is an arbitrary constant, gives a good fit. Notice that there are three unknowns in the solution — namely, a, b, and C. Unlike the examples involving exponential growth and decay, there is no way to manipulate (2.7) to construct a linear function involving three arbitrary constants. (Why?) However, from an intermediate step of the solution, that is,

$$\ln\left[\frac{y(x)}{b - y(x)}\right] = abx + c,$$

where $e^c = C$, we see that if we can estimate b, the carrying capacity, we could plot $\ln[y/(b-y)]$ against x to see whether the data are approximately linear. If they are, we can then estimate ab and c.

There are two ways we can estimate b: directly, by estimating the value of y where the logistic curve levels off, in which case b equals this value; or indirectly, by estimating the value of y where the curve has an inflection point, in which case b equals twice this value. (Remember, for the logistic equation, the point of inflection occurs at $b/2$.) Looking at Figure 2.19 on , we can try to estimate b based on these two estimates. It appears that $b \approx 9100$ is reasonable from both points of view. Figure 2.20 shows a plot of $\ln[y/(9100-y)]$ against x, together with a straight line fit. From this we estimate $ab \approx 0.396821$, because ab is the slope. Similarly, c is the y-intercept so $c \approx -7.061236$ and thus, $C = e^c \approx e^{-7.061236} \approx 8.5772 \times 10^{-4}$. In this case, the number of deaths over time is given by

$$y(x) = \frac{(9100)(8.5772 \times 10^{-4})}{e^{-0.396821x} + 8.5772 \times 10^{-4}} = \frac{78052}{10^4 e^{-0.396821x} + 8.5772}. \tag{2.8}$$

Figure 2.21 shows the data set plotted against time with the function (2.8).

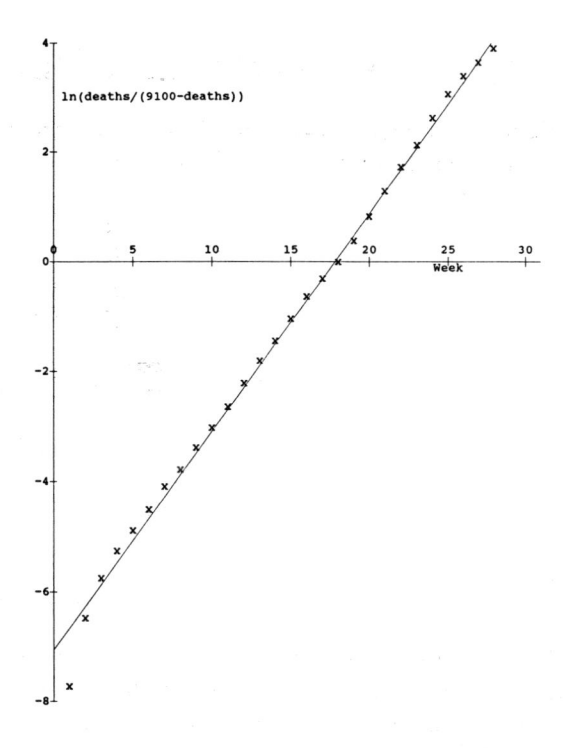

Figure 2.20 $\ln[y/(9100-y)]$ versus time

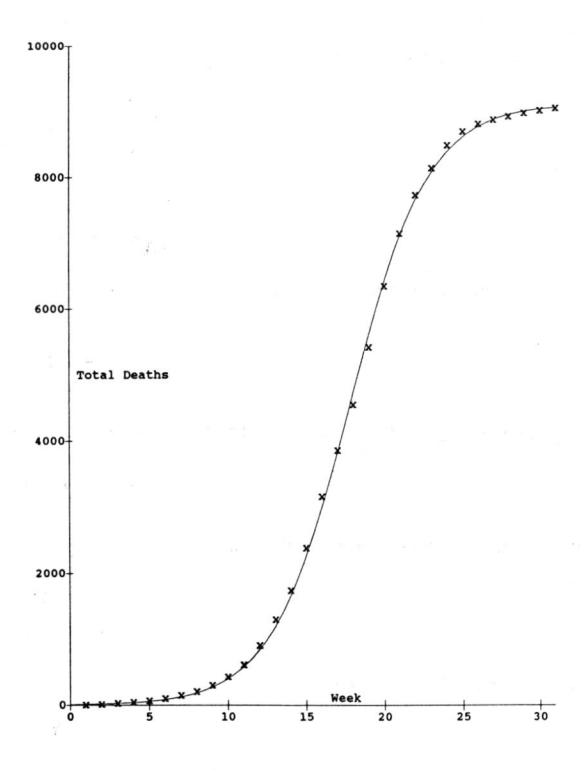

Figure 2.21 The number of deaths due to the Bombay Plague with logistic fit

ADDITIONAL MATERIALS

Parameters in ODEs — Rescaling

Differential equations frequently contain parameters, such as a and b. As far as the differential equation is concerned, these are constants. We have seen examples of this in various places:

$$\frac{dy}{dx} = ax, \tag{2.9}$$

$$\frac{dy}{dx} = ay, \tag{2.10}$$

$$\frac{dy}{dx} = ay(b-y), \tag{2.11}$$

$$\frac{dy}{dx} = ky \ln \frac{y}{a}, \tag{2.12}$$

$$\frac{dT}{dt} = k(T - T_a), \tag{2.13}$$

$$\frac{d^2 y}{dx^2} \pm a^2 y = 0 \tag{2.14}$$

While these parameters are usually significant because of the associated application, they frequently complicate the mathematics unnecessarily. In fact in many cases, it is possible to rescale and/or translate the dependent or independent variable, to simplify the mathematics. For example, let's think about rescaling (2.9), from y and x, to Y and X where

$$\begin{aligned} x &= \alpha X, \\ y &= \beta Y. \end{aligned} \tag{2.15}$$

Here α and β are constants that we want to select to free the differential equation of parameters. Substituting (2.15) into (2.9) gives

$$\frac{\beta}{\alpha} \frac{dY}{dX} = a\alpha X,$$

or

$$\frac{dY}{dX} = \frac{a\alpha^2}{\beta} X.$$

We can eliminate the parameters in this equation by choosing α and β so that

$$\frac{a\alpha^2}{\beta} = 1,$$

for given $a \neq 0$. This can be done in various ways, one of the simplest being

$$\alpha = 1 \quad \beta = a.$$

With this choice the transformation

$$\begin{aligned} x &= X, \\ y &= aY, \end{aligned} \tag{2.16}$$

converts the differential equation

$$\frac{dy}{dx} = ax,$$

into the differential equation

$$\frac{dY}{dX} = X.$$

The later has solution
$$Y = \frac{1}{2}X^2 + C,$$
which, in terms of the original variables, is
$$\frac{1}{a}y = \frac{1}{2}x^2 + C,$$
or
$$y = \frac{1}{2}ax^2 + c,$$
where $c = aC$ is an arbitrary constant.

The ability to rescale is also frequently useful when using computer programs. Usually the best picture is obtained when the horizontal and vertical axes have approximately the same scale.

Exercises

1. Show that rescaling according to (2.15) does not affect the general properties of the solutions if α and β are positive. In particular, show that extreme values in the XY plane correspond to extreme values in the xy plane. Show that similar comments apply to concavity and points of inflection.

Parameters in ODEs — Units Analysis

It is frequently possible to check the reasonableness of an answer by checking its units, or dimensions. If y is a quantity, then we will use the notation $[y]$ to denote the units of y. It is easy to see that $[y^n] = [y]^n$. Thus the units of dy/dx are $[y]/[x]$ or $[yx^{-1}]$.

Let's return to
$$\frac{dy}{dx} = ax,$$
and its solution
$$y = \frac{1}{2}ax^2 + c,$$
to see whether the units make sense.

From the differential equation we have
$$\frac{[y]}{[x]} = [a][x],$$
so that
$$[y] = [a][x]^2.$$
Thus y should have the same units as ax^2, which is in agreement with the solution.

Now let's look at the logistic equation
$$\frac{dy}{dx} = ay(b - y),$$
where b and y have the same units, that is, $[y] = [b]$. From this we conclude that
$$\frac{[y]}{[x]} = [a][y][y],$$
or
$$[y] = [a]^{-1}[x]^{-1} = [ax]^{-1}.$$
The analytical solution we obtained was
$$y(x) = \frac{y_0}{(1 - y_0/b)\,e^{-abx} + y_0/b}.$$

Now $[y] = [b] = [y_0] = [ax]^{-1}$, so $[abx]$ and $[y_0/b]$ are dimensionless. Thus the denominator is also dimensionless, whereas the numerator has dimension $[y]$.

Example 2.1 *The solution of $y'' + a^2 y = 0$ is either $y(x) = A\sin(ax) + B\cos(ax)$ or $y(x) = A\sin(x/a) + B\cos(x/a)$. Use dimensionality arguments to identify the correct solution.*

The units of y'' are $[yx^{-2}]$ and these must be the same as the units of $[a^2 y]$, that is, $[y][x]^{-2} = [a]^2[y]$. Thus $[ax]$ is dimensionless, so the first equation is the correct solution, where A and B have the units of $[y]$. □

The Doomsday Model

Table 2.36 shows the world population in billions from 1650 to 1990.[46] This data set is sketched in Figure 2.22. It is claimed that the differential equation

$$\frac{dP}{dt} = aP^2, \tag{2.17}$$

where a is a constant, models this situation very well. This exercise investigates this claim.

Table 2.36 The world population in billions from 1650

Year	Population
0	0.510
50	0.625
100	0.710
150	0.910
200	1.130
250	1.600
300	2.525
310	3.307
315	3.354
320	3.696
325	4.066
330	4.432
335	4.822
340	5.318

We assume that $P(0) = P_0$. Dividing (2.17) by P^2 and integrating we find

$$-\frac{1}{P(t)} = at + C.$$

Using $P(0) = P_0$, gives $C = -1/P_0$, gives

$$\frac{1}{P(t)} = \frac{1}{P(0)} - at, \tag{2.18}$$

so the solution of (2.17) subject to $P(0) = P_0$ is

$$P(t) = \frac{P(0)}{1 - aP(0)t}. \tag{2.19}$$

In order to compare (2.20) with the data set in Table 2.36 we need to estimate the parameters $P(0)$ and a. There are various ways of doing this, but perhaps the easiest is to notice that if we plot $1/P(t)$ against t then (2.18) suggests we should obtain a line. If this is so, then we can read off the values of $-a$ and $1/P(0)$, because these will be the slope and the intercept of the line. This is plotted in Figure 2.23, from which we estimate $a = 0.005098$ and $1/P(0) = 1.904528$. This gives $P(0) \approx 0.525$ and $aP(0) \approx 0.00267$. Notice that the exact value for $P(0)$ is 0.510. Figure 2.24 shows the data set in Table 2.36 and the function

$$P(t) = \frac{0.525}{1 - 0.00267t}, \tag{2.20}$$

which seems to fit well.

---

[46] "Introduction to the Mathematics of Population" by N. Keyfitz, Addison-Wesley (1968), as reported in "Growth and Diffusion Phenomena" by R. T. Banks, Springer-Verlag, 1991, page 17.

ADDITIONAL MATERIALS

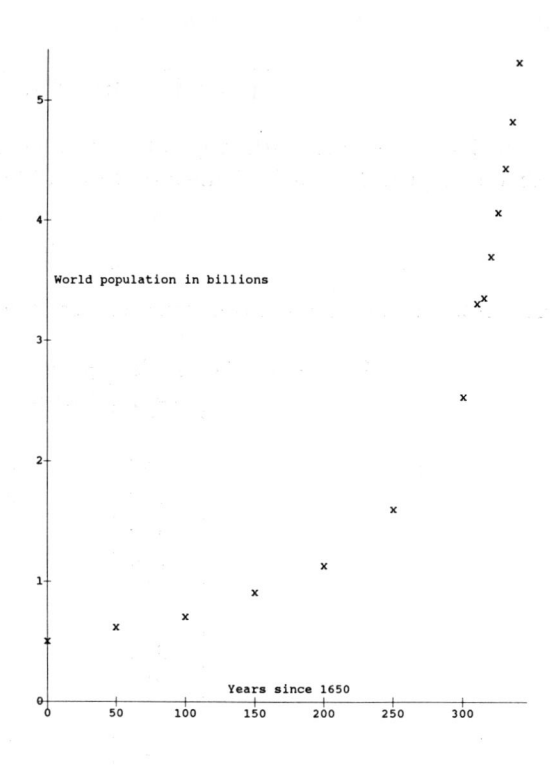

Figure 2.22   The world population in billions from 1650

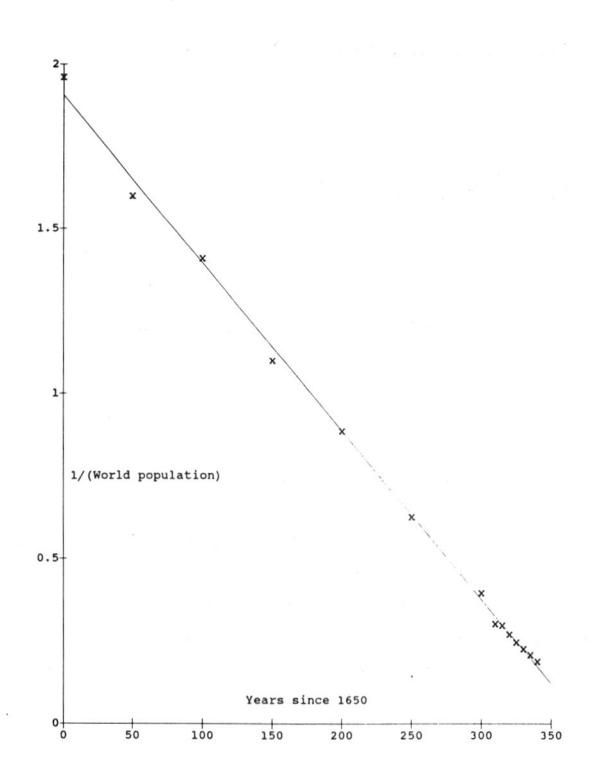

Figure 2.23   The inverse of the world population in billions from 1650 and the line $0.005098t + 1.904528$

IDEAS FOR CHAPTER 2 — "AUTONOMOUS DIFFERENTIAL EQUATIONS"

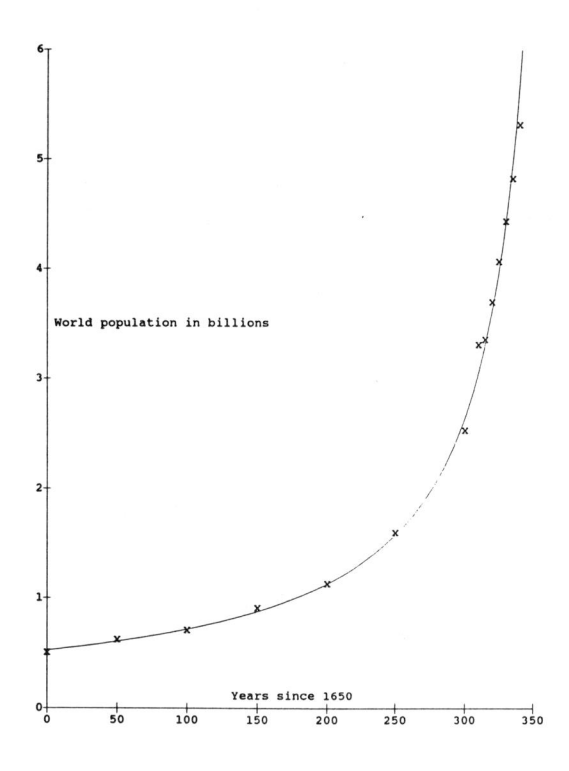

Figure 2.24 The world population in billions from 1650 and the function $P(t) = 0.525/(1 - 0.00267t)$

As $t \to 1/0.00267 \approx 376$ then $P(t) \to \infty$. Because $t = 0$ corresponds to the year 1650, 376 corresponds to the year 2026. Why is this called the doomsday model?

The manner in which we estimated the parameters a and $P(0)$ by using (2.20), is similar to the way we estimated the parameters in the exponential case (see text page 38) and the logistic case (see text page 48). This is a general property of differential equations of the type
$$\frac{dy}{dx} = ag(y),$$
because the general (nonequilibrium) solution of this equation is
$$\int \frac{1}{g(y)} \, dy = ax + C.$$

By plotting the left hand side against x we should see a straight line with slope a and intercept C.

Chemical Kinetics

When a substance $\mathcal{A}$ decomposes it is said to be an nth order reaction if it satisfies the differential equation

$$\frac{dA}{dt} = -kA^n,$$

where A is the concentration of $\mathcal{A}$ at time t and k and n are positive constants. If a_0 is the initial concentration of $\mathcal{A}$ and x is the concentration of $\mathcal{A}$ that has disappeared at time t, then $A = a_0 - x$. The differential equation then becomes

$$\frac{dx}{dt} = k(a_0 - x)^n. \qquad (2.21)$$

The constant k is called the rate constant of the reaction. The initial condition for this equation is $x(0) = 0$. (Why?)

(a) Show that the solution of (2.21) is

$$t = \frac{1}{k} \ln\left(\frac{a_0}{a_0 - x}\right) \qquad \text{if } n = 1 \qquad (2.22)$$

and

$$t = \frac{1}{k}\frac{1}{n-1}\left[\frac{1}{(a_0 - x)^{n-1}} - \frac{1}{a_0^{n-1}}\right] \qquad \text{if } n \neq 1. \qquad (2.23)$$

If $n = 2$ then (2.23) can be written in the form

$$t = \frac{1}{k}\left[\frac{x}{a_0(a_0 - x)}\right]. \qquad (2.24)$$

(b) Show that the half-life t_h — the time it takes for one half of the original substance to disappear — is given by

$$t_h = \frac{1}{k}\ln 2 \qquad \text{if } n = 1 \qquad (2.25)$$

and

$$t_h = \frac{1}{k}\frac{2^{n-1} - 1}{n - 1}\frac{1}{a_0^{n-1}} \qquad \text{if } n \neq 1. \qquad (2.26)$$

(c) There are various ways of using these equations to decide on the order of a reaction n, and to estimate the rate constant k if experimental results of x versus t, with known a_0 are available.[47] In the following we describe two of these ways — the Tabular method and the Graphical method. Use both these methods to estimate n and k from the experimental data given in Tables 2.37, 2.38, and 2.39,[48] also shown in Figures 2.25, 2.26, and 2.27.[49] [Tables 2.37, 2.38. and 2.39 are in TWIDDLE. They are called PEROXIDE.DTA, DIOXIDE.DTA, and NITRO.DTA and are in the subdirectory DD-02.]

  i. The first method — called the Tabular Method — is to rewrite (2.22) and (2.23) as

$$k = \frac{1}{t}\ln\left(\frac{a_0}{a_0 - x}\right) \qquad \text{if } n = 1 \qquad (2.27)$$

---

[47] "Chemical Kinetics, 2nd Edition" by K. J. Laidler, McGraw-Hill, 1965, Chapter 1.
[48] "Physical Chemistry" by P.W. Atkins, W.H. Freeman, 1982, 2nd edition. See page 933.
[49] We would like to thank Gabe Aldaz and Shane Sickafoose, University of Arizona, for this information.

Table 2.37 Decomposition of Hydrogen Peroxide as a function of time

Time (minutes)	Concentration (moles/liter)
0	0.0200
200	0.0160
400	0.0131
600	0.0106
800	0.0086
1000	0.0069
1200	0.0056
1600	0.0037
2000	0.0024

Table 2.38 Decomposition of Nitrogen Dioxide as a function of time

Time (minutes)	Concentration (moles/liter)
0	0.0200
30	0.0150
60	0.0120
90	0.0100
120	0.0087

Table 2.39 Decomposition of Ethyl m-Nitrobenzoate as a function of time

Time (sec)	Concentration (moles/liter)
0	0.0500
100	0.0355
200	0.0275
300	0.0225
400	0.0185
500	0.0160
600	0.0148

ADDITIONAL MATERIALS

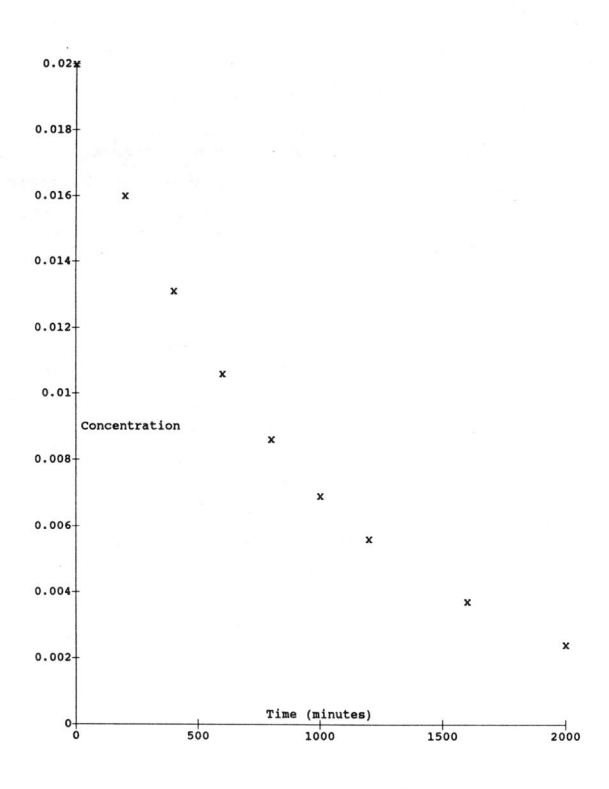

Figure 2.25 Decomposition of Hydrogen Peroxide as a function of time

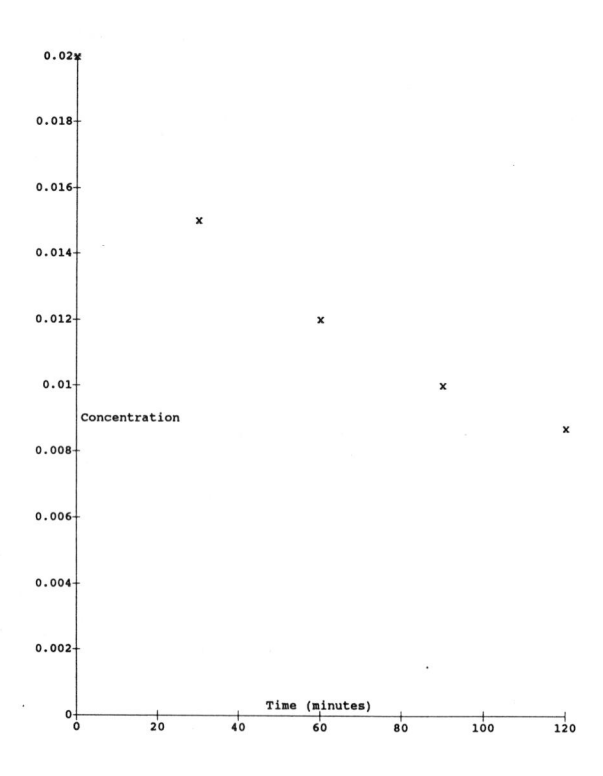

Figure 2.26 Decomposition of Nitrogen Dioxide as a function of time

IDEAS FOR CHAPTER 2 — "AUTONOMOUS DIFFERENTIAL EQUATIONS"

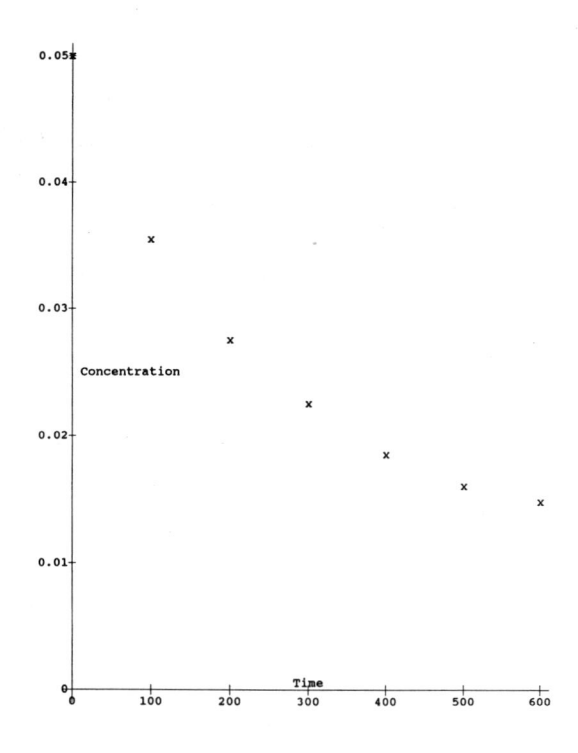

Figure 2.27 Decomposition of Ethyl m-Nitrobenzoate as a function of time

and
$$k = \frac{1}{t}\frac{1}{n-1}\left[\frac{1}{(a_0-x)^{n-1}} - \frac{1}{a_0^{n-1}}\right] \qquad \text{if } n \neq 1. \qquad (2.28)$$

Now the quantities on the right-hand sides of (2.27) and (2.28) — for various n — are computed for each data point (t,x). Select the n for which k is approximately constant.

ii. The second method — called the Graphical Method — is to rewrite (2.22) and (2.23) in the form
$$t = \frac{1}{k}X$$
where
$$X = \ln\left(\frac{a_0}{a_0-x}\right) \qquad \text{if } n = 1$$
and
$$X = \frac{1}{n-1}\left[\frac{1}{(a_0-x)^{n-1}} - \frac{1}{a_0^{n-1}}\right] \qquad \text{if } n \neq 1.$$

If a reaction is of nth order, then a plot of t versus X should give a line through the origin with slope k.

ADDITIONAL MATERIALS

The Generalized Logistic Equation

The generalized logistic equation is

$$\frac{dy}{dx} = ay(b - y^n), \qquad (2.29)$$

where a, b, and n are positive constants.[50]

(a) Show that the equilibrium solutions are $y(x) = 0$, and $y(x) = b^{1/n}$. Use the Derivative Test for Stable and Unstable Equilibrium Solutions to decide which equilibrium solutions are stable.

(b) Part (a) suggests that we introduce a new constant, B, defined by $B = b^{1/n}$, so that (2.29) can be written

$$\frac{dy}{dx} = ay(B^n - y^n). \qquad (2.30)$$

What is the physical significance of B? If you wanted to compare solutions of the generalized logistic equation for different values of n, why is it better to use the form (2.30) rather than (2.29)? [Hint: Think about the long-term behavior of the solutions.]

(c) Show that rescaling y and x according to

$$\begin{aligned} x &= X/(aB^n), \\ y &= BY, \end{aligned} \qquad (2.31)$$

converts (2.30) to

$$\frac{dY}{dX} = Y(1 - Y^n). \qquad (2.32)$$

(d) Differentiate (2.32) and show that the point of inflection (X_i, Y_i) occurs at

$$Y_i = \frac{1}{(n+1)^{1/n}}. \qquad (2.33)$$

(e) Show that

$$\frac{1}{Y(1-Y^n)} = \frac{1}{Y} + \frac{Y^{n-1}}{1-Y^n},$$

and use this to integrate (2.32) in the form

$$\frac{1}{n} \ln\left(\frac{Y^n}{1-Y^n}\right) = x + c.$$

Solve this for Y to find

$$Y(X) = \frac{1}{(1 + Ce^{-nX})^{1/n}},$$

where $C = e^{-nc}$.

---

[50]This exercise is based on "A family of generalized logistic curves and inhibited population growth" by S. Gordon, Int. J. Math. Educ. Sci. Technol., **22**, 1991, pages 919 – 925, and "Modelling with Ordinary Differential Equations" by T. Dreyer, CRC Press, 1993, page 126.

(f) Use (2.31) to change the scale from the X and Y variables back to the original x and y variables. Show from (2.33) that the point of inflection (x_i, y_i) occurs at

$$y_i = \frac{B}{(n+1)^{1/n}}. \tag{2.34}$$

Confirm that with $n = 1$ we get $y_i = 0.5B$, which is consistent with the logistic equation. Show that if $n = 1/2$, $y_i = (4/9)B \approx 0.444B$; if $n = 2$, $y_i = B/\sqrt{3} \approx 0.577B$; and if $n = 3$, $y_i = B/4^{1/3} \approx 0.630B$. Based on the interpretation of y_i and B, what do you expect to happen to $(n+1)^{1/n}$ as $n \to \infty$. Prove it.

(g) Show that if $y_i = B/2$, as occurs in the logistic equation, then (2.34) is satisfied by only one positive n — namely, $n = 1$. Thus the logistic equation is the only general logistic equation for which the inflection point occurs vertically half way between the origin and the ultimate behavior of y.

(h) Use (2.31) on (2.32) to change the scale from the X and Y variables back to the original x and y variables. Thus, show that the solution of (2.29) is

$$y(x) = \frac{B}{[1 + C\exp(-aB^n n x)]^{1/n}},$$

where $\exp(x) = e^x$. Confirm that this is consistent with the logistic equation. If initially $y(0) = y_0$ show that the solution of (2.29) may be written as

$$y(x) = \frac{y_0}{[1 - (y_0/B)^n \exp(-aB^n n x) + (y_0/B)^n]^{1/n}}. \tag{2.35}$$

Figure 2.28 shows three graphs of this function — using the same values for a, B, and y_0 — corresponding to $n = 1/2$, $n = 1$, and $n = 2$. Which is which? What do you observe? Based on (2.30), is this behavior surprising? We know that the logistic equation ($n = 1$) is symmetric about its inflection point. What can you say about the general logistic equation?

(i) Use (2.34) and (2.35) to show that the point of inflection (x_i, y_i) occurs at

$$x_i = \frac{1}{aB^n} \ln\left[\frac{1 - (y_0/B)^n}{n(y_0/B)^n}\right]. \tag{2.36}$$

(j) Comment on the following statement. "If you have a data set that you think satisfies (2.35), you can estimate the values of a, B, n, and y_0 by estimating the location of the inflection point, the location of the initial point, and the long-term behavior of y." [Hint: Look at (2.34) and (2.36).] Will this determine (2.35) completely?

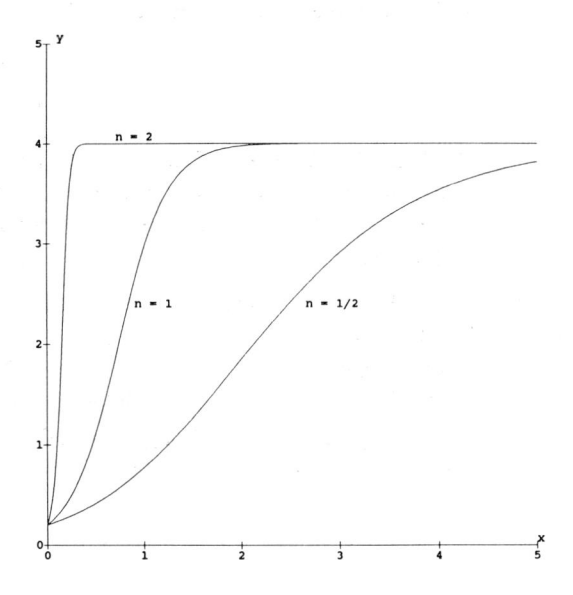

Figure 2.28 Three solutions of the generalized logistic equation with $n = 1/2$, $n = 1$, and $n = 2$

The Gompertz Equation

Table 2.40 shows the total area of the leaves of a plant as a function of time.[51] This data set is sketched in Figure 2.29. It is claimed that the Gompertz equation

$$\frac{dy}{dx} = -ky \ln \frac{y}{b}, \qquad (2.37)$$

where k and b are constants, models this situation very well. This exercise investigates this claim. [Table 2.40 is in TWIDDLE. It is called PLANT.DTA and is in the subdirectory DD-02.]

Table 2.40 The area of the leaves of a plant as a function of time

Time (days)	Area (cm^2)
0	9.0
20	39.7
40	92.5
60	142.7
80	186.6
100	209.7
120	230.5
140	235.4

[51] "Growth and Diffusion Phenomena" by R. T. Banks, Springer-Verlag, 1991, page 153.

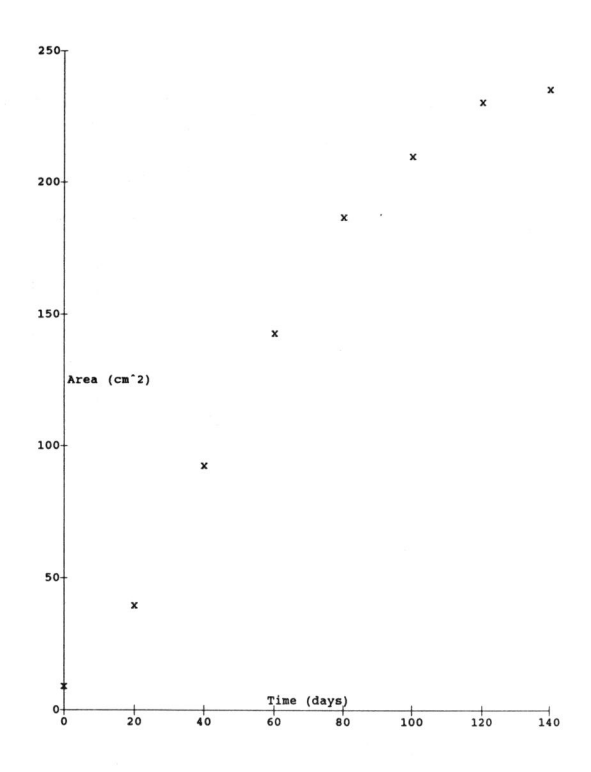

```
                 250┐
                                                          x    x
                                                      x
                 200┤
                                                  x
                 150┤
                                          x
       Area (cm^2)
                 100┤
                                  x
                  50┤
                          x
                   0┤x    Time (days)
                    └────┬────┬────┬────┬────┬────┬────┬
                    0   20   40   60   80   100  120  140
```

Figure 2.29 The area of the leaves of a plant as a function of time

To make the algebra easier, we make the change of scale

$$\begin{aligned} y &= bY, \\ x &= X/k, \end{aligned} \qquad (2.38)$$

in which case (2.37) becomes

$$\frac{dY}{dX} = -Y \ln Y. \qquad (2.39)$$

There are two equilibrium solutions of this equation — namely, $Y = 0$ and $Y = 1$ (which correspond to $y = 0$ and $y = b$). From the previous phase line analysis, we know that $y = 0$ is unstable, and $y = b$ is stable, so b has the properties of a carrying capacity.

If we differentiate (2.39) with respect to X we find

$$\frac{d^2Y}{dX^2} = -\frac{dY}{dX} \ln Y - \frac{dY}{dX} = -\frac{dY}{dX}(\ln Y + 1).$$

Thus, the point of inflection occurs when $\ln Y = -1$ — that is, when $Y = e^{-1}$ (which corresponds to $b/e \approx 0.368b$). By way of comparison, the point of inflection of the logistic equation occurs at $0.5b$.

If we write (2.39) in the form

$$\frac{1}{Y \ln Y} \frac{dY}{dX} = -1$$

and integrate with respect to X, we find

$$\int \frac{1}{Y \ln Y} dY = -X + C.$$

In order to integrate the left-hand side, we make the substitution $u = \ln Y$, in which case we have

$$\int \frac{1}{u} du = -X + C,$$

ADDITIONAL MATERIALS

which yields
$$\ln|u| = -X + C,$$
or
$$\ln|\ln Y| = -X + C. \tag{2.40}$$
In terms of the original variables, this corresponds to
$$\ln\left|\ln\left(\frac{y}{b}\right)\right| = -kx + C. \tag{2.41}$$

Solving (2.40) for Y gives
$$Y(X) = \exp\left(\pm e^{-X+C}\right) = \exp\left(\pm e^C e^{-X}\right),$$
or, putting $c = \pm e^C$,
$$Y(X) = \exp\left(ce^{-X}\right).$$
In terms of the original variables this corresponds to
$$y(x) = b\exp\left(ce^{-kx}\right). \tag{2.42}$$

Notice that $c > 0$ implies that y is increasing, while $c < 0$ implies that y is decreasing. Thus, $c > 0$ for $0 < y < b$.

In order to compare (2.42) with the data set in Table 2.40 we need to estimate the parameters b, c and k. There are various ways of doing this, but perhaps the easiest is to notice that if we plot $\ln|\ln(y/b)|$ against x then (2.41) suggests we should obtain a line. If this is so, then we can read off the values of $-k$ and C, because these will be the slope and the intercept of the line, from which we can estimate $c = e^C$. (Why don't we consider $c = -e^C$?) In order to do this, we first need to estimate b. We can do this by realizing that b is the carrying capacity. From Figure 2.29 we estimate $b = 250$. In Figure 2.30 we plot $\ln|\ln(y/250)|$ against x, from which we estimate $-k = -0.029484$ and $C = 1.181128$. This gives $c \approx 3.258$. Figure 2.31 shows the data set in Table 2.40 and the function
$$y(x) = 250\exp\left(-3.258e^{-0.029484x}\right), \tag{2.43}$$
which seems to fit well.

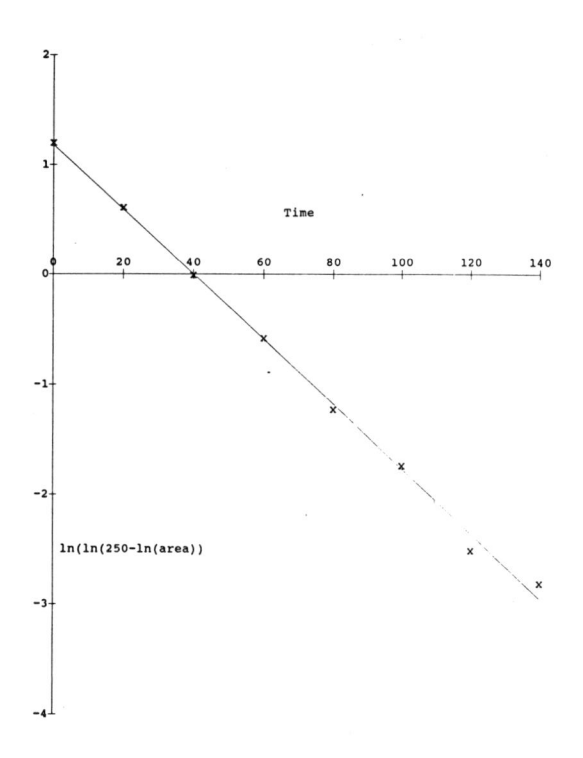

Figure 2.30 The adjusted data and the line $-0.029484x + 1.81128$

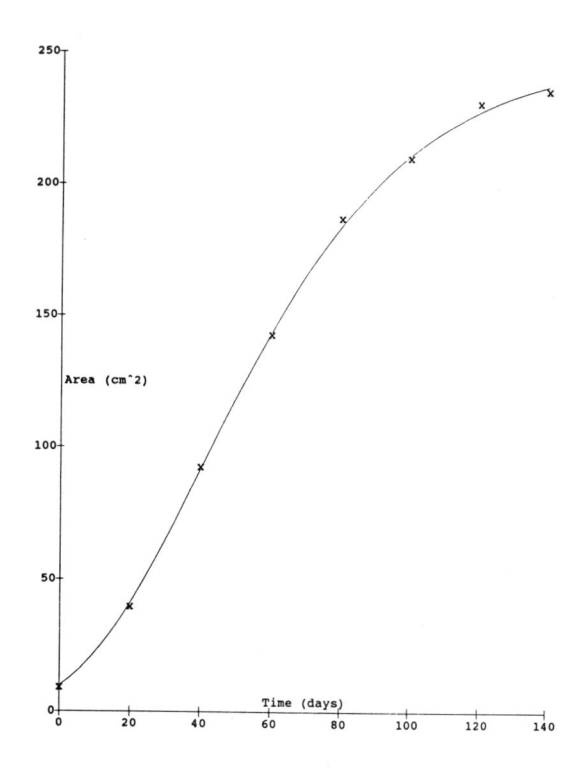

Figure 2.31 The area of the leaves of a plant as a function of time and the function $y(x) = 250\exp\left(-3.258e^{-0.029484x}\right)$

ADDITIONAL MATERIALS

3. IDEAS FOR CHAPTER 3 — "FIRST ORDER DIFFERENTIAL EQUATIONS — QUALITATIVE AND QUANTITATIVE ASPECTS"

3.1 Page by Page Comments

Chapter 3 requires 4 to 6 class meetings of 50 minutes each.

Section 3.4, *Comparing Solutions of Differential Equations*, is not used in later sections, and so may be selectively omitted.

Section 3.5, *Finding Power Series Solutions,* is not extensively used in later sections, and so may be selectively omitted. If you plan to do series solutions (Chapter 12, *Using Power Series*), here is an elementary introduction, so students can get their feet wet.

Page 82. The differential equation in Example 3.1, namely, $y' = x - y$, will be used a number of times later in the book. (Exercise 1, page 91; Example 3.5, page 99; and Example 5.2, page 192.)

Page 84. We give the students copies of Figure 3.4, and have them draw something similar to Figure 3.5.

Page 85. We find it useful to discuss the difficulties associated with trying to plot implicit functions.

Page 85. If you can display slope fields in class, students enjoy discovering the graphs of $x^n + y^n = 2^n$, for various positive integer n.

Page 87. We give the students copies of Figure 3.6, and have them draw something similar to Figures 3.7 and 3.8.

Page 90. We give the students copies of Figure 3.10, and have them draw something similar to Figure 3.12.

Page 91. If you used Example 3.1 then you should assign Exercise 1 of Section 3.1. This differential equation, namely, $y' = x - y$, will be used a number of times later in the book. (Example 3.5, page 99; and Example 5.2, page 192.)

Page 91. If you assign Exercise 6 of Section 3.1., stress that the student is not supposed to solve for y and then substitute into the differential equation. Rather the student is supposed to mimic the process used to obtain (3.6) from (3.4).

Page 92. The differential equation in Exercise 10 of Section 3.1, namely, $y' = 1 - 2xy$, will be seen a number of times later in the book. (Exercise 2, page 117; Example 5.8, page 205; Exercise 27, page 211; and — in its differentiated form, $y'' + 2xy' + 2y = 0$ — in Example 12.3, page 580; and Example 12.5, page 589.)

Page 92. Exercise 13(b). The second suggestion here is a means to observe two different slope fields on the same graphics screen.

Page 94. There are additional exercises for Section 3.1 on page 91 of this manual.

Page 94. It is possible to introduce symmetry in a purely geometrical way. See page 93 of this manual, SYMMETRY.

Page 95. There is a big distinction between a slope field having a certain symmetry and a solution curve having that symmetry. Here we are concerned with the symmetry of the slope field.

Page 96. Exercises 6 and 7 of Section 3.2 reveal whether students understand the power of symmetry. We hand out Figures 3.19 and 3.20 as an aid to students doing Exercises 6 and 7. These figures are repeated on page 97 of this manual — with the other three quadrants added — so that you can duplicate them to hand out to students.

Page 97. There are additional exercises for Section 3.2 on page 92 of this manual.

Page 98. Many students have seen Euler's method in other classes. Check with your students before spending a lot of time on it.

Page 99. The differential equation in Example 3.5, namely, $y' = x - y$, will be used a number of times in the text. (Example 3.1, page 82; Exercise 1, page 91; and Example 5.2, page 192.)

Page 102. We point out that what Euler's method generally does is to jump from one exact solution with a particular initial value to another exact solution with a different initial value.

Page 103. The section *Improvements on Euler's Method* may be omitted, if students are willing to take the Runge-Kutta 4 method on faith.

Page 106. The section *Period Doubling and Chaos* is optional. Our students enjoy it.

Page 112. There are additional exercises for Section 3.3 on page 92 of this manual.

Page 112. Section 3.4, *Comparing Solutions of Differential Equations*, is optional.

Page 115. Section 3.5, *Finding Power Series Solutions,* is optional. If you plan to do series solutions later, here is a further elementary introduction, so students can get their feet wet. The solution of $y' = x - y$ is also obtained in a different way in Chapter 5 (Example 5.2).

Page 117. The differential equation in Exercise 2 of Section 3.5, namely, $y' = 1 - 2xy$, will be seen a number of times later in the book. (Exercise 10, page 92; Example 5.8, page 205; Exercise 27, page 211; and — in its differentiated form, $y'' + 2xy' + 2y = 0$ — in Example 12.3, page 580; and Example 12.5, page 589.)

3.2 Additional Exercises

Section 3.1

Page 94. Add exercises after the current Exercise 18.

19. The equation $ax^2+bxy+cy^2+dx+ey+f = 0$ characterizes conic sections. (If $b^2-4ac < 0$ the curve is an ellipse, circle, point, or no curve; if $b^2 - 4ac = 0$ the curve is a parabola, two parallel lines, one line, or no curve, and if $b^2 - 4ac > 0$ the curve is a hyperbola or two intersecting lines.) Show that a conic section satisfies the differential equation

$$\frac{dy}{dx} = \frac{-2ax - by - d}{bx + 2cy + e}.$$

20. We wish to find all curves $y = y(x)$ with the property that the tangent line at each point on the curve is perpendicular to the line connecting that point to the origin (see Figure 3.1). Use the fact that if the line is perpendicular to the curve, then the product of the slopes of the line and the curve must be -1 to arrive at the differential equation $y' = -x/y$. Use graphical analysis to sketch solutions of this equation. Based on what you see, guess the equation of the curve $y = y(x)$. Check that your guess satisfies the differential equation.

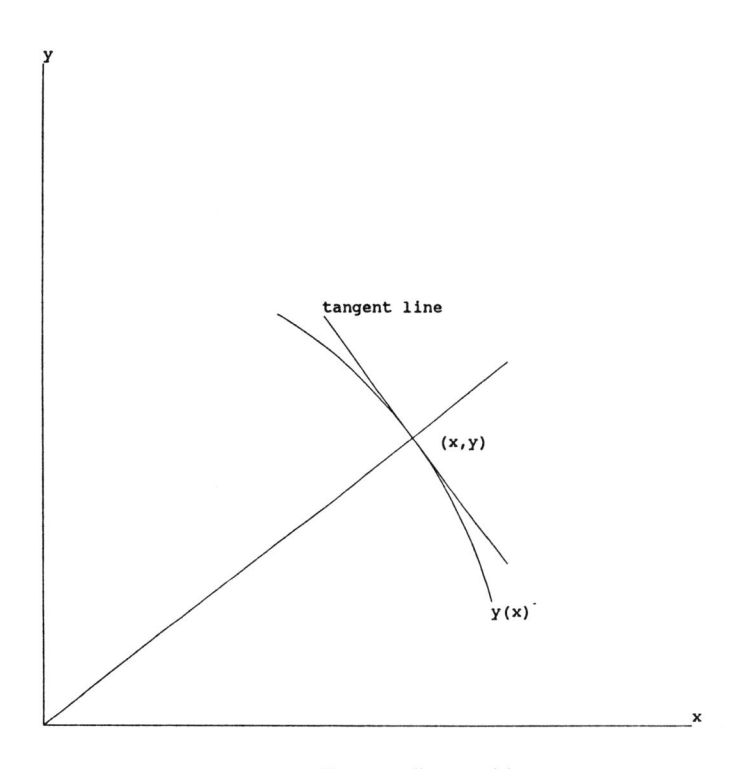

Figure 3.1 Tangent line problem

21. Use slope fields to discuss the graphs of $x^n + y^n = 2^n$ for $n = 1, 2, 3, \cdots$. Make some general comments as to the behavior of these graphs as n increases.

Section 3.2

Page 97. Add exercise after the current Exercise 12.

13. Consider the differential equation

$$y' = \frac{ax + by}{cx + dy},$$

where a, b, c, and d are constants. What conditions must these constants satisfy if this differential equation is symmetric (a) across the x-axis, (b) across the y-axis, (c) across both the x-axis and the y-axis? Confirm your answers by constructing the slope field for the appropriate differential equation.

Section 3.3

Page 112. Add exercise after the current Exercise 14.

15. Computer Experiment: Use a computer program to analyze the numerical solution of Example 2.8 on page 58 of the text — namely, $y' = (y-1)^{2/3}$ in the window $-5 \leq x \leq 5$, $-5 \leq y \leq 5$. After displaying the slope field, plot the numerical solution which starts at $(1, 2)$. Explain its behavior. Repeat for the initial points $(1, 1)$ and $(1, 0)$.

3.3 Additional Materials

Symmetry

Symmetry can also be introduced in a geometrical way.[1] We start with the slope field corresponding to

$$\frac{dy}{dx} = g(x, y) \tag{3.1}$$

and a particular point $\mathcal{P}$ with coordinates (x_0, y_0) in the first quadrant. At $\mathcal{P}$ the slope field will have slope $m = g(x_0, y_0)$ which we depict by drawing a short line with slope m at $\mathcal{P}$ on a piece of paper. This is shown in Figure 3.2.

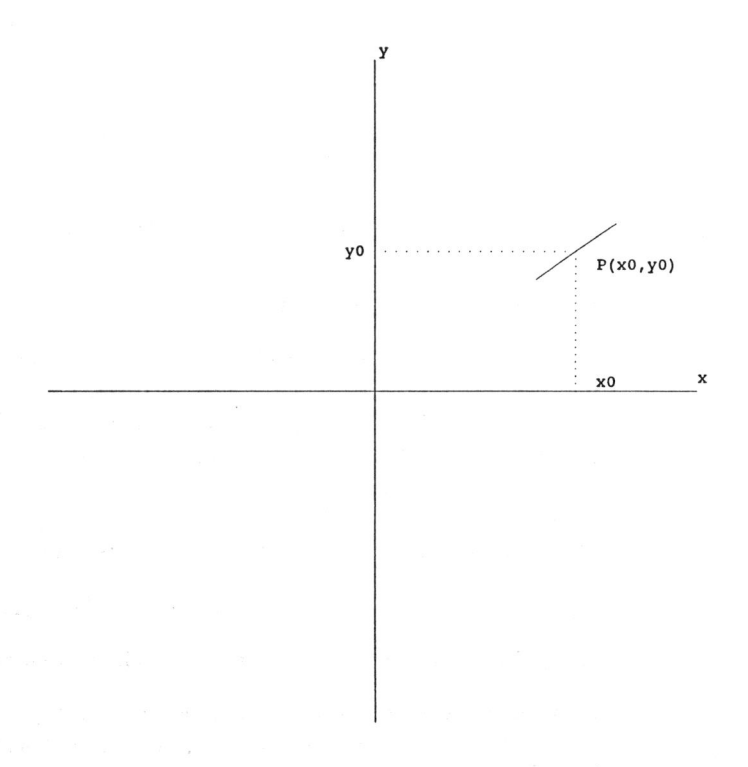

Figure 3.2 The short line with slope $g(x_0, y_0)$

1. **Symmetry across the x-axis.** If we fold the paper along the x-axis, and trace the image of the short line it will be in the fourth quadrant. If we unfold the paper we will see two short lines which are symmetric across the x-axis. This is shown in Figure 3.3. The second short line has slope $-m$ and is located at the point $\mathcal{Q}$ with coordinates $(x_0, -y_0)$. Thus, returning to (3.1), if the function $g(x, y)$ has the value $-m$ at the point $(x_0, -y_0)$, that is, if $g(x_0, -y_0) = -m = -g(x_0, y_0)$, then the slope field through $\mathcal{P}$ will be symmetric across the x-axis. If this happens for every point (x, y), that is, **if**

$$g(x, y) = -g(x, -y),$$

---

[1] We thank our colleague, Maciej Wojtkowski, for this suggestion.

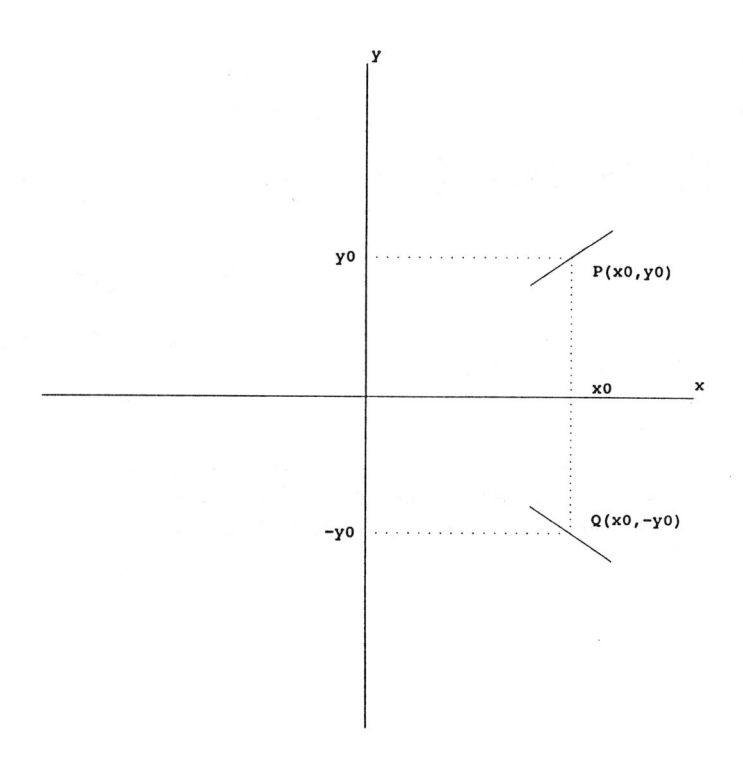

Figure 3.3 Symmetry across the x-axis

then the slope field will be symmetric across the x-axis.

2. **Symmetry across the y-axis.** If we fold the paper along the y-axis, and trace the image of the short line it will be in the second quadrant. If we unfold the paper we will see two short lines which are symmetric across the y-axis. The second short line has slope $-m$ and is located at the point $\mathcal{R}$ with coordinates $(-x_0, y_0)$. This is shown in Figure 3.4. Thus, returning to (3.1), if the function $g(x, y)$ has the value $-m$ at the point $(-x_0, y_0)$, that is, if $g(-x_0, y_0) = -m = -g(x_0, y_0)$, then the slope field through $\mathcal{P}$ will be symmetric across the y-axis. If this happens for every point (x, y), that is, **if**

$$g(x, y) = -g(-x, y),$$

then the slope field will be symmetric across the y-axis.

3. **Symmetry about the origin.** If we fold the paper along the x-axis and then along the y-axis, and trace the image of the short line it will be in the third quadrant. If we unfold the paper we will see two short lines which are symmetric about the origin. The second short line has slope m and is located at the point $\mathcal{S}$ with coordinates $(-x_0, -y_0)$. This is shown in Figure 3.5. Thus, returning to (3.1), if the function $g(x, y)$ has the value m at the point $(-x_0, -y_0)$, that is, if $g(-x_0, -y_0) = m = g(x_0, y_0)$, then the slope field through $\mathcal{P}$ will be symmetric about the origin. If this happens for every point (x, y), that is, **if**

$$g(x, y) = g(-x, -y),$$

then the slope field will be symmetric about the origin.

4. **Symmetry across the line $y = x$.** If we fold the paper along the line $y = x$, and trace the image of the short line it will be in the first quadrant. If we unfold the paper we will see two short lines which are symmetric about the line $y = x$. The second short line has slope $1/m$ and is located at the point $\mathcal{T}$ with coordinates (y_0, x_0). This

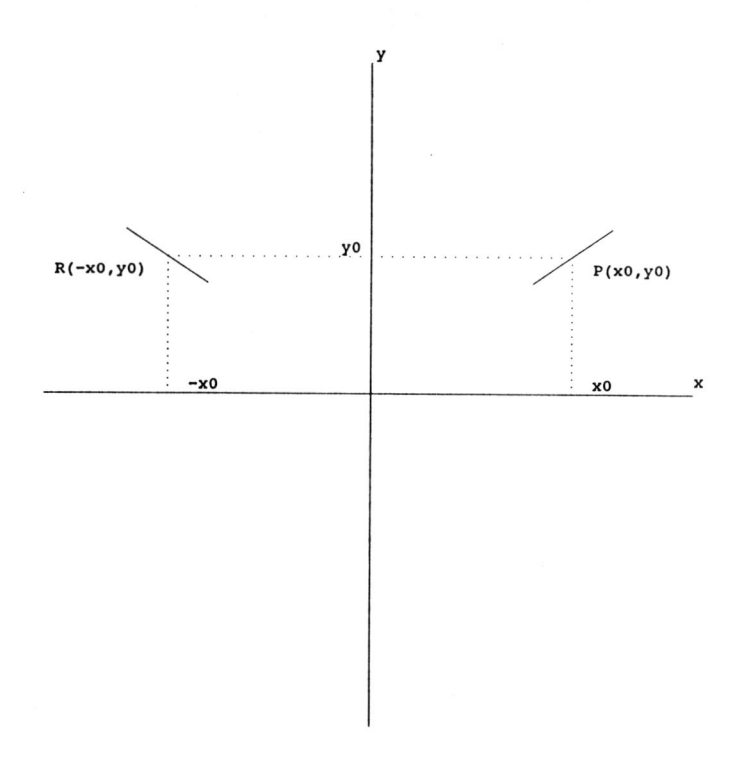

Figure 3.4 Symmetry across the y-axis

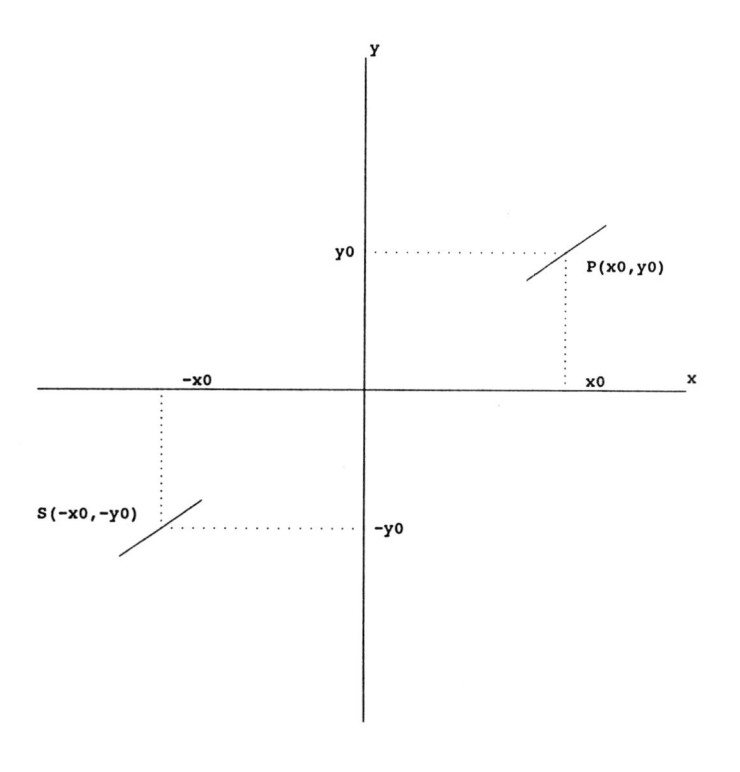

Figure 3.5 Symmetry about the origin

ADDITIONAL MATERIALS

is shown in Figure 3.6. Thus, returning to (3.1), if the function $g(x,y)$ has the value $1/m$ at the point (y_0, x_0), that is, if $g(y_0, x_0) = 1/m = 1/g(x_0, y_0)$, then the slope field through $\mathcal{P}$ will be symmetric across the line $y = x$. If this happens for every point (x, y), that is, **if**

$$g(x,y) = 1/g(y,x),$$

then the slope field will be symmetric across the line $y = x$.

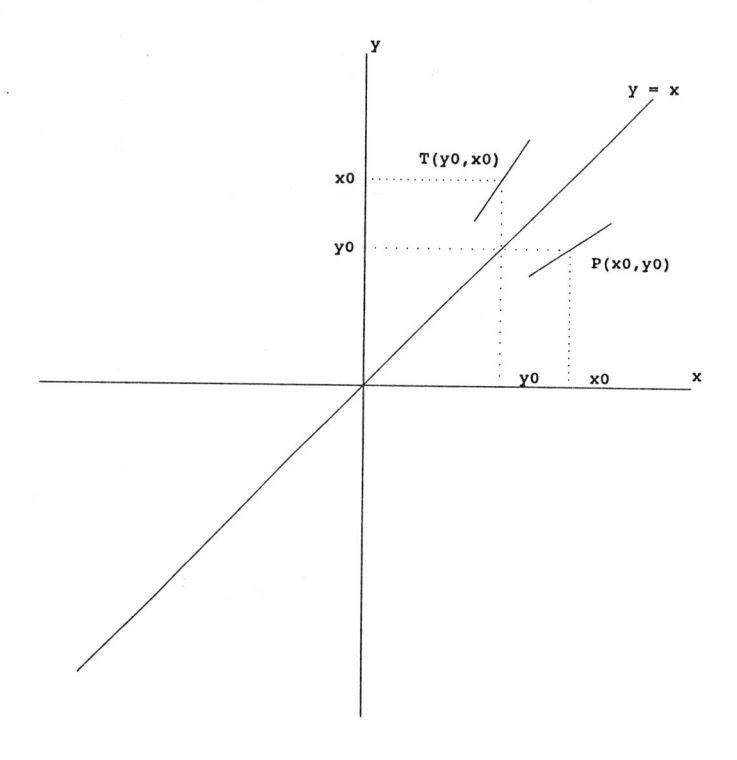

Figure 3.6 Symmetry across the line $y = x$

Figures 3.19 and 3.20

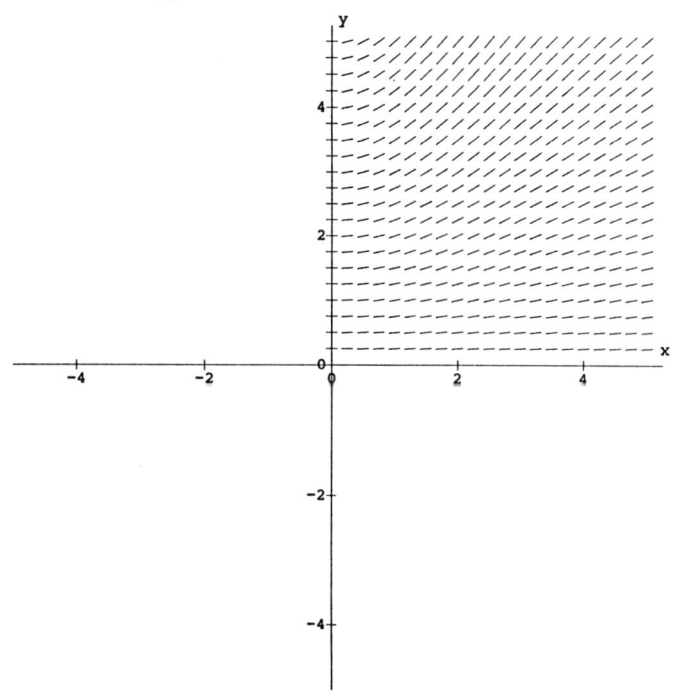

Figure 3.7 Slope field for Exercise 6, Section 3.2

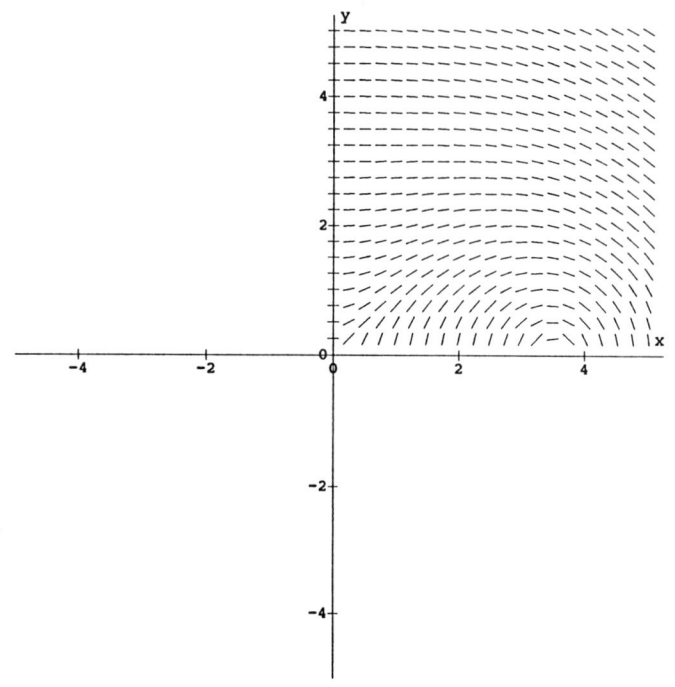

Figure 3.8 Slope field for Exercise 7, Section 3.2

ADDITIONAL MATERIALS

4. IDEAS FOR CHAPTER 4 — "MODELS AND APPLICATIONS LEADING TO NEW TECHNIQUES"

4.1 Page by Page Comments

Chapter 4 requires 6 or 7 class meetings of 50 minutes each.

Sections 4.5, *Applications — Orthogonal Trajectories*, and 4.6, *Piecing Together Differential Equations*, are not extensively used in later sections, and so may be selectively omitted.

Page 120. Mixture problems, such as Example 4.1, will occur again, such as Example 4.2. The key to solving these is to pay attention to the units.

Page 121. We give the students copies of Figure 4.1 (without the hand-drawn curve) and have them draw the solution curve for $y(0) = 50$.

Page 122. The equation $y' = xy + x + y + 1$ is separable because $xy + x + y + 1 = (x+1)(y+1)$. Sometimes it is difficult to identify whether a function $f(x,y)$ is separable. An easy way to test for separability is to select any two points $x = a$ and $y = b$ for which $f(a,b) \neq 0$. A function $f(x,y)$ is separable if and only if $f(x,y) = f(x,b)f(a,y)/f(a,b)$. For example, in the case of $f(x,y) = xy + x + y + 1$, with $a = b = 0$ we have $f(0,0) = 1$, $f(x,0) = x+1$, $f(0,y) = y+1$, and $f(x,y) = f(x,0)f(0,y)/f(0,0) = (x+1)(y+1)$.

Page 126. Example 4.3 is very rich. It is possible to obtain the implicit solution (4.12) in a few lines. However, with the slope field we can make and prove a number of conjectures about the solution curve. It is also possible to make a conjecture which turns out to be false! This example demonstrates the interplay between the graphical and analytical approaches.

Page 129. The section on 'Sketching Solutions of Separable Equations' may be new to you. It is easier to demonstrate than explain. It is used in Chapter 10, when discussing the predator-prey problem, as is previewed in Exercise 13 on page 135.

Page 132. We assign Exercise 5(c) of Section 4.1 to see if the lessons of Section 2.4 have been fully comprehended.

Page 133. Table 4.1 is in TWIDDLE. It is called PLAICE.DTA and is in the subdirectory DD-04.

Page 133. Exercise 8 of Section 4.1 is used to motivate Bernoulli's equation in Section 5.3, page 212.

Page 134. Table 4.2 is in TWIDDLE. It is called TOLUENE.DTA and is in the subdirectory DD-04.

Page 134. Table 4.3 is in TWIDDLE. It is called NICOTINE.DTA and is in the subdirectory DD-04.

Page 135. There are additional exercises for Section 4.1 on page 103 of this manual.

Page 136. In Section 4.2 we avoid calling these homogeneous differential equations to reduce confusion with the term second order homogeneous differential equations.

Page 144. Exercise 4 of Section 4.2 is instructive.

Page 144. Exercise 6 of Section 4.2 illustrates that the slope field of a differential equation with homogeneous coefficients is unchanged if the x- and y-axes are rescaled by equal amounts. This is important when we consider linear autonomous systems in Chapter 9.

Pages 145 – 149. Example 4.7 could be done by either you or your students working individually, or in groups, to gather fresh data.
Simple Experiment. A simple experiment to obtain data from a cooling body was suggested by one of our colleagues.[1] You require a thermometer and a stopwatch. Heat the thermometer until it is well above room temperature. Suspend the thermometer by its top and then, when the mercury is lined up with a graduation mark on the thermometer, start the stopwatch. Then every time the mercury reaches another graduation mark record the time. (It is easier if there are two of you so one can call and the other note the time.)

Page 145. Table 4.4 is in TWIDDLE. It is called PROBE.DTA and is in the subdirectory DD-04.

Pages 145 – 149. Table 4.5 is in TWIDDLE in two different ways. The first way has $\Delta_C T/\Delta t$ versus t. It is called PROBE1.DTA and is in the subdirectory DD-04. The second way has $\Delta_C T/\Delta t$ versus T. It is called PROBE2.DTA and is in the subdirectory DD-04.

Pages 145 – 149. Example 4.7 of the text could be replaced by THE LENGTH OF A STURGEON on page 113 of this manual. This is Exercise 7 of Section 4.3 on page 155.

Page 147. We give the students copies of Figure 4.25 so they can construct Figure 4.27.

Page 148. If we use nonlinear regression to find the parameters k, T_a and T_0 that best fit the Heating of a Probe (starting from the ones we found, namely, $k = -0.135$, $T_a = 34.14$, and $T_0 = 32.78$, and using the program Kalkulator) we find $k = -0.1361$, $T_a = 34.132$, and $T_0 = 32.791$. The graph with these revised values of k, T_a and T_0 does not differ appreciably from Figure 4.29 of the text.

Page 149. Table 4.6 is in TWIDDLE. It is called KENYA.DTA and is in the subdirectory DD-04.

Pages 150 – 153. Table 4.7 is in TWIDDLE in two different ways. The first way has $(1/P)\Delta_C P/\Delta t$ versus P. It is called KENYA1.DTA and is in the subdirectory DD-04. The second way has $(1/P)\Delta_C P/\Delta t$ versus t. It is called KENYA2.DTA and is in the subdirectory DD-04.

Page 151. If we use nonlinear regression to find the parameters $P(0)$, a and b that best fit the Population of Kenya (starting from the ones we found, namely, $P(0) = 6.265$,

---

[1] Jim Cushing, Department of Mathematics, University of Arizona

$a = 0.00406$, and $b = 0.026434$, and using the program Kalkulator) we find $P(0) = 6.280$, $a = 0.000439$, and $b = 0.02589$. The graph with these revised values of $P(0)$, a and b does not differ appreciably from Figure 4.34 of the text.

Page 152. Exercises 1 and 2 of Section 4.3 are very rich.

Page 152. Table 4.8 is in TWIDDLE. It is called COFFEE.DTA and is in the subdirectory DD-04.

Page 154. Exercise 4(c) of Section 4.3 shows another way we could estimate the ambient temperature T_a.

Page 155. Exercise 7 of Section 4.3 is used to motivate Bernoulli's equation in Section 5.3, page 212.

Page 155. Table 4.10 is in TWIDDLE. It is called PUPILS1.DTA and is in the subdirectory DD-04.

Page 155. Table 4.11 is in TWIDDLE. It is called STURGEON.DTA and is in the subdirectory DD-04.

Page 156. Table 4.12 is in TWIDDLE. It is called CAR.DTA and is in the subdirectory DD-04.

Page 157. Table 4.13 is in TWIDDLE. It is called AIDS.DTA and is in the subdirectory DD-04.

Page 157. Exercises 15 and 16 of Section 4.3 show that if the data set is exponential then using the right-hand or central difference quotient to approximate the derivative may not be the best thing to do.

Page 158. There are additional exercises for Section 4.3 on page 105 of this manual.

Page 159. Table 4.14 is in TWIDDLE. It is called FREEFALL.DTA and is in the subdirectory DD-04.

Page 161. The steps from the initial value problem (4.51), namely,

$$\frac{dx}{dt} = V\frac{e^{\alpha t} - 1}{e^{\alpha t} + 1} \qquad x(0) = 0, \tag{4.1}$$

to (4.52), namely,

$$x(t) = \frac{2V}{\alpha} \ln\left[\frac{1}{2}\left(e^{\alpha t} + 1\right)\right] - tV \tag{4.2}$$

are as follows.

Integrating (4.1), using $u = e^{\alpha t} + 1$ and $du = \alpha e^{\alpha t} dt = \alpha(u-1)\, dt$, gives

$$x(t) = V \int \frac{e^{\alpha t} - 1}{e^{\alpha t} + 1}\, dt = \frac{V}{\alpha} \int \frac{u - 2}{u(u-1)}\, du.$$

By partial fractions

$$\int \frac{u-2}{u(u-1)}\, du = \int \left(\frac{2}{u} - \frac{1}{u-1}\right) du = 2\ln u - \ln(u-1) + C,$$

so
$$x(t) = \frac{V}{\alpha}\left[2\ln u - \ln(u-1) + C\right] = \frac{V}{\alpha}\left[2\ln\left(e^{\alpha t}+1\right) - \ln e^{\alpha t} + C\right] = \frac{V}{\alpha}\left[2\ln\left(e^{\alpha t}+1\right) - \alpha t + C\right].$$

The initial condition $x(0) = 0$ requires that $C = -\frac{V}{\alpha}2\ln 2$, so that
$$x(t) = \frac{V}{\alpha}\left[2\ln\left(e^{\alpha t}+1\right) - 2\ln 2 - \alpha t\right] = x(t) = \frac{2V}{\alpha}\ln\left[\frac{1}{2}\left(e^{\alpha t}+1\right)\right] - tV.$$

Page 161. Equation (4.51) can be expressed in the form
$$\frac{dx}{dt} = V\frac{e^{\alpha t}-1}{e^{\alpha t}+1} = V\frac{e^{\alpha t/2}-e^{-\alpha t/2}}{e^{\alpha t/2}+e^{-\alpha t/2}} = V\frac{\sinh(\alpha t/2)}{\cosh(\alpha t/2)},$$

which can be integrated subject to the initial condition $x(0) = 0$ to give
$$x(t) = \frac{2V}{\alpha}\ln\left|\cosh\frac{\alpha t}{2}\right|.$$

This is another way of writing (4.52).

Page 161. The shape of the graph in Figure 4.42 is not surprising and can be estimated from the differential equation (4.47), namely,
$$\frac{V^2}{g}\frac{dv}{dt} = V^2 - v^2.$$

On the one hand, if v is small, so $v \approx 0$, the equation becomes $dv/dt = g$ with solution $v(t) = gt$ (using $v(0) = 0$) which is a line through the origin. On the other hand, if v is large, so that $v \approx V$, the equation becomes $dv/dt = 0$, with solution $v(t) = C$ which is a horizontal line.

Page 161. If we use nonlinear regression to find the parameter V that best fits Sky Diving (starting from the one we found, namely, $V = 182$, and using the program Kalkulator) we find $V = 182.5$. The graph with this revised values of V does not differ appreciably from Figure 4.43 of the text.

Page 162. Table 4.15 is in TWIDDLE. It is called BRAKE.DTA and is in the subdirectory DD-04.

Page 163. The parameter h in $y = hx^2$ can also be estimated by plotting y/x^2 against x.

Page 163. If we use nonlinear regression to find the parameter h that best fits Braking in an Emergency Stop (starting from the one we found, namely, $h = 0.054$, and using the program Kalkulator) we find $h = 0.0539$. The graph with this revised values of h does not differ appreciably from Figure 4.47 of the text.

Page 164. Exercise 3 of Section 4.4 is worth doing.

Page 164. Exercises 5 and 9 of Section 4.4 are very interesting.

Page 165. Exercise 7 of Section 4.4 shows that air-resistance proportional to the velocity is a poor model.

Page 165. Table 4.16 is in TWIDDLE. It is called SPHERE.DTA and is in the subdirectory DD-04.

Page 166. Table 4.17 is in TWIDDLE. It is called BULLET.DTA and is in the subdirectory DD-04.

Page 166. Exercise 10 of Section 4.4 is very rich.

Page 167. There are additional exercises for Section 4.4 on page 109 of this manual.

Page 167. Section 4.5, *Applications — Orthogonal Trajectories*, is not used subsequently, and may be omitted.

Page 175. There are additional exercises for Section 4.5 on page 112 of this manual.

Page 175. Section 4.6, *Piecing Together Differential Equations*, is not used subsequently, and may be omitted.

Page 176. Table 4.18 is in TWIDDLE. It is called IRELAND.DTA and is in the subdirectory DD-04.

Page 178. Exercises 3 and 5 of Section 4.6 have important applications. Exercise 3 shows that a constant drug dose that is taken at equal time intervals has a limiting concentration, if the concentration in the body decay exponentially. Exercise 5 shows that a constant drug dose that is taken at equal time intervals does not have a limiting concentration, if the concentration in the body decays linearly. Many prescribed drugs decay exponentially (approximately) and so have a limiting concentration. Alcohol decays linearly (approximately) for a certain time interval, and then decays exponentially (approximately). Thus, if someone takes alcohol at equal time intervals while it decays linearly, the concentration in the body is unlimited. It leads to alcohol poisoning. However, if someone takes alcohol at equal time intervals while it decays exponentially, the concentration in the body is limited.[2]

Page 181. The sections on FUNCTIONAL EQUATIONS (on page 120 of this manual) and CALCULUS (on page 127 of this manual) show some nonstandard applications of differential equations to mathematics. Students do not find these sections easy.

Page 181. Applications of $y' = g(x)$ — given on pages 27 and 29 of this manual, in sections called WATER-SKIER PROBLEM and the TAPE DECK COUNTERS — could be included here. Students find these sections interesting, but not easy.

4.2 Additional Exercises

Section 4.1

Page 135. Add exercise after the current Exercise 14.

15. A particular function $f(x)$ is defined in the first quadrant, where it is decreasing and concave up. A tangent line is drawn through the point $(x_0, f(x_0))$ where $x_0 > 0$. It intersects the x- and y-axes at $(a, 0)$ and $(0, b)$. See Figure 4.1.

---

[2]We would like to thank our colleague, Michael Mayersohn, Department of Pharmaceutical Sciences, University of Arizona, for this information.

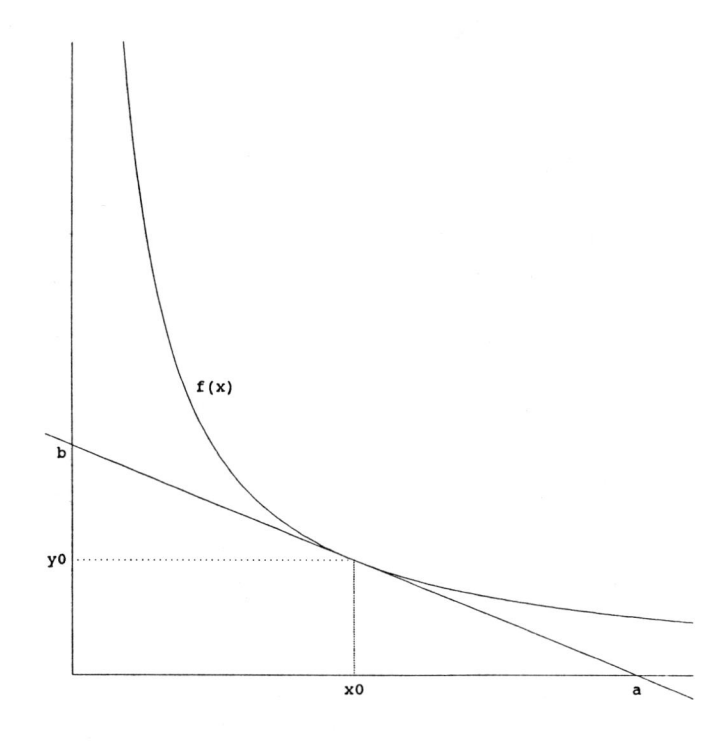

Figure 4.1 The tangent line to $f(x)$

Show that
$$a = x_0 - \frac{f(x_0)}{f'(x_0)}$$
and
$$b = f(x_0) - x_0 f'(x_0).$$

Now show that the area A of the triangle with vertices $(0,0)$, $(a,0)$ and $(0,b)$ — that is, $A = ab/2$ — can be written as
$$A = -\frac{1}{2f'(x_0)} \left[f(x_0) - x_0 f'(x_0)\right]^2.$$

For a given $f(x)$, the area usually depends on the point x_0. A particular $f(x)$ has the property that the area A is the same for all points $x_0 > 0$. That is the quantity
$$A = -\frac{1}{2f'(x)} \left[f(x) - x f'(x)\right]^2$$
is a constant. Differentiate this quantity with respect to x to find
$$f''(x) \left[f(x) - x f'(x)\right] \left[x f'(x) + f(x)\right] = 0.$$

Explain why this implies that
$$x f'(x) + f(x) = 0.$$

Solve this separable differential equation to show that the hyperbola is the only function that has the property that the area A of the triangle as drawn in Figure 4.1 is independent of x_0.

Section 4.3

Page 158. Add exercises after the current Exercise 17.

18. **The Running Lizard:** Experiments have been performed where a lizard is encouraged to run as fast as possible, starting from rest.[3] The distance run is then measured as a function of time. The data gathered are given in Table 4.1 and shown in Figure 4.2. [Table 4.1 is in TWIDDLE. It is called LIZARD.DTA and is in the subdirectory DD-04.]

Table 4.1 Distance traveled by accelerating lizard as a function of time

Time (sec)	Distance (meters)
0.000	0.000
0.044	0.040
0.076	0.100
0.104	0.160
0.150	0.275
0.252	0.520
0.336	0.760
0.416	1.020
0.500	1.280
0.576	1.520
0.661	1.780
0.750	2.050

(a) It is believed that the differential equation governing the acceleration of a lizard is
$$\frac{d^2x}{dt^2} = ae^{-bt} \qquad (4.3)$$
where a and b are positive constants and x is the distance traveled in time t. Explain what happens to the acceleration as t increases. Is this a reasonable model?

(b) Integrate (4.3) once, using the fact that the lizard starts from rest at $t = 0$, to find
$$\frac{dx}{dt} = \frac{a}{b}\left(1 - e^{-bt}\right). \qquad (4.4)$$
What happens to dx/dt as $t \to \infty$? Explain what this means physically, and show how this can be used, in conjunction with Figure 4.2 to estimate a/b.

(c) Integrate (4.4) again, using the fact that the lizard starts from $x = 0$ at $t = 0$, to find
$$x(t) = \frac{a}{b}\left[t - \frac{1}{b}\left(1 - e^{-bt}\right)\right]. \qquad (4.5)$$

(d) Use one of the points from Table 4.1 — say $x(0.661) = 1.780$ — and your estimate for a/b from part (b) to estimate the value of b. Using these estimates for a and b plot (4.5) on Figure 4.2. Figure 4.3 is one possibility. How good is the differential equation (4.3) at modeling an accelerating lizard?

---

[3] "Effects of body size and slope on acceleration of a lizard (*Stellio Stellio*)" by R. B. Huey and P. E. Hertz, J. Exp. Biol., **110**, 1984, pages 113 – 123.

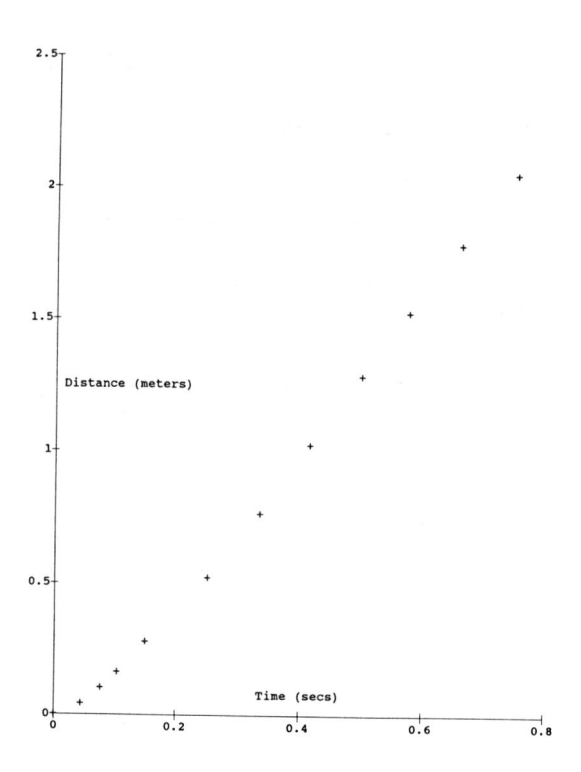

Figure 4.2 The distance run (in meters) as a function of time (in seconds) by an accelerating lizard

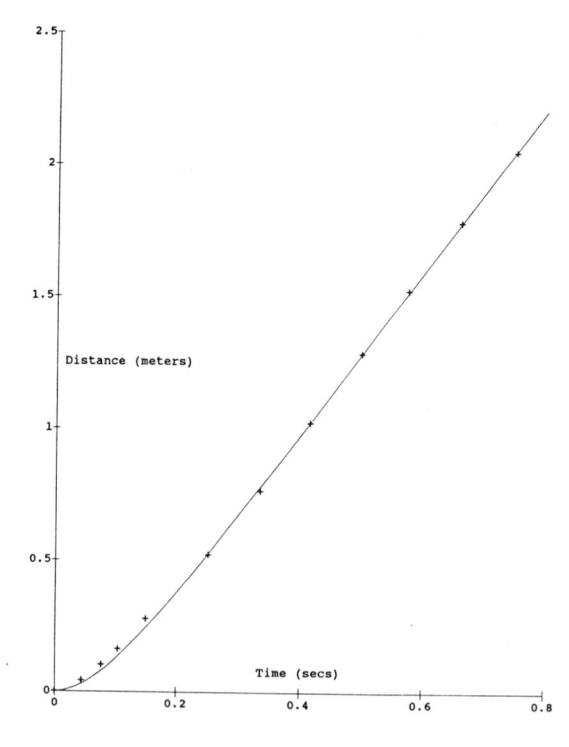

Figure 4.3 The distance run by a lizard and the function $x(t) = (a/b)[t - (1 - e^{-bt}/b)]$

(e) Table 4.2 is a data set corresponding to the times achieved every 10 meters by Carl Lewis in the 100 m final of the World Championship in Rome in 1987.[4] This data set is shown in Figure 4.4. Does the model (4.3) also apply to world-class sprinters? According to this model does Carl Lewis attain his maximum speed while running the 100 m race? [Table 4.2 is in TWIDDLE. It is called LEWIS.DTA and is in the subdirectory DD-04.]

Table 4.2 Distance traveled by Carl Lewis as a function of time

Time (sec)	Distance (meters)
0.00	0
1.94	10
2.96	20
3.91	30
4.78	40
5.64	50
6.50	60
7.36	70
8.22	80
9.07	90
9.93	100

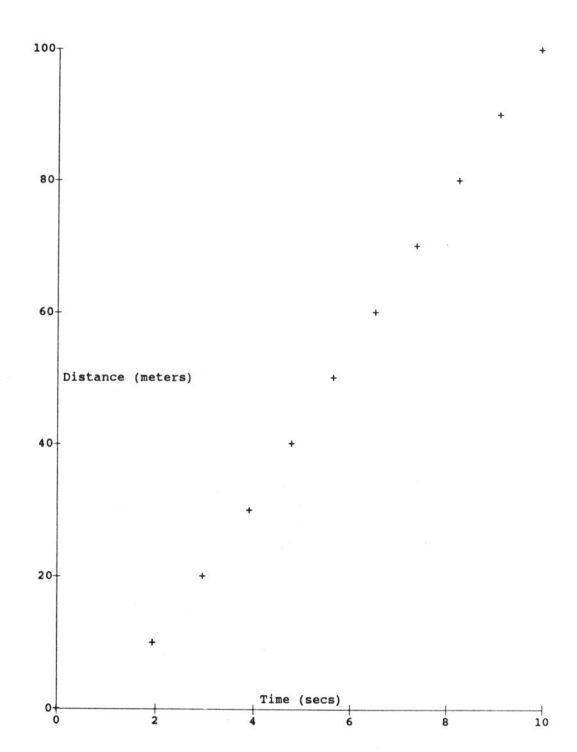

Figure 4.4 The distance run (in meters) as a function of time (in seconds) by Carl Lewis

19. **The Population of the USA - Part 1.** See THE POPULATION OF THE USA - PART 1 on page 131 of this manual.

20. **Animal Tumors.** Observations on the growth of animal tumors indicate that the size y obeys the differential equation $dy/dt = me^{-ht}y$, where h and m are positive constants.

---

[4] "Mathematical Models of Running" by W. G. Pritchard, SIAM Review, **35**, 1993, pages 359 – 379.

This differential equation is sometimes called the Gompertz growth law. Note that physical considerations require $y > 0$ and $t > 0$.

(a) Check this differential equation for equilibrium solutions and label any you find.

(b) Look at the slope field for $y > 0$, $t > 0$ for this equation to obtain some idea on how solution curves behave. What sort of behavior does this remind you of?

(c) Determine regions where the solutions of the differential equation are increasing, decreasing, concave up, or concave down.

(d) Just by looking at the differential equation, what do you expect to happen as $t \to \infty$?

(e) Solve this differential equation subject to the initial condition $y(0) = y_0$, $y_0 > 0$.

(f) Find a relationship between y_0, h, and m such that the graph of y versus t will have no inflection point.

(g) Show that the carrying capacity b for the Gompertz growth law depends on the initial condition. Is it possible to find an initial condition $y(0)$ for which $y(0) > b$?

(h) The Gompertz growth law is sometimes written $dy/dt = -ky \ln \frac{y}{b}$, where k and b are positive constants. (See Exercise 11 on page 50.) What is the relationship between h, m, k and b?

21. **Simple Experiment.** Perform the following experiments related to Newton's law of cooling.

    (a) Place a hot liquid in a cup, and measure the temperature of the liquid as a function of time. The appropriate time interval will depend on the insulating properties of your container. Plot this data set as in Exercise 1 (a) and (b) to see which gives a better fit.

    (b) Repeat this experiment with hot liquid at the same initial temperature as in your experiment in part (a), only this time cover the top of the liquid with whipping cream (or a substitute). Compare the two data sets and explain any differences.

22. **Contracting Pupil.** Table 4.3 shows the result of an experiment where 10 subjects where kept in a darkened room and then brought into light, causing the pupils to contract, and the diameter d in millimeters of the pupils was measured every 1/3 sec.[5]

Table 4.3 The average diameter D in millimeters of a dilating pupil at time t in seconds

t	0.33	0.67	1.00	1.33	1.67	2.00	2.33	2.67	3.00
D	6.56	5.58	4.91	4.51	4.19	3.91	3.59	3.52	3.42

(a) A model proposed for this phenomena is $dD/dt = a + bD$, where t is the time in seconds, and a and b are constants.

   i. What is the equilibrium solution of this differential equation? Show that this differential equation can be rewritten in terms of its equilibrium solution, D_e, as $dD/dt = a(D - D_e)$. Explain why this differential equation is a reasonable model. What physical meaning can be attached to D_e? What is the sign of a?

   ii. What other phenomena does this equation model?

   iii. Solve this differential equation. How many parameters are there in this solution? Using the ideas from Exercise 4 estimate these parameters for the data set in Table 4.3. How well does this model fit the data set?

---

[5] Adapted from "Quantitative Methods in Psychology" by D. Lewis, McGraw-Hill, New York, 1960.

iv. What does this model predict about the ultimate diameter of the pupil?

(b) Another model proposed for this phenomena is $dA/dt = a + bA$, where $A = \pi D^2/4$ is the area of the pupil, t is the time in seconds, and a and b are constants.

i. What is the equilibrium solution of this differential equation? Show that this differential equation can be rewritten in terms of its equilibrium solution, A_e, as $dA/dt = a(A - A_e)$. Explain why this differential equation is a reasonable model. What physical meaning can be attached to A_e? What is the sign of a?

ii. What other phenomena does this equation model?

iii. Solve this differential equation. How many parameters are there in this solution? Using the ideas from Exercise 4 estimate these parameters for the data set in Table 4.3. How well does this model fit the data set?

iv. What does this model predict about the ultimate diameter of the pupil?

(c) Which of the models proposed in parts (a) and (b) do you think is more reasonable?

23. The Chanter Growth Equation. The Chanter growth equation is $dy/dt = \alpha y (b - y) e^{-\beta t}$, where α, b, and β are positive constants.

(a) It is claimed that this equation is a hybrid of the logistic equation and the Gompertz growth law described in Exercise 20. Justify this claim.

(b) Look at the slope field for this equation to obtain some idea on how solution curves behave. What sort of behavior does this remind you of?

(c) Determine regions where the solutions of the differential equation are increasing, decreasing, concave up, or concave down.

(d) Solve this differential equation subject to the initial condition $y(0) = y_0$, $y_0 > 0$.

Section 4.4

Page 167. Add exercises after the current Exercise 10.

11. Free Fall.[6] Table 4.4 and Figure 4.5 show the results of an experiment in which people fell freely from rest from heights up to 31,400 feet. It shows the distance fallen in feet as a function of time in seconds. How well does this fit the model given by (4.45) in the text? [Table 4.4 is in TWIDDLE. It is called JUMPS.DTA and is in the subdirectory DD-04.]

Table 4.4 Freefall distance against time

Time (sec)	Distance (feet)	Time (sec)	Distance (feet)	Time (sec)	Distance (feet)
0.0	0	46.7	10850	88.1	22000
9.9	620	51.3	12000	92.7	23260
14.5	1200	55.9	13300	97.3	24320
19.1	2550	60.5	14800	101.9	25700
23.7	3700	65.1	16250	106.5	26950
28.3	5250	69.7	17330	111.1	28230
32.9	6800	74.3	18600	116.0	29300
37.5	8200	78.9	19750		

---

[6] "The Physiology of Free Fall Through the Air: Delayed Parachute Jumps" by A. J. Carlson, A. C. Ivy, L. R. Krasmo, and A. H. Andrews, Quarterly Bulletins, Northwestern University Medical School, **16**, 1942, page 254, as reported in "Introduction to Nonlinear Differential and Integral Equations" by H. T. Davis, Dover, 1962, page 62.

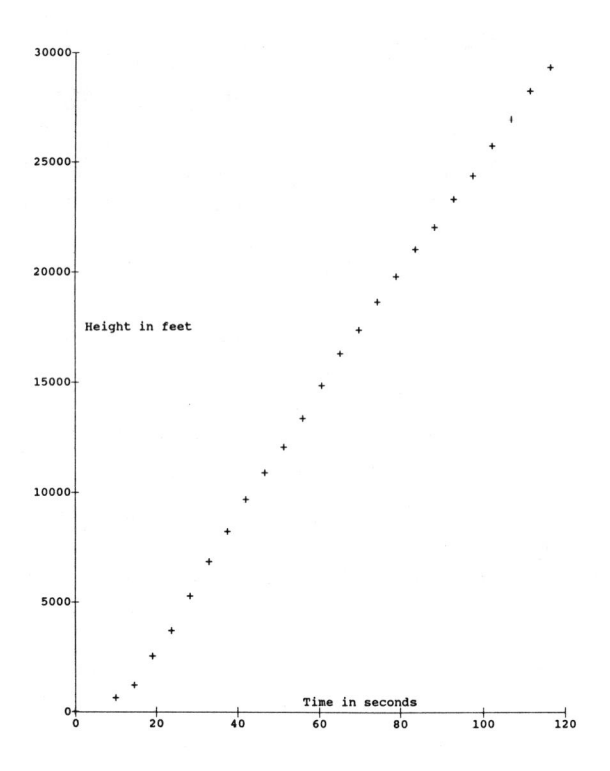

[Graph: Height in feet vs Time in seconds, showing freefall data points]

Figure 4.5 Freefall distance against time

12. **Terminal Velocity of a Shuttlecock.**[7] The data in Table 4.5 and shown in Figure 4.6 was obtained by dropping a badminton shuttlecock and recording its distance as a function of time. Which of the two differential equations

$$m\frac{dv}{dt} = mg - kv$$

and

$$m\frac{dv}{dt} = mg - kv^2$$

best models the motion of the shuttlecock? [Table 4.5 is in TWIDDLE. It is called BADMIN.DTA and is in the subdirectory DD-04.]

13. **Investigating Traffic Accidents**[8] When an investigating officer arrives at the scene of a traffic accident, one of the most important pieces of information is to estimate the speed of the vehicles involved. In a car accident where a vehicle has left straight skidmarks on the road, the formula that investigators use to determine the speed V_0, in miles per hour, at which the vehicle was traveling to cause skidmarks of length ℓ is

$$V_0 = 5.5\sqrt{f\ell}.$$

Here ℓ is measured in feet, $g = 32.2$ ft/sec^2 is the gravitational constant, and f is a constant (the coefficient of friction) that can be determined from test skids at the scene of the accident or from tables. This formula is used under the assumption that there is no drag. Where does this formula come from? This is the same problem as Example 4.10 on page 161 of the text with $k = fg$, and so formula (4.57) gives

$$V_0 = \sqrt{2fg\ell} = 8.025\sqrt{f\ell}$$

---

[7] "Terminal velocity of a shuttlecock in vertical fall" by M. Peastrel, R. Lynch, and A. Armenti, Jr., The American Journal of Physics, **48**, 1980, pages 511 through 513.

[8] "Physics in accident investigations" by M. L. Brake, The Physics Teacher, January 1981, pages 26 through 29.

Table 4.5 Vertical distance fallen by a shuttlecock as a function of time

Time (seconds)	Distance (meters)
0.347	0.61
0.470	1.00
0.519	1.22
0.582	1.52
0.650	1.83
0.674	2.00
0.717	2.13
0.766	2.44
0.823	2.74
0.870	3.00
1.031	4.00
1.193	5.00
1.354	6.00
1.501	7.00
1.726	8.50
1.873	9.50

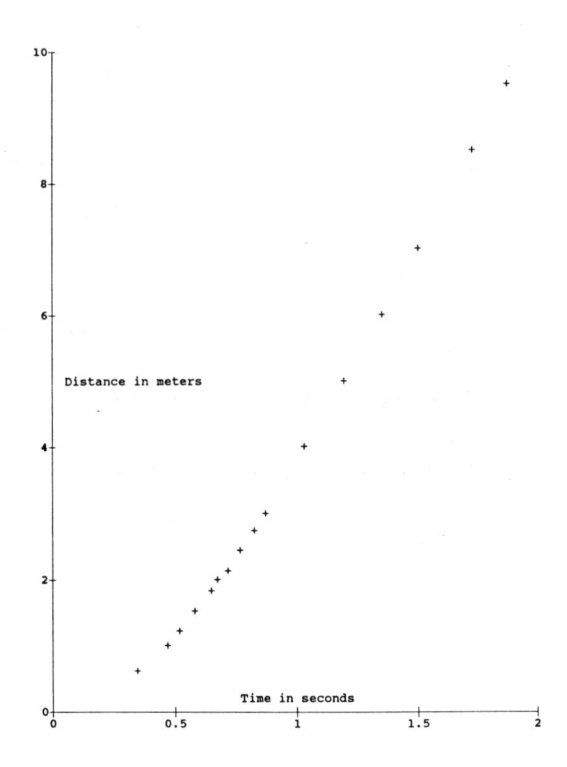

Figure 4.6 Vertical distance fallen by a shuttlecock as a function of time

ADDITIONAL EXERCISES

in ft/sec. Converting this to miles per hour gives

$$V_0 = 5.47\sqrt{f\ell}.$$

14. **Exceeding Terminal Velocity.** In all the examples involving terminal velocity, the parachutist started from rest and approached the terminal velocity from below, never actually reaching it. Is it possible to select a function $f(v)$ so that the initial value problem

$$\frac{dv}{dt} = g - f(v)$$

    $v(0) = 0$, has an equilibrium solution $v = V > 0$, and also has a solution $v(t)$ so that for some time interval the velocity exceeds the terminal velocity? Explain fully.[9]

15. (Because the differential equations encountered in this exercise can be solved in different ways, this same exercise is repeated — with appropriate change of notation and extensions — on page 231 of this manual.)

    Projectiles. See the section called PROJECTILES on page 139 of this manual.

16. **Allometric Growth.** See page 134 of this manual on ALLOMETRIC GROWTH.

Section 4.5

Page 175. Add exercise after the current Exercise 13.

14. **Oblique Trajectories.** Oblique (or isogonal) trajectories occur when one family of curves intersects another family of curves at some constant angle θ, $\theta \neq \pi/2$. If one family is given by the solution of the differential equation $y' = f(x, y)$, the other family, which intersects these curves at an angle of θ, is given by the solution of the differential equation

$$\frac{dy}{dx} = \frac{f(x,y) + \tan\theta}{1 - f(x,y)\tan\theta}.$$

    (a) Derive this formula. [Hint: Use the identity for the tangent of the sum of two angles.]
    (b) Find the family of lines that intersects the lines $y = x + \lambda$ at an angle of $\pi/3$.
    (c) Find the family of curves that intersects the hyperbolas $y^2 = 2x^2 + \lambda$ at an angle of $\arctan(1/2)$.
    (d) Find the family of curves that intersects the circles $x^2 + y^2 = \lambda^2$ at an angle of $\pi/4$.
    (e) Find the family of curves that intersects the parabolas $y^2 = \lambda x$ at an angle of $\pi/4$.

---

[9]Based on "Can Terminal Velocity be Exceeded" by J. Fink, J. H. Freeman, and C. Hampton, CoODEoE, Winter 1994, pages 5 through 7.

4.3 Additional Materials

The Length of a Sturgeon

Example 4.1 : The Length of a Sturgeon

A biologist measured the length of a sturgeon over a 21 year period.[10] Table 4.6 shows the resulting data set where the length is in centimeters and the time is in years. This data set is plotted in Figure 4.7. Our goal is to determine a formula that will allow us to find the length at other times. [Table 4.6 is in TWIDDLE. It is called STURGEON.DTA and is in the subdirectory DD-04.]

Table 4.6 Length of sturgeon as a function of time

Time (years)	Length) (cm)	Time (years)	Length) (cm)
0	21.1	11	107.6
1	32.0	12	112.7
2	42.3	13	117.7
3	51.4	14	122.2
4	60.1	15	126.5
5	68.0	16	130.9
6	75.3	17	135.3
7	82.3	18	140.2
8	89.0	19	145.0
9	95.3	20	148.6
10	101.6	21	152.0

Because the length is clearly changing with time we ask what is the relationship between the length of the sturgeon L and the time t. To answer this question, we try to find a differential equation governing this process, which means we want to relate the derivative dL/dt to a function of t and L. We can obtain an approximate numerical value for dL/dt from Table 4.6 by using one of several approximate forms of the derivative. Here we list three such forms, the right-hand difference quotient

$$\frac{dL}{dt} \approx \frac{\Delta_R L}{\Delta t} = \frac{L(t+h) - L(t)}{h}, \qquad (4.6)$$

the left-hand difference quotient

$$\frac{dL}{dt} \approx \frac{\Delta_L L}{\Delta t} = \frac{L(t-h) - L(t)}{-h}, \qquad (4.7)$$

and the average of the last two, the central difference quotient,

$$\frac{dL}{dt} \approx \frac{\Delta_C L}{\Delta t} = \frac{L(t+h) - L(t-h)}{2h}. \qquad (4.8)$$

Of these, the central difference quotient is the approximation most commonly used. Table 4.7 shows the central difference quotient calculations $\Delta_C L/\Delta t$ for Table 4.6.

---
[10] "General System Theory" by L. von Bertalanffy, Braziller, 1968, page 177, Table 7.5.

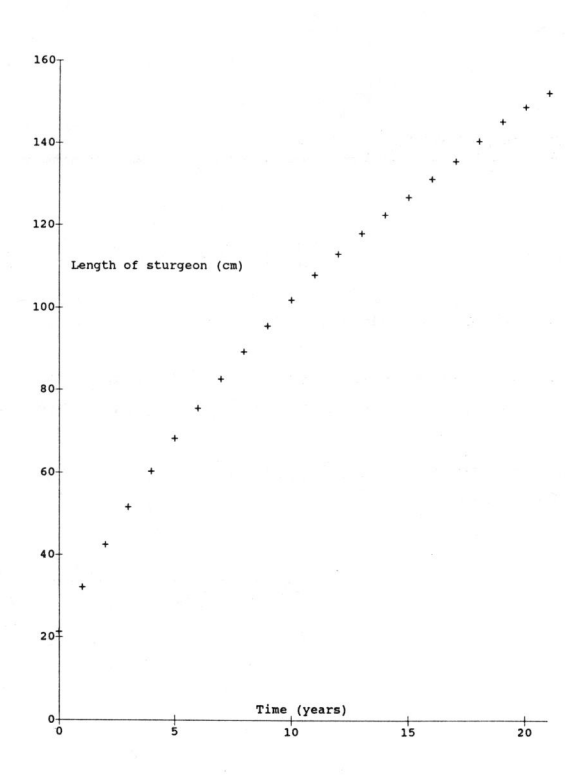

Figure 4.7 Length of sturgeon versus time

Table 4.7 Length of sturgeon as a function of time and $\Delta_C L/\Delta t$

Time (years)	Length (cm)	$[L(t+1) - L(t-1)]/2$
0	21.1	
1	32.0	10.60
2	42.3	9.70
3	51.4	8.90
4	60.1	8.30
5	68.0	7.60
6	75.3	7.15
7	82.3	6.85
8	89.0	6.50
9	95.3	6.30
10	101.6	6.15
11	107.6	5.55
12	112.7	5.05
13	117.7	4.75
14	122.2	4.40
15	126.5	4.35
16	130.9	4.40
17	135.3	4.65
18	140.2	4.85
19	145.0	4.20
20	148.6	3.50
21	152.0	

IDEAS FOR CHAPTER 4 — "MODELS AND APPLICATIONS LEADING TO NEW TECHNIQUES"

In Figure 4.8 we plot this numerical approximation for dL/dt against the time t. We do not see any obvious relation between these variables. In Figure 4.9 we plot the numerical approximation for dL/dt against the length L, and it appears that the relationship is approximately linear with a negative slope. Figure 4.10 is Figure 4.9 with the addition of straight line with slope -0.055 and vertical intercept 11.66.

This suggests that the differential equation governing the change in length might be of the form
$$\frac{dL}{dt} = a + kL,$$
where a and k are constants, with $k < 0$. This differential equation has the equilibrium solution $-a/k$, which we label L_a, so $L_a = -a/k$. We can then rewrite this differential equation in terms of L_a as
$$\frac{dL}{dt} = k(L - L_a). \tag{4.9}$$

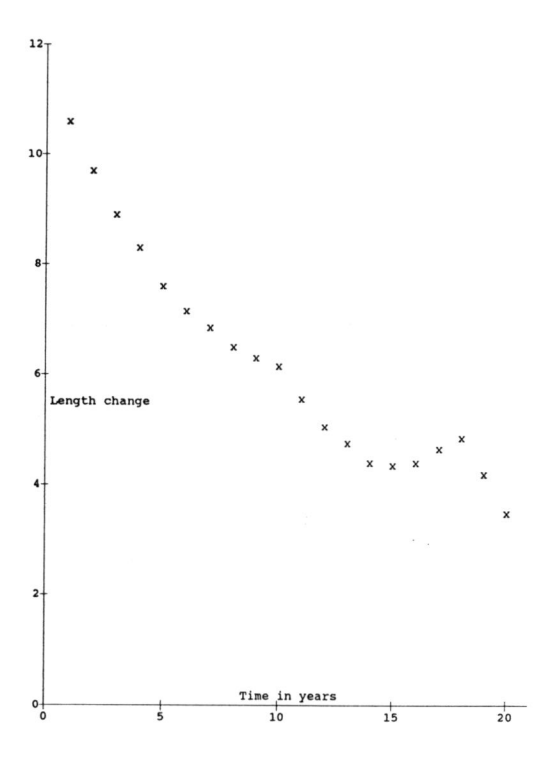

Figure 4.8 Numerical approximation for dL/dt versus time t

Before trying to solve (4.9), we will use graphical arguments to analyze this model. For this to be a reasonable model for the growth of a sturgeon, we expect the solution curves to be increasing, concave down, and tend to the greatest length as time increases. This agrees with the slope field for (4.9), as shown in Figure 4.11 for $L_a = 4$ and $k = -0.5$.

From (4.9) we see that if L, the length of the sturgeon, is less than L_a, then $k(L - L_a)$ is positive (remember $k < 0$), so that L is an increasing function. Furthermore, the nearer L gets to L_a, the smaller the increase. If we take the derivative of (4.9) we find
$$\frac{d^2L}{dt^2} = k\frac{dL}{dt} = k^2(L - L_a).$$

This equation tells us that for $L < L_a$, the solution curves are concave down. All of this seems to suggest that (4.9) is a reasonable model.

ADDITIONAL MATERIALS

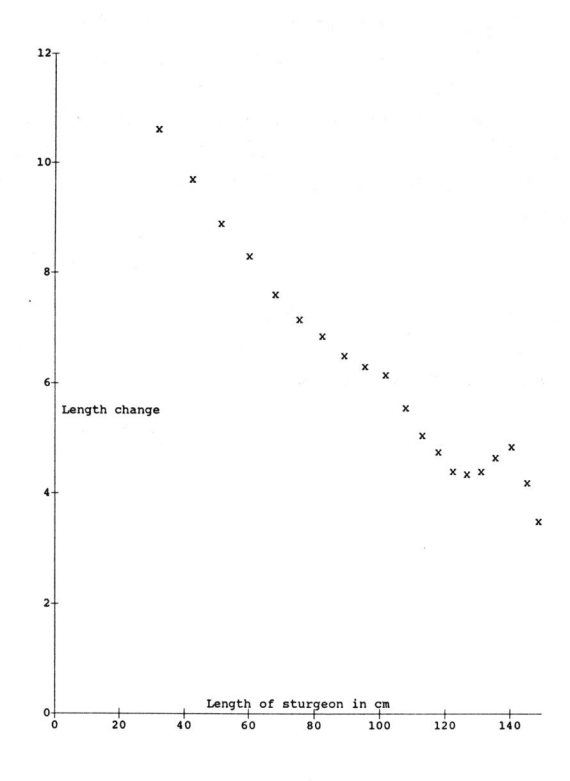

Figure 4.9 Numerical approximation for dL/dt versus length L

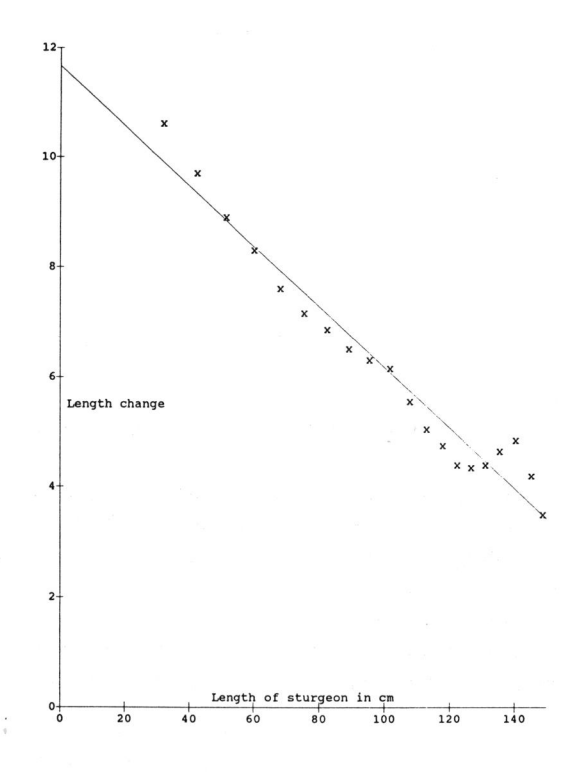

Figure 4.10 The line $11.66 - 0.055L$ and the numerical approximation for dL/dt versus length L

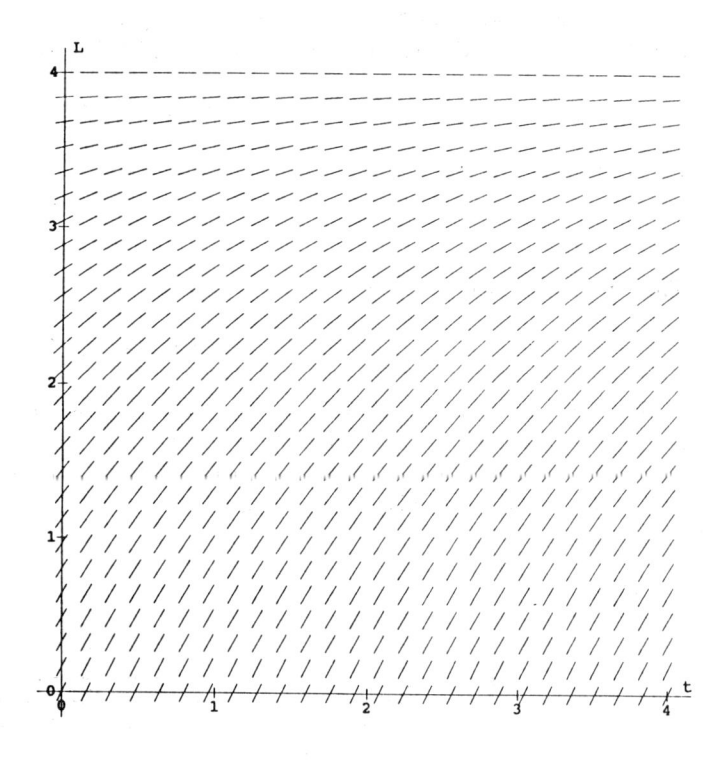

Figure 4.11 Slope field for $dL/dt = k(L - L_a)$

Because (4.9) is an autonomous differential equation (the right-hand side is independent of t), a phase line analysis can be performed on (4.9), and we find from Figure 4.12 that L_a is a stable equilibrium. This means that the length will tend to the quantity L_a as time increases, which means that L_a is the ultimate length of the sturgeon.

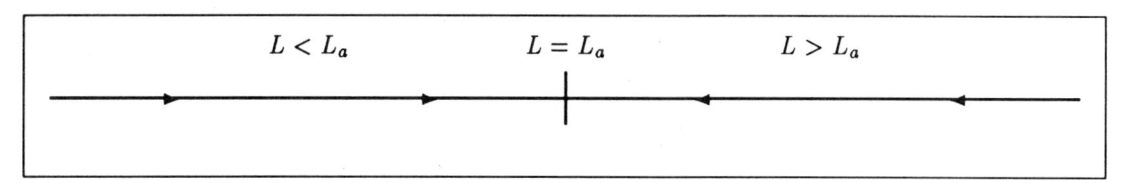

Figure 4.12 Phase line analysis for $dL/dt = k(L - L_a)$

We now turn to finding the explicit solution of (4.9), which has the equilibrium solution

$$L(t) = L_a. \tag{4.10}$$

By rewriting (4.9) in the form

$$\frac{1}{L - L_a} \frac{dL}{dt} = k$$

and then integrating, we find

$$\ln|L - L_a| = kt + c.$$

We isolate the length by taking the exponential of this equation,

$$|L - L_a| = e^{kt+c},$$

which can be written as

$$L - L_a = Ce^{kt},$$

ADDITIONAL MATERIALS

where the arbitrary constant C may be either positive or negative. (Why?) If we permit $C = 0$, we can absorb the equilibrium solution (4.10) into this solution. Rearranging gives

$$L(t) = L_a + Ce^{kt},$$

where the value of C is determined by requiring the value of the length at $t = 0$ to equal the initial length L_0

$$L_0 = L_a + C.$$

Thus, the final form of the solution is

$$L(t) = L_a + (L_0 - L_a)e^{kt}. \qquad (4.11)$$

A brief look at this equation leads us to the conclusion that for large values of time, the exponential term will be insignificant (remember that $k < 0$) and the length of the sturgeon will approach the ultimate length This agrees with the slope field for (4.9) as shown in Figure 4.13, and it also agrees with our common sense.

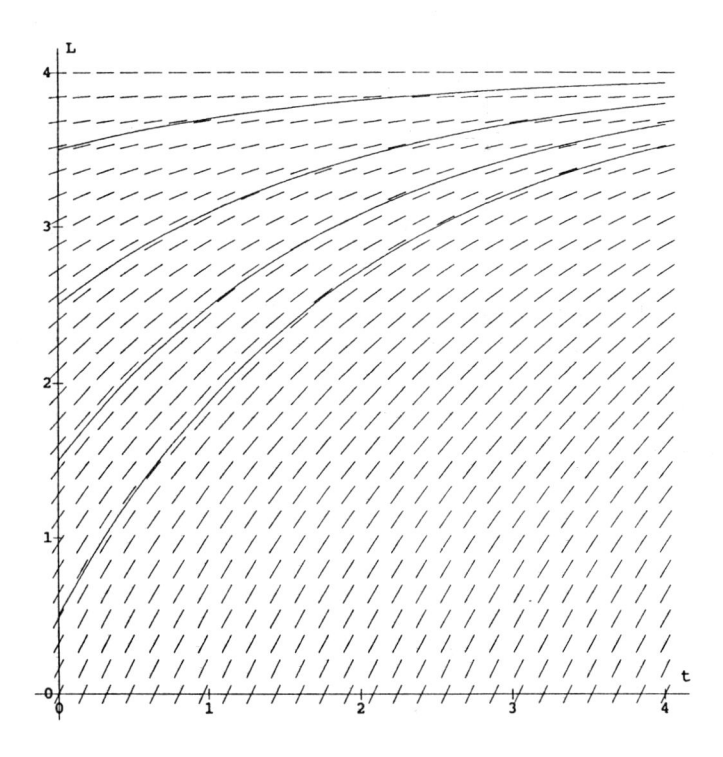

Figure 4.13 Some solution curves and slope field for $dL/dt = k(L - L_a)$

We now return to the original experimental data set (Table 4.6) to see how well it compares to our mathematical model. It is clear that Figure 4.10 gives a crude estimate (based on the differential equation) of $k \approx -0.055$ and $kL_a = -a \approx 11.66$, from which we can estimate the ultimate length as $L_a \approx 212$. However, we expect to get a more accurate estimate of the parameters by comparing the data with the exact solution (4.11). If we rewrite (4.11) in the form

$$L_a - L(t) = (L_a - L_0)e^{kt}$$

and take the logarithm of both sides, we find

$$\ln[L_a - L(t)] = \ln(L_a - L_0) + kt.$$

Thus, if we use the estimate $L_a \approx 212$ and plot $\ln[212 - L(t)]$ against t and then find a straight line, its slope will determine k, and its vertical intercept will determine $\ln(L_a - L_0)$. Figure 4.14 shows this plot, together with the straight line $5.246 - 0.054t$. Thus, we estimate $k \approx -0.054$ and $L_a - L_0 \approx e^{5.246}$, which gives $L_0 \approx 22.19$, in good agreement with the experimental value of L_0 and our crude initial estimate of $k \approx -0.055$. When these estimates, together with $L_a \approx 212$, are substituted in (4.11), we find

$$L(t) = 212 - e^{5.246 - 0.054t}.$$

The agreement between the graph of this function and the original data set may be seen in Figure 4.15. [The function $L(t)$ is in TWIDDLE. It is called STURGEON.TWD and is in the subdirectory DD-04.]□

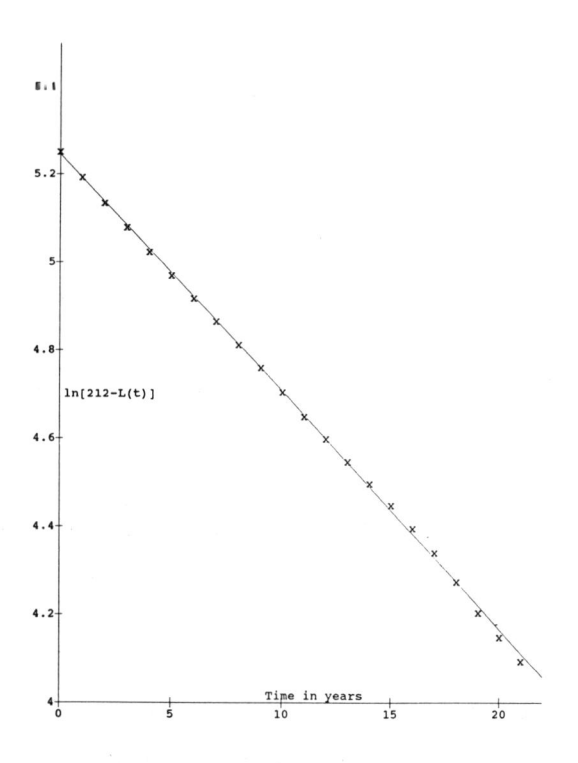

Figure 4.14 The line $5.246 - 0.054t$ and $\ln[212 - L(t)]$ versus time

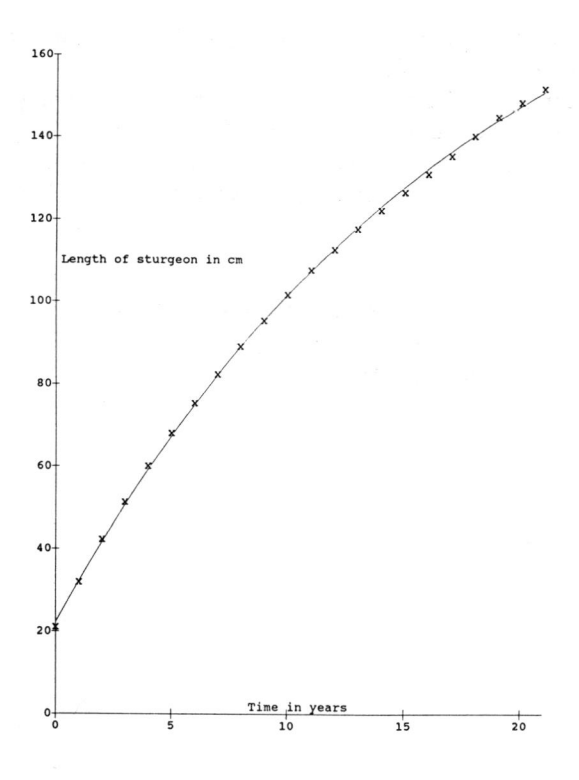

Figure 4.15 The function $212 - e^{5.246-0.054t}$ and length of sturgeon versus time

Applications — Functional Equations

In this section we show how differential equations can be used to solve functional equations.

Example 4.2 : A Functional Equation

We now look at a type of problem that arises in various branches of mathematics — namely, solving functional equations. A simple example of this problem is to find all differentiable functions $f(x)$ that satisfy the condition

$$f(u+v) = f(u) + f(v) \text{ for all } u, v. \tag{4.12}$$

At first sight this problem appears unrelated to differential equations.

By setting $v = 0$ in (4.12) we see that

$$f(u) = f(u) + f(0),$$

which tells us that $f(0) = 0$.

If (4.12) is rewritten in the form

$$\frac{f(u+v) - f(u)}{v} = \frac{f(v)}{v} = \frac{f(v) - f(0)}{v}$$

and we let $v \to 0$, the left-hand side becomes $f'(u)$ and the right-hand side becomes $f'(0)$, which is independent of u, and so is a constant, say, m. Thus, we find the differential equation

$$f'(u) = m,$$

subject to $f(0) = 0$, with solution
$$f(u) = mu. \tag{4.13}$$

That (4.13) satisfies (4.12) is easily seen. Thus, *the most general differentiable function $f(u)$ satisfying (4.12) is the linear function mu.* □

Example 4.3 : Another Functional Equation

The function $\tan x$ satisfies the identity
$$\tan(x+y) = \frac{\tan x + \tan y}{1 - \tan x \tan y},$$

which is the cornerstone of many applications of the tangent function. A natural problem is to find the most general differentiable function, defined at $x = 0$, that satisfies the functional equation
$$f(x+y) = \frac{f(x) + f(y)}{1 - f(x)f(y)}. \tag{4.14}$$

If we put $x = y = 0$ in (4.14), we obtain
$$f(0) = \frac{2f(0)}{1 - [f(0)]^2}.$$

This is true if $[f(0)]^2 = -1$ or $f(0) = 0$, so we have
$$f(0) = 0.$$

If we consider the definition of $f'(0)$ and use this last fact, we have
$$f'(0) = \lim_{y \to 0} \frac{f(y) - f(0)}{y} = \lim_{y \to 0} \frac{f(y)}{y}. \tag{4.15}$$

From (4.14) we see that
$$f(x+y) - f(x) = \frac{f(x) + f(y)}{1 - f(x)f(y)} - f(x) = \frac{[1 + f^2(x)] f(y)}{1 - f(x)f(y)}$$

so that
$$\frac{df}{dx} = \lim_{y \to 0} \frac{f(x+y) - f(x)}{y} = \lim_{y \to 0} \frac{[1 + f^2(x)]}{[1 - f(x)f(y)]} \frac{f(y)}{y},$$

from which, by (4.15) and the fact that $f(y) \to 0$ as $y \to 0$, we obtain the differential equation
$$\frac{df}{dx} = [1 + f^2(x)] f'(0).$$

This equation can be written in separated form
$$\int \frac{df}{1 + f^2(x)} = \int f'(0) \, dx,$$

giving
$$\tan^{-1}(f(x)) = f'(0)x + C.$$

Using the initial condition $f(0) = 0$, gives $C = 0$, and
$$f(x) = \tan(f'(0)x).$$

Thus, the most general function satisfying (4.14) is $\tan ax$. If it is further specified that the derivative of the function we seek has the value 1 at the origin, then the most general function is $\tan x$.

ADDITIONAL MATERIALS

Exercises

1. Imagine we conduct the following experiment using a spring scale that measures weight (that is, force). We want to see how the distance the spring pointer moves down is related to the weight of the object hung on the scale. With nothing on the scale it reads 0, so the distance displaced is 0 when there is 0 weight on the scale. Put a weight w_1 on the scale and measure the displacement of the pointer, namely, $d(w_1)$. (See Figure 4.16.) Now we replace the weight w_1 with a weight w_2 and measure $d(w_2)$. Finally we put both w_1 and w_2 on the scale and measure $d(w_1 + w_2)$. We discover that

$$d(w_1 + w_2) = d(w_1) + d(w). \qquad (4.16)$$

We repeat this for various w_1 and w_2 and discover that (4.16) is always valid (unless of course we stretch the spring too much by hanging a very heavy weight on the scale!). What does (4.16) tell us about the relationship between d and w? [Hint: Look at (4.12).] What does this tell us about the relationship between w and d? This relationship is called HOOKE'S LAW.

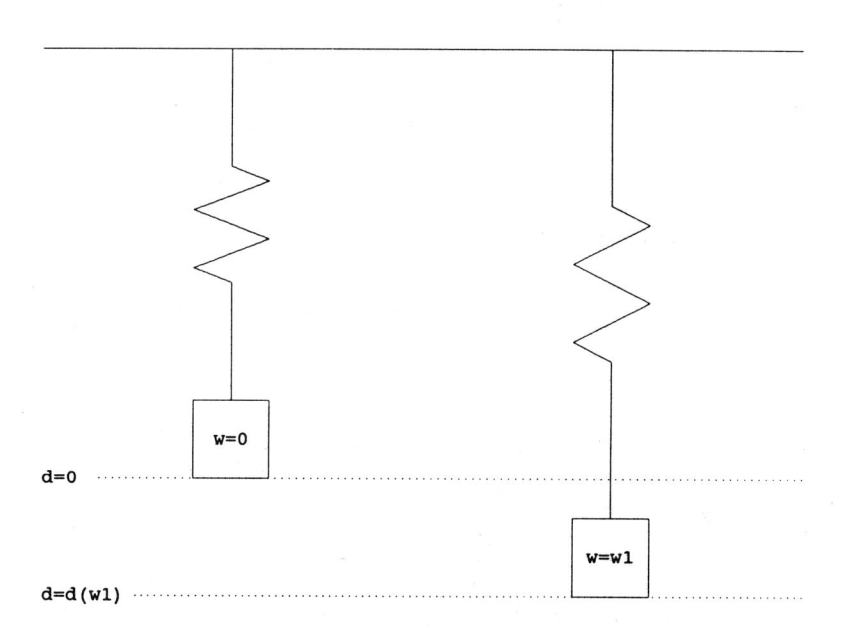

Figure 4.16 Spring with and without weight

2. Show that the only functions $f(x)$ that satisfy the condition

$$f(u + v) = f(u)$$

for all u, v, are $f(x) = C$, where C is a constant.

   (a) Use this result to show that if

   $$\frac{dy}{dx} = g(x, y)$$

   is invariant under the interchange of x with $x + a$ for every constant a —so that if x is replaced by $x + a$ in $y' = g(x, y)$ the resulting equation is identical to $y' = g(x, y)$ —then $g(x, y)$ is a function of y alone.

(b) Show that if
$$\frac{dy}{dx} = g(x,y)$$
is invariant under the interchange of y with $y+b$ for every constant b, then $g(x,y)$ is a function of x alone.

(c) Show that if
$$\frac{dy}{dx} = g(x,y)$$
is invariant under the interchange of x with $x+a$ and y with $y+b$, for every a and b, then $g(x,y)$ is constant.

3. The purpose of this exercise is to find all differentiable functions $f(x)$ that satisfy the condition
$$f(u+v) = f(u)f(v) \text{ for all } u, v,$$
where $f(x) \neq 0$.

(a) Show that
$$f(0) = 1$$
and
$$f'(u) = af(u),$$
where a is constant, namely, $a = f'(0)$.

(b) For the function $f(u)$ and the constant a obtained in part (a), define
$$g(u) = f(u)e^{-au}.$$
Prove that
$$g(0) = 1$$
and that
$$g'(u) = 0.$$

(c) Integrate the last equation in part (b) subject to $g(0) = 1$, and explain how this shows that $f(x) = e^{ax}$ is the only nonzero differentiable function that has the property that
$$f(u+v) = f(u)f(v) \text{ for all } u, v.$$

4. Repeat the technique you used in Exercise 3 to show that the only function $f(x)$ that satisfies
$$f(u+v) = \frac{1}{k}f(u)f(v) \text{ for all } u, v,$$
is $f(x) = ke^{ax}$.

5. This problem deals with two functions, $f(x)$ and $g(x)$, which have the following properties:

   i. $f(0) = 0$, and $g(0) = 1$.
   ii. $f(x)$ and $g(x)$ are differentiable for all x.
   iii. $f'(x) = g(x)$, and $g'(x) = -f(x)$.

(a) Construct the function
$$h(x) = f^2(x) + g^2(x),$$
and show that $h'(x) = 0$; deduce that
$$f^2(x) + g^2(x) = 1$$
for all x.

ADDITIONAL MATERIALS

(b) Let $F(x)$ and $G(x)$ be another two functions that have exactly the same properties as $f(x)$ and $g(x)$ (that is, properties i through iii). Construct the function

$$k(x) = [F(x) - f(x)]^2 + [G(x) - g(x)]^2,$$

and show that $k'(x) = 0$. Now show that $F(x) = f(x)$, and $G(x) = g(x)$, for all x.

(c) Can you think of any functions $f(x)$ and $g(x)$ that satisfy all of the preceding conditions?

(d) Explain how parts (b) and (c) guarantee that $f(x) = \sin x$ and $g(x) = \cos x$.

(e) Use the preceding ideas to show that $\sinh x$ and $\cosh x$ are the only functions that satisfy the following conditions:

   i. $f(0) = 0$, and $g(0) = 1$.
   ii. $f(x)$ and $g(x)$ are differentiable for all x.
   iii. $f'(x) = g(x)$, and $g'(x) = f(x)$.

Applications of Functional Equations — Law of Thermal Expansion

Consider a rod that has length L when the temperature is t. If the temperature of the rod is changed by an amount h, then the length of the rod will change. The simplest assumption to make is that the new length is the old length multiplied by a function that depends only on the change in temperature h, that is,

$$L(t+h) = L(t)f(h). \tag{4.17}$$

If we set $h = 0$ we have $f(0) = 1$. We also see that the left-hand side is invariant under the interchange of t and h, so we must have

$$L(t)f(h) = L(h)f(t). \tag{4.18}$$

If we put $h = 0$ in this and use $f(0) = 1$, we find that

$$L(t) = L(0)f(t).$$

Substituting this back into (4.17) yields

$$L(0)f(t+h) = L(0)f(t)f(h),$$

or

$$f(t+h) = f(t)f(h).$$

This is a functional equation with solution

$$f(t) = e^{kt},$$

where k is a constant. If we substitute this into (4.18) we find

$$L(t) = L(0)e^{kt},$$

which is the usual Law of Thermal Expansion.

Applications of Functional Equations — The Change of Pressure with Altitude

Consider the air pressure P at altitude a. If the altitude is changed by an amount h, then the pressure will change. The simplest assumption to make is that the new pressure is the old pressure multiplied by a function that depends only on the change in pressure h, that is,

$$P(a+h) = P(a)f(h). \tag{4.19}$$

If we set $h = 0$ we have $f(0) = 1$. We also see that the left-hand side is invariant under the interchange of a and h, so we must have

$$P(a)f(h) = P(h)f(a). \tag{4.20}$$

If we put $h = 0$ in this and use $f(0) = 1$, we find that

$$P(a) = P(0)f(a).$$

Substituting this back into (4.19) yields

$$P(0)f(a+h) = P(0)f(a)f(h),$$

or

$$f(a+h) = f(a)f(h).$$

This is a functional equation with solution

$$f(a) = e^{ka},$$

where k is a constant. If we substitute this into (4.20) we find

$$P(a) = P(0)e^{ka}.$$

This is the standard assumption that is made concerning the change in pressure with altitude.

Applications — Calculus

In this section we show how differential equations arise when answering questions from elementary calculus.

Example 4.4 : An Application Based on the Mean Value Theorem

We now discuss an application based on the MEAN VALUE THEOREM, as follows:

Theorem 4.1 *If $f(x)$ is continuous for $u \leq x \leq v$ and is differentiable for $u < x < v$, then there is at least one point c between u and v, $u < c < v$, for which*

$$f'(c) = \frac{f(u) - f(v)}{u - v}.$$

In general, for an arbitrary function $f(x)$, it is impossible to find c explicitly in terms of the end points u and v. However, if $f(x)$ is quadratic, that is,

$$f(x) = a_0 + a_1 x + a_2 x^2,$$

then c is always the average of u and v —that is,

$$c = \frac{u+v}{2}.$$

In other words the value c in the conclusion of the mean value theorem when applied to a quadratic function always occurs at the average of the endpoints of the interval. (See Exercise 1, page 130.)

A natural question to ask is what functions have this property—that is, what functions satisfy

$$f'\left(\frac{u+v}{2}\right) = \frac{f(v) - f(u)}{v - u} \tag{4.21}$$

for every u and v? For our purposes we will assume that the function $f(u)$ we seek is defined for all u and is twice differentiable.

We know that a general quadratic satisfies this condition, so let us exploit this by defining the function $g(u)$ as the difference between $f(u)$ and the first three terms in its Taylor series, namely,

$$g(u) = f(u) - f(0) - f'(0)u - \frac{1}{2}f''(0)u^2. \tag{4.22}$$

Thus, if we can show that

$$g(u) = 0,$$

then the general function $f(u)$ is a quadratic, otherwise it is not. It is easy to see from (4.22) that

$$g(0) = g'(0) = g''(0) = 0, \tag{4.23}$$

whereas substituting (4.22) into (4.21) shows that $g(u)$ must also satisfy

$$g'\left(\frac{u+v}{2}\right) = \frac{g(v) - g(u)}{v - u} \tag{4.24}$$

for every u and v. If we can find the most general $g(u)$ satisfying (4.23) and (4.24), then we can substitute it into (4.22) to find $f(u)$. So we now turn to (4.23) and (4.24).

If we put $v = -u$ into (4.24) and use (4.23), we find that

$$g(u) = g(-u) \tag{4.25}$$

for every u. If we differentiate (4.24) with respect to u (keeping v constant) we find

$$\frac{1}{2}g''\left(\frac{u+v}{2}\right) = \frac{-g'(u)(v-u) + [g(v) - g(u)]}{(v-u)^2} \tag{4.26}$$

for every u and v. If we set $v = -u$ in (4.26) and use (4.23) and (4.25), we find the differential equation

$$g'(u) = 0,$$

which implies that

$$g(u) = a,$$

where a is an arbitrary constant. But from (4.23), $g(0) = 0$, which means that $a = 0$, so

$$g(u) = 0.$$

From this and (4.22) we conclude that

$$f(u) = f(0) + f'(0)u + f''(0)u^2.$$

In other words, *the most general (twice differentiable) function that satisfies (4.21) is a quadratic.* □

Example 4.5 : The Failure of Newton's Method

Newton's method finds the roots of a specified function $f(x)$ by an iterative procedure. In essence, we make an initial approximation, x_0, for the root of $f(x)$, and obtain the next approximation, x_1, from

$$x_1 = x_0 - \frac{f(x_0)}{f'(x_0)}.$$

Then we take x_1 as the new approximation and use it in place of x_0 in the preceding expression, to produce

$$x_2 = x_1 - \frac{f(x_1)}{f'(x_1)}.$$

The whole process is repeated until either a root, c, is found or the technique fails. One way in which this technique fails [assuming $f'(x_0) \neq 0$] is with a poor choice for x_0, which leads to $x_1 = c - x_0$, and then, using x_1, we end up back at x_0. Thus, the process oscillates indefinitely between $c - x_0$ and x_0. Usually a better choice for the initial approximation x_0 avoids this problem.

However, it can happen that no matter what initial approximation x_0 is selected, the process oscillates indefinitely between $c - x_0$ and x_0. This leads to the following problem (where, without loss of generality, we assume $c = 0$).

Find all functions with the following property: For all x ($x \neq 0$), $f(x)$ satisfies

$$-x = x - \frac{f(x)}{f'(x)}, \tag{4.27}$$

where $f'(x) \neq 0$.

If we put $y = f(x)$, we can rewrite (4.27) in the form

$$\frac{dy}{dx} = \frac{y}{2x}. \tag{4.28}$$

Before solving this separable differential equation, let's use our graphical techniques to try to identify properties of the graph of f. The differential equation (4.28) implies that the slope field is symmetric about the x-axis, about the y-axis, and about the origin. The slope field for (4.28) is shown in Figure 4.17 and has these properties. The isoclines with slope m are characterized by the straight lines

$$y = 2mx,$$

which is consistent with Figure 4.17. The concavity is obtained from d^2y/dx^2, which in this case is

$$\frac{d^2y}{dx^2} = \frac{1}{2}\left(\frac{xy' - y}{x^2}\right) = -\frac{1}{4}\left(\frac{y}{x^2}\right).$$

Thus, the function we seek will be concave down for $y > 0$ and concave up for $y < 0$. This is consistent with Figure 4.17.

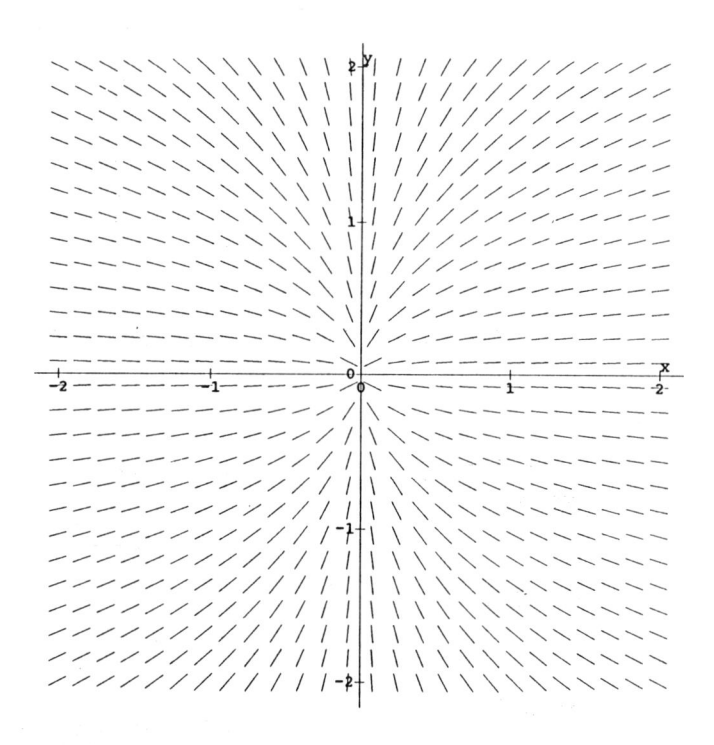

Figure 4.17 Slope field for failure of Newton's method

Now let's turn to solving (4.28). We first note that $y(x) = 0$ is the equilibrium solution. If we rewrite (4.28) as

$$\frac{1}{y}\frac{dy}{dx} = \frac{1}{2x},$$

we may integrate to find

$$\ln|y| = \frac{1}{2}\ln|x| + C.$$

In our usual manner, we write this solution in the form

$$y(x) = A\sqrt{|x|},$$

which includes the equilibrium solution. This is the most general function for which Newton's method fails in the manner just described. The graph with the $A = 1$ is shown in Figure 4.18. Notice that this solution is not symmetric about the x-axis, even though the slope field is. □

ADDITIONAL MATERIALS

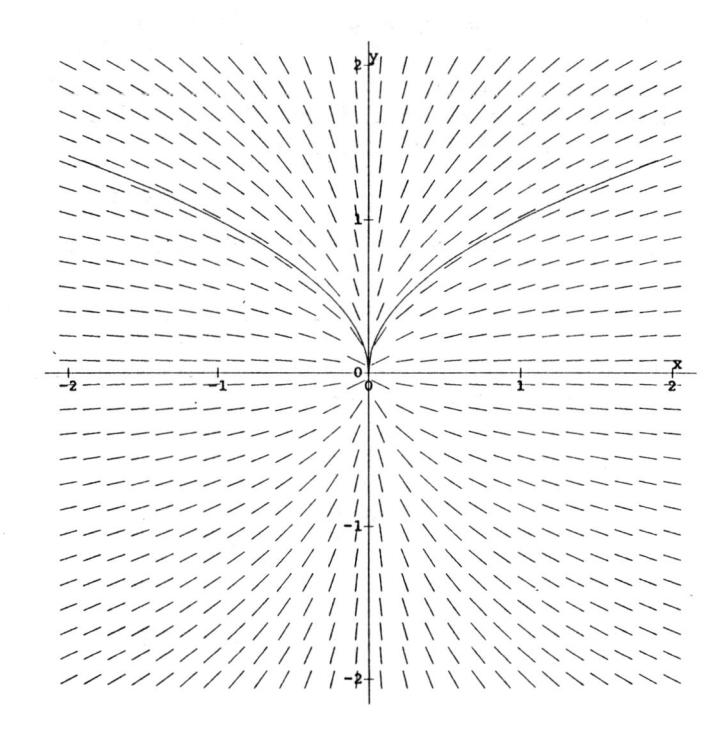

Figure 4.18 Solution and slope field for failure of Newton's method

Exercises

1. Show that if $f(x)$ is a quadratic function—that is, $f(x) = a_0 + a_1 x + a_2 x^2$ —then solving
$$f'(c) = \frac{f(u) - f(v)}{u - v}$$
for c (for every u and v) gives
$$c = \frac{u + v}{2}.$$

2. Show that if
$$\frac{dT}{dt} = \frac{T(t + h) - T(t)}{h},$$
then $T(t)$ is a linear function of t.

3. Show that if
$$\frac{dT}{dt} = \frac{T(t + h) - T(t - h)}{2h},$$
then $T(t)$ is a linear function of t.

4. Show that if
$$\frac{1}{T}\frac{dT}{dt} = \frac{\ln T(t + h) - \ln T(t)}{h},$$
then $T(t)$ is an exponential function of t.

5. Show that if
$$\frac{1}{T}\frac{dT}{dt} = \frac{\ln T(t + h) - \ln T(t - h)}{2h},$$
then $T(t)$ is an exponential function of t.

The Population of the USA - Part 1

The population $P(t)$ of the USA from 1790 to 1990 is shown in Table 4.8 where $t = 0$ corresponds to 1790.[11] We have also computed the central difference quotient $\Delta_C P/\Delta t$ divided by P, namely $(1/P)\Delta_C P/\Delta t$. In Figure 4.19 we have plotted the population as a function of time. In Figure 4.20 we have plotted $(1/P)\Delta_C P/\Delta t$ as a function of time and in Figure 4.21 we have plotted $(1/P)\Delta_C P/\Delta t$ as a function of population. [Table 4.8 is in TWIDDLE. It is called USAPOP.DTA and is in the subdirectory DD-04.]

Table 4.8 USA population since 1790

Year	Time	Population (millions)	$(1/P)\Delta_C P/\Delta t$
1790	0	3.929	
1800	10	5.308	0.0312
1810	20	7.240	0.0299
1820	30	9.638	0.0292
1830	40	12.861	0.0289
1840	50	17.063	0.0303
1850	60	23.192	0.0310
1860	70	31.443	0.0244
1870	80	38.558	0.0243
1880	90	50.189	0.0243
1890	100	62.980	0.0207
1900	110	76.212	0.0192
1910	120	92.228	0.0162
1920	130	106.022	0.0146
1930	140	123.203	0.0106
1940	150	132.165	0.0106
1950	160	151.326	0.0156
1960	170	179.323	0.0145
1970	180	203.302	0.0116
1980	190	226.546	0.0104
1990	200	248.710	

(a) One interpretation that might be drawn from Figure 4.20 is that the first 6 points lie approximately on a horizontal line. What does that suggest about the value of P'/P? Solve this differential equation, and try to find parameters so that this model gives a reasonable fit to the years $0 \leq t \leq 65$. What is the long-term consequence of this model?

(b) Another interpretation that might be drawn from Figure 4.20 is that the first 15 points lie approximately on a line. What does that suggest about the value of P'/P? Solve this differential equation, and try to find parameters so that this model gives a reasonable fit to the years $0 \leq t \leq 155$. What is the long-term consequence of this model?

(c) A further interpretation that might be drawn from Figure 4.20 is that the last 4 points lie approximately on a line. What does that suggest about the value of P'/P? Solve this differential equation, and try to find parameters so that this model gives a reasonable fit to the years $155 \leq t \leq 200$. What is the long-term consequence of this model?

---

[11] "The Universal Almanac 1995" by J. W. Wright, Andrews and McMeel, page 291.

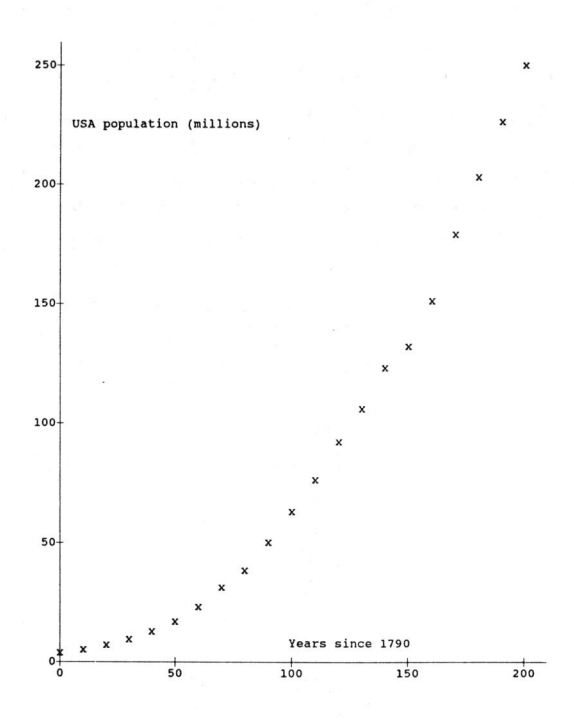

Figure 4.19   USA population since 1790

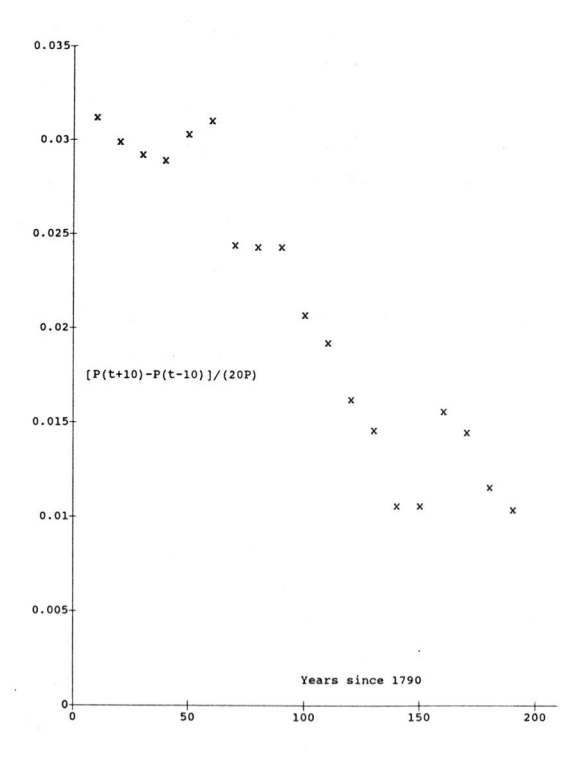

Figure 4.20 $(1/P)\Delta_C P/\Delta t$ as a function of time

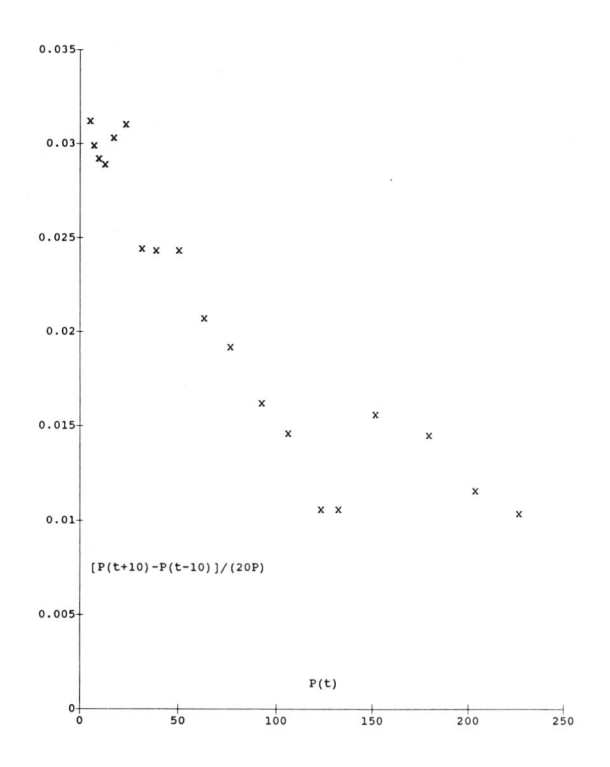

Figure 4.21 $(1/P)\Delta_C P/\Delta t$ as a function of population

(d) Comments similar to those made in parts (a) through (c) also apply to Figure 4.21. Repeat parts (a) through (c) using Figure 4.21. Compare the models considered in this part to their counterparts in parts (a) through (c). Which have the more realistic predictions?

ADDITIONAL MATERIALS

Allometric Growth

Allometry is the study of the relative size of different parts of an organism as a consequence of growth.[12] One of the simplest models of allometry is one in which it is assumed that the relative growth rates of the two components are proportional. Thus, if the size of the two components at time t are $x(t)$ and $y(t)$ then, because the relative growth rates of x and y are $(1/x)\,dx/dt$ and $(1/y)\,dy/dt$, we have

$$\frac{1}{y}\frac{dy}{dt} = a\frac{1}{x}\frac{dx}{dt}$$

where a is the constant of proportionality. We may eliminate t in this equation to find

$$\frac{dy}{dx} = a\frac{y}{x}.$$

Solving this differential equation for nonequilibrium solutions yields

$$\ln y = a \ln x + c, \tag{4.29}$$

or

$$y = Cx^a, \tag{4.30}$$

where $C = e^c$. This equation is often called the allometry equation, and it is a simple power law. It says that the size of y is related to the size of x raised to the power a. Notice that if we have a real data set which we think obeys the allometry equation, we can estimate the constants a and C by using (4.29). This tells us that if we plot $\ln y$ versus $\ln x$ we should get a straight line with slope a and intercept $c = \ln C$.

This supports an argument that suggests that the weight $w(t)$ of a fish is related to its length $L(t)$ by $w(t) = bL^3(t)$ where b is a constant. The argument is based on units. If L is a typical unit of length then L^3 is a typical unit of volume, and volume and weight are proportional. Thus, $w(t)$ is proportional to L^3.

We can see whether this is accurate. In Table 4.9 and Figure 4.22 we show the results of an experiment relating the weight of plaice $w(t)$ to its length $L(t)$.[13] [Table 4.9 is in TWIDDLE. It is called PLAICE.DTA and is in the subdirectory DD-04.] Figure 4.23 plots $\ln w(t)$ versus $\log L(t)$ and the line of best fit $w(t) = 2.952605L(t) - 4.534074$. This gives the estimate of $a \approx 2.952605$ and $C \approx e^{-4.534074} = 0.0107368$. Figure 4.24 shows the weight of plaice as a function of length together with the curve $w(t) = 0.0107368L(t)^{2.952605}$. Because 2.952605 is so close to 3 we also tried to fit the curve $w(t) = CL(t)^3$ to the data set. Figure 4.25 shows the weight of plaice as a function of length together with the curve $w(t) = 0.00892L(t)^3$. Figure 4.26 shows the weight of plaice as a function of length together with the curves $w(t) = 0.0107368L(t)^{2.952605}$ and $w(t) = 0.00892L(t)^3$. It is almost impossible to tell them apart. This supports the suggestion that $w(t)$ is proportional to L^3. [The function $w(t) = 0.00892L(t)^3$ is in TWIDDLE. It is called PLAICE.TWD and is in the subdirectory DD-04.]

Exercises

1. **Allometric Growth of Fiddler-Crabs.** Table 4.10 and Figure 4.27 show the results of an experiment in which the relative weights of the body and the claw of fiddler-crabs were obtained.[14] The body weight is the total weight of the crab with the weight of the claw subtracted. Show that allometric growth can account for these results. [Table 4.10 is in TWIDDLE. It is called CRAB.DTA and is in the subdirectory DD-04.]

---

[12] "Problems of Relative Growth" by J. S. Huxley, Dover, 1972.

[13] "On the Dynamics of Exploited Fish Populations" by R. J. H. Beverton and S. J. Holt, Fishery Investigations, Series II, **19**, 1957, page 281.

[14] "Problems of Relative Growth" by J. S. Huxley, Dover, 1972, page 12, Table 1.

Table 4.9 Weight of plaice as a function of length

Length (cm)	Weight (gm)	Length (cm)	Weight (gm)
23.5	124	37.5	455
24.5	146	38.5	500
25.5	155	39.5	538
26.5	174	40.5	574
27.5	190	41.5	623
28.5	213	42.5	674
29.5	236	43.5	724
30.5	259	44.5	808
31.5	284	45.5	812
32.5	308	46.5	909
33.5	332	47.5	1039
34.5	363	48.5	1124
35.5	391	49.5	1163
36.5	419		

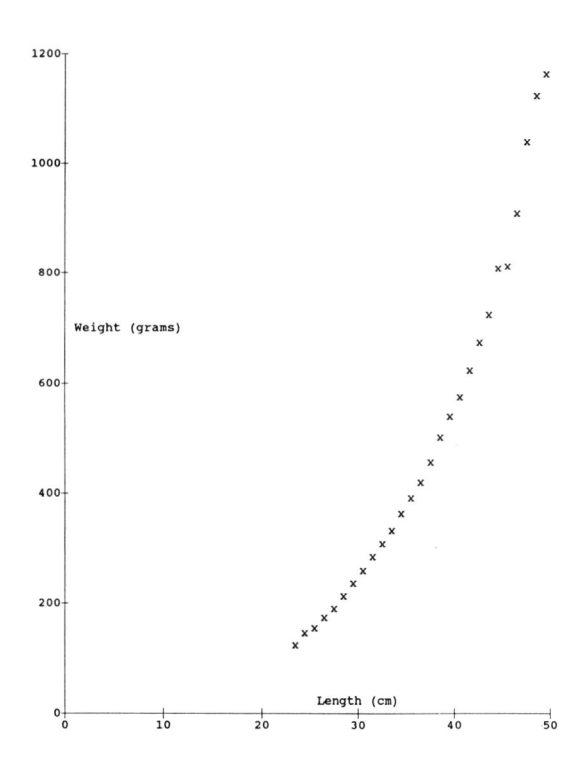

Figure 4.22 Weight of plaice as a function of length

ADDITIONAL MATERIALS

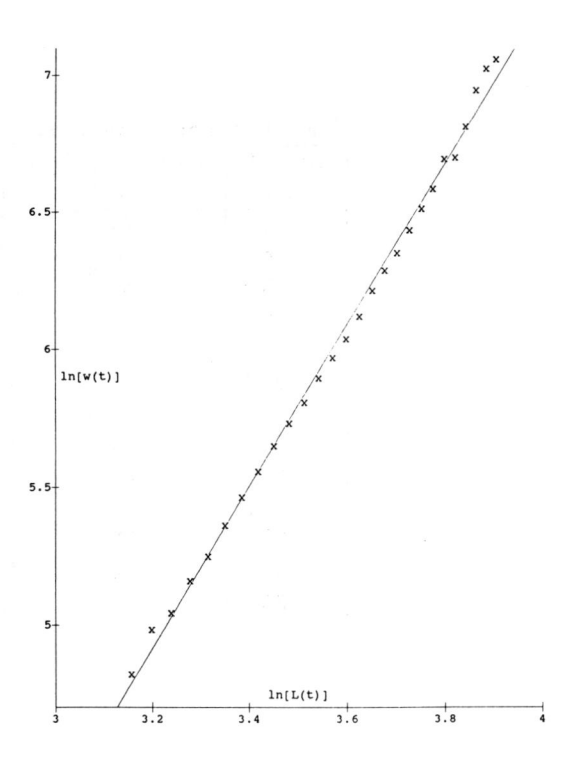

Figure 4.23 $\ln w(t)$ versus $\ln L(t)$ and the line $w(t) = 2.952605L(t) - 4.534074$

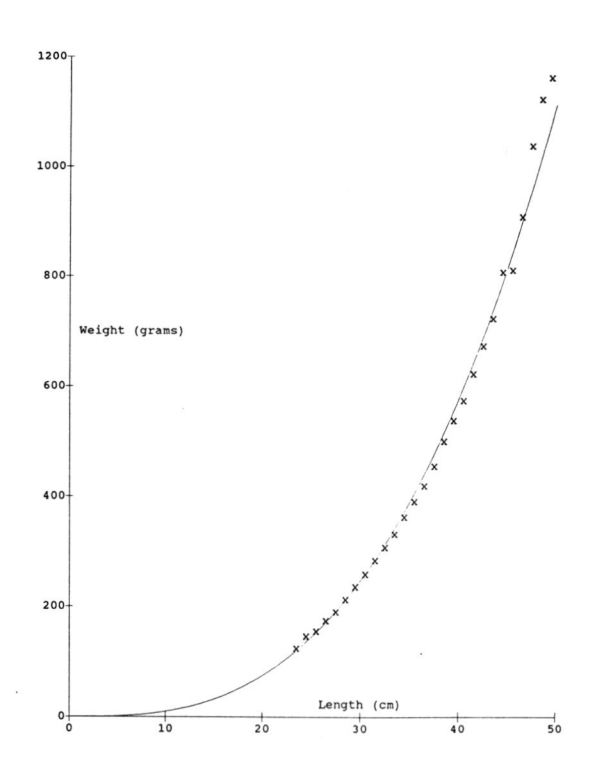

Figure 4.24 Weight of plaice as a function of length and the curve $w(t) = 0.0107368L(t)^{2.952605}$

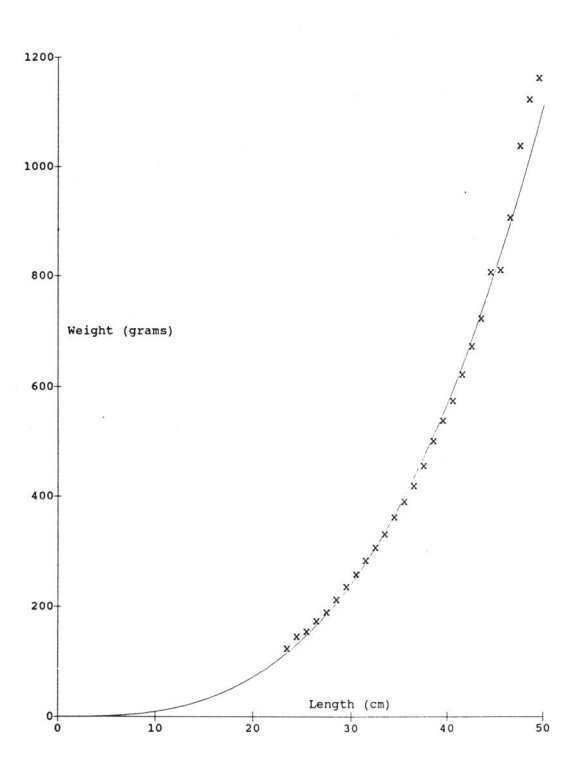

Figure 4.25 Weight of plaice as a function of length and the curve $w(t) = 0.00892L^3$

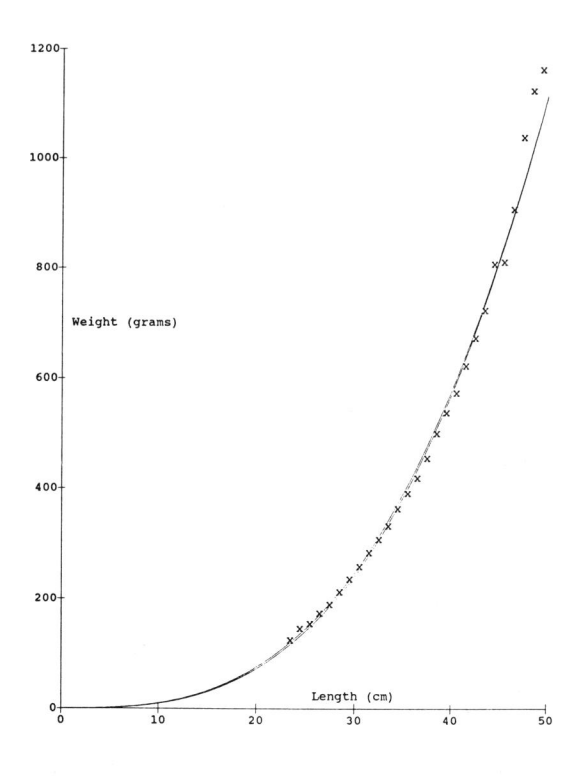

Figure 4.26 Weight of plaice as a function of length and the curves $w(t) = 0.0107368L(t)^{2.952605}$ and $w(t) = 0.00892L^3$

ADDITIONAL MATERIALS

Table 4.10 Claw weight versus body weight of fiddler-crabs

Body Weight (mg)	Claw Weight (mg)	Body Weight (mg)	Claw Weight (mg)
57.6	5.3	680.6	271.6
80.3	9.0	743.3	319.2
109.2	13.7	872.4	417.6
156.1	25.1	983.1	460.8
199.7	38.3	1079.9	537.0
238.3	52.5	1165.5	593.8
270.0	59.0	1211.7	616.8
300.2	78.1	1291.3	670.0
355.2	104.5	1363.2	699.3
420.1	135.0	1449.1	777.8
470.1	164.9	1807.9	1009.1
535.7	195.6	2235.0	1380.0
617.9	243.0		

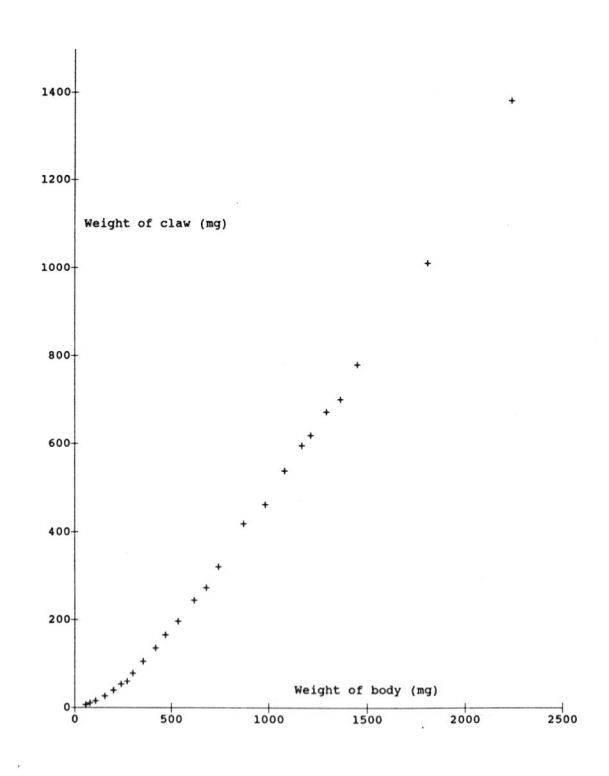

Figure 4.27 Claw weight versus body weight of fiddler-crabs

Projectiles

We have dealt with the trajectories of particles in one-dimension. Here we generalize this to two dimensions where a typical position of a particle of mass m has coordinates $(x(t), y(t))$ in the xy-plane. Two typical applications of this are to firing projectiles[15] and to skydiving.

No Drag (No Air Resistance)

If gravity is the only force acting on the particle — and it acts downward, parallel to the y-axis — then the particle's trajectory is governed by the two differential equations

$$mu' = 0$$

and

$$mv' = -mg,$$

where

$$u = \frac{dx}{dt}$$

and

$$v = \frac{dy}{dt}.$$

These equations are independent of each other and are solved separately giving

$$x'(t) = u(t) = u_0,$$
$$x(t) = u_0 t + x_0,$$
$$y'(t) = v(t) = -gt + v_0,$$

and

$$y(t) = -\frac{1}{2}gt^2 + v_0 t + y_0,$$

where u_0, x_0, v_0, and y_0 are constants. There is a simple interpretation for these constants. They correspond to the position (x_0, y_0) and velocity (u_0, v_0) of the particle at time $t = 0$.

Comments about this motion:

- Notice that neither $x(t)$ nor $y(t)$ depend on m. This means that if we take two particles with different masses at the same initial height y_0 and simultaneously release them from rest (so the initial velocities are $u_0 = v_0 = 0$) then they will fall the same distance y in the same time. **A simple experiment** to demonstrate this consists of taking two tennis balls, one loaded with lead shot, and dropping them from the same height. They hit the floor at the same time.[16]

- Notice that $y(t)$ does not depend on x_0 or u_0. This means that if we take two particles at the same initial height y_0 and simultaneously release them with different initial horizontal velocities u_0 then they will fall the same distance y in the same time. A **simple experiment** to demonstrate this consists of taking two large marbles and holding them side-by-side between thumb and first finger of one hand. Make sure that the balls are the same height above the ground. Now strike one of the balls horizontally with the other hand. If done correctly, the other marble drops vertically. They both

---

[15] "The Mathematics of Projectiles in Sport" by N. de Mestre, Australian Mathematical Society Lecture Series 6, Cambridge University Press, 1990.

[16] "Doing physics" by J. Bozovsky, The Physics Teacher, December 1983, pages 611 through 612.

have the same initial vertical velocities — namely, zero — but different initial horizontal velocities. They hit the floor at the same time.[17] A consequence of this is if a bullet is dropped at the same time that a rifle is fired horizontally then the two bullets will strike the earth at the same time.

The speed of the particle $w(t) = \sqrt{u^2 + v^2}$ at any time t is

$$w(t) = \sqrt{u_0^2 + (-gt + v_0)^2}.$$

The initial speed is $w_0 = w(0) = \sqrt{u_0^2 + v_0^2}$. There is no terminal speed because $\lim_{t \to \infty} w(t) = \infty$.

If we define the angle α by $u_0 = w_0 \cos \alpha$ and $v_0 = w_0 \sin \alpha$, where $-\pi/2 \leq \alpha \leq \pi/2$ — which means that the particle initially has a speed of w_0 in the direction α — then we can write

$$x(t) = (w_0 \cos \alpha) t + x_0,$$

$$y(t) = (w_0 \sin \alpha) t - \frac{1}{2} g t^2 + y_0,$$

and

$$w(t) = \sqrt{w_0^2 - 2gt w_0 \sin \alpha + g^2 t^2}.$$

It is sometimes convenient to consider the situation where initially the particle is at the origin of the coordinate system. We can do this by using a translation from (x, y) to (X, Y) which places the particle's initial position at the origin of the XY-plane — namely,

$$X = x - x_0,$$

and

$$Y = y - y_0.$$

This gives the trajectory

$$X(t) = (w_0 \cos \alpha) t, \tag{4.31}$$

$$Y(t) = (w_0 \sin \alpha) t - \frac{1}{2} g t^2. \tag{4.32}$$

A natural question to ask is what is the shape of the trajectory followed by the particle in the XY-plane assuming that $u_0 \neq 0$. (What happens if $u_0 = 0$?) This we can obtain by eliminating t between the $X(t)$ and $Y(t)$ equations. Thus,

$$t = \frac{X}{w_0 \cos \alpha},$$

and so

$$Y = (\tan \alpha) X - \frac{1}{2} g \left(\frac{1}{w_0 \cos \alpha} \right)^2 X^2.$$

Thus, Y is a quadratic function of X and so the trajectory is a parabola opening downwards in the XY-plane. Typical trajectories are shown in Figure 4.28 for $\alpha > 0$, $\alpha = 0$, and $\alpha < 0$ — namely, $g = 32$, $w_0 = 200$, and $\alpha > 45°$, $\alpha = 0°$, and $\alpha < -45°$.

The maximum height of the particle will occur when $dY/dX = 0$ and, because $dY/dX = Y'/X'$, this occurs when $Y' = 0$, that is

$$t = \frac{w_0}{g} \sin \alpha.$$

---

[17] "A demonstration to show the independence of horizontal and vertical motion" by J. Hoskins and L. Lonney, The Physics Teacher, November 1983, page 525.

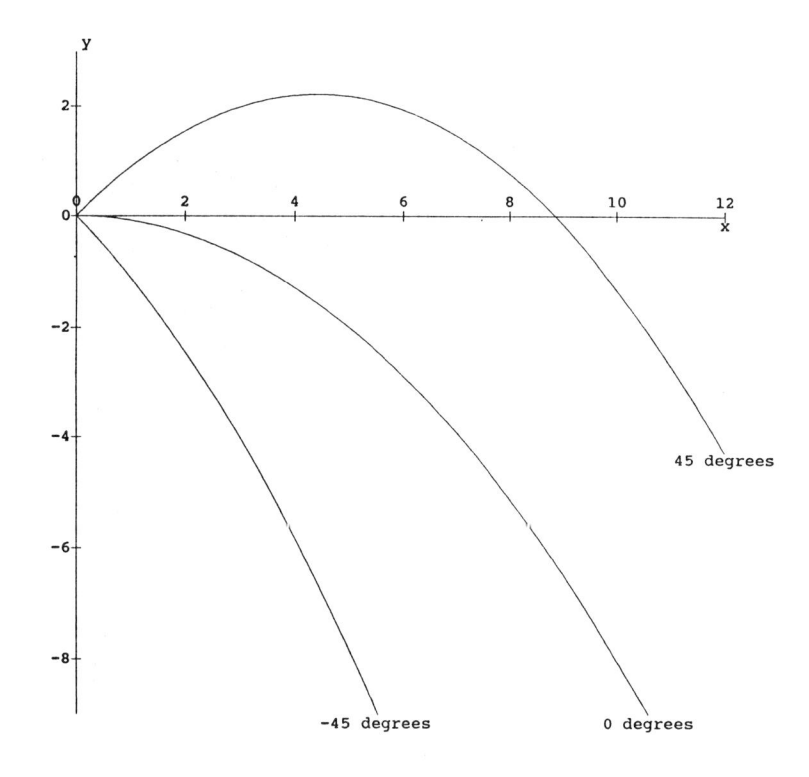

Figure 4.28 Trajectories with no drag for $\alpha > 0$, $\alpha = 0$, and $\alpha < 0$

For this to occur for $t > 0$ we need $0 < \alpha \leq \pi/2$, which agrees with intuition, and at this time the particle is at

$$X = \frac{w_0^2}{2g} \sin 2\alpha,$$

when the maximum height H is

$$H = \frac{w_0^2}{2g} \sin^2 \alpha. \tag{4.33}$$

We notice that $Y(t) = 0$ when $t = 0$ and when $t = t_h$, where

$$t_h = \frac{2w_0}{g} \sin \alpha. \tag{4.34}$$

The time t_h is when the particle hits the X-axis — if the particle were a football we would call this the **hang time**. If we want $t_h > 0$, this requires that $0 < \alpha \leq \pi/2$ which also agrees with intuition. Notice it is twice the time taken to attain the maximum height, which is to be expected on the grounds of symmetry. In this case the particle hits the X-axis at a distance

$$R = \frac{w_0^2}{g} \sin 2\alpha \tag{4.35}$$

from where is was released, called the **range**.

Table 4.11 and Figure 4.29 show the results of an experiment recording the range R as a function of the initial angle of elevation α of a projectile.[18] Is this consistent with (4.35)? [Table 4.11 is in TWIDDLE. It is called RANGE.DTA and is in the subdirectory DD-04.]

---

[18] "A study of the trajectories of projectiles" by A. Ruari Grant, Physics Education, **25**, 1990, pages 288 through 292. Figure 4, page 289.

ADDITIONAL MATERIALS

Table 4.11 The range R as a function of the elevation α

Angle (degrees)	Range (cm)
10.2	17.60
16.8	26.95
22.7	33.28
30.0	42.35
32.2	45.24
35.1	47.44
40.2	48.95
45.1	49.78
48.9	48.95
52.0	47.30
57.8	42.63
65.1	35.48
68.0	31.63
76.8	20.21

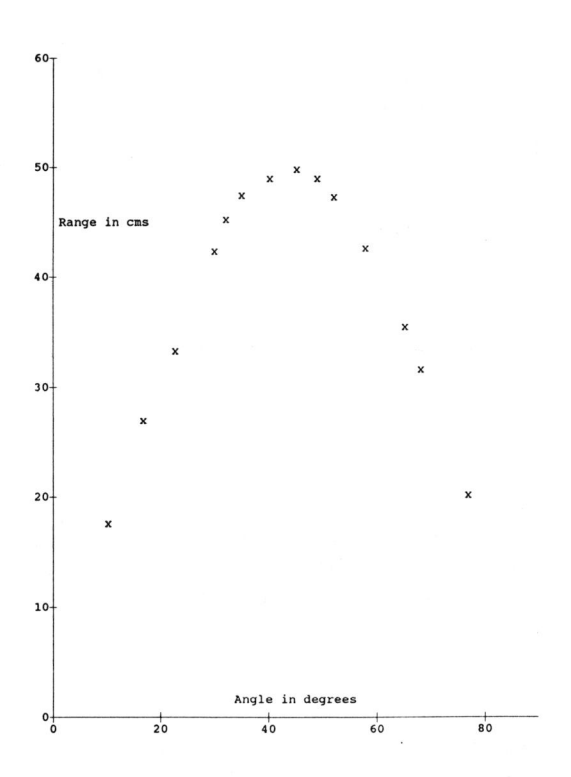

Figure 4.29 The range R as a function of the elevation α

For a given positive range R and an initial velocity w_0 the initial angle α can be computed from (4.35). Apart for $\alpha = \pi/4$, this always has two solutions for $0 < \alpha \leq \pi/2$ — namely

$$\alpha_1 = \frac{1}{2} \arcsin\left(\frac{Rg}{w_0^2}\right)$$

and $\alpha_2 = \pi/2 - \alpha_1$. Thus, there are two different initial angles that have identical ranges for the same w_0 but different trajectories. The angle α_1 is less than $\pi/4$ while α_2 is greater. We can calculate the hang times t_1 and t_2 for each of these trajectories from (4.34), and we find

$$t_1 = \frac{2w_0}{g}\sin\alpha_1 \qquad t_2 = \frac{2w_0}{g}\sin\left(\frac{\pi}{2} - \alpha_1\right) = \frac{2w_0}{g}\cos\alpha_1.$$

The difference in flight time is

$$t_2 - t_1 = \frac{2w_0}{g}(\cos\alpha_1 - \sin\alpha_1).$$

This means that if we release two particles — say, snowballs — with the same w_0, one at an initial angle of α_2 and the other at an angle α_1 at time $t_2 - t_1$ later, they will both arrive at the range at the same time. Imagine you were throwing snowballs with an initial speed w_0 of 72 ft/sec at a person 100 feet away. In this case $\alpha_1 \approx 19°$ and $\alpha_2 \approx 71°$, and $t_2 - t_1 \approx 2.75$ sec. Thus throwing a high snowball followed by a low one 2.75 seconds later will cause confusion![19]

The particle which was launched with an initial angle of α will land at an angle of $\pi - \alpha$. One way to confirm this is to compute $dY/dX = Y'/X'$ at $t = (2w_0/g)\sin\alpha$, which will give the slope of the trajectory at the time of impact. In this way we find

$$\frac{dY}{dX} = -\tan\alpha.$$

If w_0 (the initial speed) is fixed then changing α will change the range. The maximum range occurs when $\sin 2\alpha = 1$, that is, when $\alpha = \pi/4$.

If w_0 (the initial speed) is fixed then changing α will change the maximum height of the projectile. The maximum height occurs when $\sin\alpha = 1$, that is, when $\alpha = \pi/2$. This occurs when the particle is propelled straight up — not a recommended practice for firing projectiles.

If we observe the trajectory of a particle, there are two natural measurements we can make — the range R and the hang time t_h. If we measure these two parameters, can we identify the trajectory? From (4.34) we have

$$w_0 \sin\alpha = \frac{1}{2}gt_h, \tag{4.36}$$

which when used in (4.35) gives

$$w_0 \cos\alpha = \frac{R}{t_h}. \tag{4.37}$$

If we substitute these into (4.31) and (4.32) we find

$$X(t) = \frac{R}{t_h}t, \tag{4.38}$$

$$Y(t) = \frac{1}{2}g(t_h - t)t, \tag{4.39}$$

---

[19] "Snowball Fighting: A Study in Projectile Motion" by P. N. Henriksen, The Physics Teacher, January 1975, page 43.

so we can express the trajectory in terms of R and t_h.

It is sometimes useful to have the main features of a trajectory expressed in terms of R and t_h. From (4.36) and (4.37) we find that the initial speed is

$$w_0 = \sqrt{\left(\frac{1}{2}gt_h\right)^2 + \left(\frac{R}{t_h}\right)^2}$$

and the initial angle is

$$\alpha = \arctan\left(\frac{gt_h^2}{2R}\right).$$

From (4.33) and (4.36) we find that the maximum height H is

$$H = \frac{1}{8}gt_h^2. \tag{4.40}$$

Thus, a knowledge of the range and hang time will determine the trajectory completely.

Example 4.6 : Simple Experiment[20]

While watching a football game estimate and record the hang time and the range of passes and kicks.[21] This will usually require two people, one with a stop watch to record the hang time, and the other to estimate the range. For each of these trajectories determine the maximum height, the initial speed, and the initial angle of the ball. For example, a punt might have a hang time of 4.9 secs and a range of 25 yards. How high did the punt go? What was its initial speed and angle? □

Example 4.7 : Kicking a Football[22]

When kicking a football the kicker usually tries to maximize the range and to maximize the hang time. What initial angle α should the kicker use to accomplish this?

The range is maximized when $\sin 2\alpha$ is maximum — that is, when $\sin 2\alpha = 1$ — so $\alpha = \pi/4$. The hang time is maximized when $\sin \alpha$ is maximum — that is, when $\sin \alpha = 1$ — so $\alpha = \pi/2$. Thus, it is not possible to simultaneously maximize the range and hang time. This is sometimes called the "kicker's dilemma." □

Example 4.8 : Investigating Traffic Accidents[23]

When an investigating officer arrives at the scene of a traffic accident, one of the most important pieces of information is to estimate the speed of the vehicles involved. In a single car accident where a vehicle has left the road and crashed some distance below the road, the formula that investigators use to determine the speed w_0, in miles per hour, at which the vehicle left the road at an angle α and fell a vertical distance d_v and a horizontal distance d_h is

$$w_0 \approx \frac{2.74 d_h}{\sqrt{d_h \tan \alpha - d_v}}.$$

Here d_v and d_h are measurements (in feet) of the center of mass of the vehicle. (They are measured to where the vehicle landed, and not to where the vehicle eventually came to rest.)

---

[20] "Television, football, and physics: Experiments in kinematics" by A. A. Bartlett, The Physics Teacher, September 1984, pages 386 through 387.

[21] The range of a football is not necessarily the same as the distance quoted by announcers. Why?

[22] "The physics of kicking a football" by P. J. Brancazio, The Physics Teacher, **23**, 1985, pages 403 through 407.

[23] "Physics in accident investigations" by M. L. Brake, The Physics Teacher, January 1981, pages 26 through 29.

This formula is used when the vehicle is not traveling very fast and does not fall a great distance, so that air resistance can be neglected, and so there is no drag. Where does this formula come from?

From (4.31) and (4.32) we have these horizontal and vertical distances given by

$$d_h = (w_0 \cos \alpha) t,$$

and

$$d_v = (w_0 \sin \alpha) t - \frac{1}{2} g t^2,$$

where t is the time of impact. Eliminating t between these equations yields

$$d_v = d_h \tan \alpha - \frac{1}{2} g \left(\frac{d_h}{w_0 \cos \alpha} \right)^2.$$

Solving for w_0 gives

$$w_0 = \frac{d_h}{\cos \alpha} \sqrt{\frac{g}{2 (d_h \tan \alpha - d_v)}}.$$

This is the exact calculation for w_0. With $g = 32.2$ ft/sec^2 this is

$$w_0 = 4.01 \frac{d_h}{\cos \alpha \sqrt{d_h \tan \alpha - d_v}}$$

in ft/sec and

$$w_0 = \frac{2.74 d_h}{\cos \alpha \sqrt{d_h \tan \alpha - d_v}}$$

in mph. If α is small, so that $\cos \alpha \approx 1$, we have

$$w_0 \approx \frac{2.74 d_h}{\sqrt{d_h \tan \alpha - d_v}}$$

in mph. □

Example 4.9 : Simple Experiment[24]

Take a large coffee can and drill 5 holes of about 5 mm diameter equally spaced down one side of the can. Place the can near a sink and cover the holes. Fill the can with water. We are going to open the holes, keep the water level constant, and see which of the 5 streams travels the greatest horizontal distance. Before opening the holes, what does your intuition tell you? Now open the holes and keep the water level constant. Look at the 5 streams, and where they intersect the plane on which the can is standing. Which stream travels the greatest horizontal distance? Does the result of this experiment agree with your intuition? Surprised? We can explain this by thinking of the water stream as a succession of particles leaving a hole with a horizontal velocity due to the pressure of the water above it.

To do this, we let the can have height ℓ and consider one of the holes which is a distance h below the water level. We can find the water stream's trajectory by thinking of a particle leaving this hole with a horizontal velocity due to the head of water h. By Toriccelli's law, this velocity is proportional to $\sqrt{h}$. Thus, we can model this situation in the following way.

A particle is fired horizontally from a height of $\ell - h$ above the ground with an initial horizontal velocity of $b\sqrt{h}$, where h, b, and ℓ are positive constants. Find the horizontal distance traveled when the particle hits the ground. What value of h makes this distance the greatest?

---

[24] "'Canned' Physics" by H. Kruglak, The Physics Teacher, 30, 1992, pages 392 through 396, and "The Water Can Paradox" by L. G. Paldy, The Physics Teacher, 1, 1963, page 126.

Here we have $u_0 = b\sqrt{h}$, $v_0 = 0$, $x_0 = 0$, $y_0 = \ell - h$, so

$$x(t) = b\sqrt{h}\,t$$

and

$$y(t) = -\frac{1}{2}gt^2 + \ell - h.$$

The time T when the particle hits the ground is determined from $y(T) = 0$ and is

$$T = \frac{1}{g}\sqrt{2(\ell - h)}.$$

The horizontal distance traveled in this time is

$$x(T) = \frac{b}{g}\sqrt{2(\ell - h)h}.$$

If we think of h as a parameter and plot $\sqrt{2(\ell - h)h}$ versus h, we see that this has a maximum when $h = \ell/2$. Thus, $x(T)$ is maximum when $h = \ell/2$. □

Now repeat the above experiment, but first place the coffee can on a raised platform. Does the result of this experiment agree with your intuition? Is it possible to adjust the height of the platform so that the water streams from middle and lowest holes travel the same horizontal distance? Use the above model to verify your conclusions. Based on this, write a paragraph explaining this phenomena — and in particular why it does not violate intuition — in language that a nonscientist could understand.

For applications to other areas, see

- Basketball. "Physics of basketball" by P. J. Brancazio, The American Journal of Physics, **49**, 1981, pages 356 through 365. "Kinematics of the free throw in basketball" by A. Tan and G. Miller, The American Journal of Physics, **49**, 1981, pages 542 through 544.

- Shot Put. "Maximizing the range of the shot put" by D. B. Lichtenberg and J. G. Wills, The American Journal of Physics, **46**, 1978, pages 546 through 549.

- Shooting. "A puzzle in elementary ballistics" by O. A. Haugland, The Physics Teacher, April 1983, pages 246 through 248.

Linear Drag (Air Resistance Linear in Velocity)

Now let's include air resistance which is a force that opposes the motion. For slow velocities this force will be proportional to (x', y'), so our equations become

$$mx'' = -kx'$$

and

$$my'' = -mg - ky',$$

where k is a positive constant. Once more these equations are independent of each other and are solved separately giving

$$x(t) = x_0 + \frac{u_0}{\lambda}\left(1 - e^{-\lambda t}\right)$$

and
$$y(t) = y_0 - \frac{g}{\lambda}t + \left(\frac{v_0}{\lambda} + \frac{g}{\lambda^2}\right)\left(1 - e^{-\lambda t}\right)$$

where $\lambda = k/m$.

We can see whether these are reasonable answers by checking their behavior for small λ. Expanding $e^{-\lambda t}$ in powers of λt, namely,

$$e^{-\lambda t} = 1 - \lambda t + \frac{1}{2!}(\lambda t)^2 - \frac{1}{3!}(\lambda t)^3 + \cdots,$$

and collecting like terms, we find that x and y can be written

$$x(t) = x_0 + u_0 t - \lambda t^2 u_0 \left(\frac{1}{2!} - \frac{1}{3!}\lambda t + \cdots\right)$$

and

$$y(t) = y_0 + v_0 t - \lambda t^2 (v_0 + g\lambda)\left(\frac{1}{2!} - \frac{1}{3!}\lambda t + \cdots\right).$$

This reduces to the no drag case when $\lambda = 0$. Thus, these are reasonable equations.

The associated velocities are

$$x'(t) = u_0 e^{-\lambda t}$$

and

$$y'(t) = -\frac{g}{\lambda} + \left(v_0 + \frac{g}{\lambda}\right)e^{-\lambda t}.$$

As $t \to \infty$ we see that $x'(t) \to 0$ and $y'(t) \to -g/\lambda$. Also, because $x''(t) = -\lambda u_0 e^{-\lambda t} < 0$ and $y''(t) = -\lambda(v_0 + g/\lambda)e^{-\lambda t} < 0$ both x' and y' decrease with time. Thus as time increases the horizontal component of the velocity decreases to zero, while the vertical component decreases towards a terminal velocity of $-W$, where $W = g/\lambda = gm/k$. (Notice that we can obtain the same conclusion directly from the differential equations by looking for equilibrium solutions.) We can express $y(t)$ in terms of W as follows:

$$y(t) = y_0 - Wt + \frac{1}{\lambda}(v_0 + W)\left(1 - e^{-\lambda t}\right).$$

The speed of the particle $w(t) = \sqrt{(x')^2 + (y')^2}$ at any time t is

$$w(t) = \sqrt{u_0^2 e^{-2\lambda t} + [W - (v_0 + W)e^{-\lambda t}]^2}.$$

Notice that $\lim_{t \to \infty} w(t) = W$, so W is the terminal speed.

If we define the angle α by $u_0 = w_0 \cos \alpha$ and $v_0 = w_0 \sin \alpha$, where $-\pi/2 \leq \alpha \leq \pi/2$ and $w_0 = w(0)$ — which means that the particle initially has a speed of w_0 in the direction α — then we can write

$$x(t) = x_0 + \frac{w_0 \cos \alpha}{\lambda}\left(1 - e^{-\lambda t}\right),$$

and

$$y(t) = y_0 - Wt + \frac{1}{\lambda}(w_0 \sin \alpha + W)\left(1 - e^{-\lambda t}\right).$$

It is sometimes convenient to consider the situation where initially the particle is at the origin of the coordinate system. We can do this by using a translation from (x, y) to (X, Y) which places the particle's initial position at the origin of the XY-plane — namely,

$$X = x - x_0,$$

and

$$Y = y - y_0.$$

ADDITIONAL MATERIALS

This gives the trajectory
$$X(t) = \frac{w_0 \cos \alpha}{\lambda} \left(1 - e^{-\lambda t}\right),$$
$$Y(t) = -Wt + \frac{1}{\lambda}\left(w_0 \sin \alpha + W\right)\left(1 - e^{-\lambda t}\right).$$

As $t \to \infty$ we see that $X(t) \to (w_0 \cos \alpha)/\lambda$ and $Y(t) \to -\infty$. Thus, this model predicts that there is a limit to the horizontal distance the particle can travel.

A natural question to ask is what is the shape of the trajectory followed by the particle in the XY-plane assuming that $u_0 \neq 0$. (What happens if $u_0 = 0$?) This we can obtain by eliminating t between the $X(t)$ and $Y(t)$ equations. Thus, if $u_0 \neq 0$ we have
$$\left(1 - e^{-\lambda t}\right) = \frac{\lambda X}{w_0 \cos \alpha},$$
from which we find
$$t = -\frac{1}{\lambda} \ln\left(1 - \frac{\lambda X}{w_0 \cos \alpha}\right),$$
and so
$$Y = \frac{W}{\lambda} \ln\left(1 - \frac{\lambda X}{w_0 \cos \alpha}\right) + \left(\tan \alpha + \frac{W}{w_0 \cos \alpha}\right) X.$$

Remembering that $W = g/\lambda$, and that the value of g is known, we see that the previous equation has three parameters in it, namely λ, w_0, and α.

To compare this to the case of no drag we rewrite it in the form
$$Y = \frac{W}{\lambda} \left[\ln\left(1 - \frac{\lambda X}{w_0 \cos \alpha}\right) + \frac{\lambda X}{w_0 \cos \alpha}\right] + (\tan \alpha) X,$$

and expand the quantity in brackets using the Taylor series expansion for $\ln(1 + z)$, namely,
$$\ln(1 + z) = z - \frac{1}{2}z^2 + \frac{1}{3}z^3 + \cdots,$$

with $z = -\lambda X/(w_0 \cos \alpha)$. We thus find
$$Y = \frac{W}{\lambda} \left[-\frac{1}{2}\left(\frac{\lambda X}{w_0 \cos \alpha}\right)^2 - \frac{1}{3}\left(\frac{\lambda X}{w_0 \cos \alpha}\right)^3 + \cdots\right] + (\tan \alpha) X,$$

or
$$Y = -\frac{1}{2}g\left(\frac{1}{w_0 \cos \alpha}\right)^2 X^2 - \frac{1}{3}g\lambda\left(\frac{1}{w_0 \cos \alpha}\right)^3 X^3 + \cdots + (\tan \alpha) X.$$

In this form we can see that the air resistance comes into play at the X^3 and higher terms.

The maximum height of the particle will occur when $dY/dX = 0$ and, because $dY/dX = Y'/X'$, this occurs when $Y' = 0$, that is
$$e^{-\lambda t} = \frac{W}{w_0 \sin \alpha + W},$$
or
$$t = \frac{1}{\lambda} \ln\left(1 + \frac{w_0}{W} \sin \alpha\right).$$

For this to occur for $t > 0$ we need $0 < \alpha \leq \pi/2$, which agrees with intuition, and at this time the particle is at
$$X = \frac{w_0^2 \sin \alpha \cos \alpha}{\lambda (w_0 \sin \alpha + W)},$$
when the maximum height H is
$$H = -W\frac{1}{\lambda} \ln\left(1 + \frac{w_0}{W} \sin \alpha\right) + \frac{1}{\lambda} w_0 \sin \alpha.$$

We notice that $Y(t) = 0$ when $t = 0$ and $t = t_h$ where

$$Wt_h = \frac{1}{\lambda}(w_0 \sin \alpha + W)\left(1 - e^{-\lambda t_h}\right)$$

or

$$1 - e^{-\lambda t_h} = \frac{gW}{w_0 \sin \alpha + W} t_h.$$

Solving this for t_h by analytically techniques is beyond us. However, for given constants λ, g, w_0, α, and W, we can solve this either numerically or graphically by plotting the line $\{gW/(w_0 \sin \alpha + W)\} T$ and the curve $1 - e^{-T}$ as functions of T to find the point of intersection.[25] There are no values of t_h if $gW/(w_0 \sin \alpha + W) > 1$. (Why?)

Having obtained a numerical value for the time of flight t_h, the range R is computed from

$$R = \frac{gW w_0 \cos \alpha}{w_0 \sin \alpha + W} t_h,$$

while the angle of impact is determined from

$$\frac{dY}{dX} = \frac{-W + (w_0 \sin \alpha + W) e^{-\lambda t_h}}{w_0 \cos \alpha e^{-\lambda t_h}}.$$

Example 4.10 : The Flight of a Bullet with Drag

We illustrate these results by analyzing the flight of a bullet fired from a .17 Remington rifle. Table 4.12 shows the height of the bullet in inches above the muzzle of the rifle as a function of the horizontal distance in yards from the muzzle.[26]

Table 4.12 Hang time and range for different initial angles

Distance (yards)	Height (inches)
0	0.0
100	2.1
150	2.5
200	1.9
250	0.0
300	−3.4
400	−17.0
500	−44.3

According to the previous model, X and Y should be related by the equation

$$Y = \frac{W}{\lambda}\left[\ln\left(1 - \frac{\lambda X}{w_0 \cos \alpha}\right) + \frac{\lambda X}{w_0 \cos \alpha}\right] + (\tan \alpha) X.$$

Using the program Kalkulator to perform a nonlinear regression, we find $\lambda = 1.8927$, $w_0 = 46478$, and $\alpha = 0.001057$, after converting the horizontal measurements to inches. This predicts a muzzle velocity of 3873 feet/sec and that the bullet will travel a maximum horizontal distance of 2046 feet. These are plausible predictions. Figure 4.30 shows the bullet's trajectory together with the theoretical curve.

---

[25] "The Mathematics of Projectiles in Sport" by N. de Mestre, Australian Mathematical Society Lecture Series 6, Cambridge University Press, 1990, page 29.

[26] Data from "Armed and Dangerous", by M. Newton, Writer's Digest Books, 1990, page 174. (This book also contains similar data for many other rifles.)

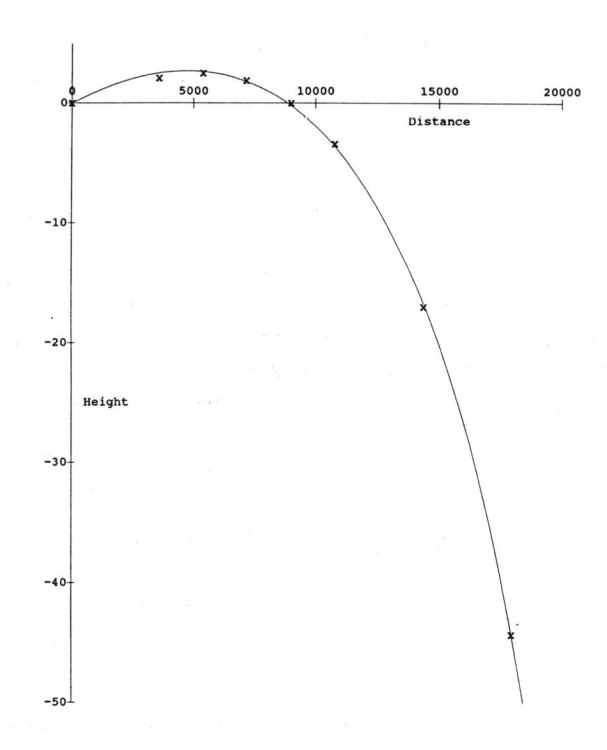

Figure 4.30 Bullet's trajectory together with the theoretical curve

Example 4.11 : The Flight of an Arrow with Drag

We also illustrate these results by analyzing the flight of an arrow fired from a Precision Shooting Equipment 30 inch Mach 7 bow with a 60 pound draw. Table 4.13 shows the height of the arrow in inches above the initial height of the arrow as a function of the horizontal distance in feet.[27]

Table 4.13 Height of arrow as a function of horizontal distance

Distance (feet)	Height (inches)
0	0.0
14	1.54
29	1.52
44	−1.35

According to the previous model, X and Y should be related by the equation

$$Y = \frac{W}{\lambda}\left[\ln\left(1 - \frac{\lambda X}{w_0 \cos \alpha}\right) + \frac{\lambda X}{w_0 \cos \alpha}\right] + (\tan \alpha) X.$$

Using the program Kalkulator to perform a nonlinear regression, we find $\lambda = 3.3212$, $w_0 = 3295$, and $\alpha = 0.01254$, after converting the horizontal measurements to inches. This predicts an initial velocity of 274.58 feet/sec and that the arrow will travel a maximum horizontal distance of 82.66 feet. These are plausible predictions. Figure 4.31 shows the arrow's trajectory together with the theoretical curve.

---

[27]We would like to thank Doug Marcoux of Precision Shooting Equipment, Tucson, Arizona, for his help in conducting this experiment.

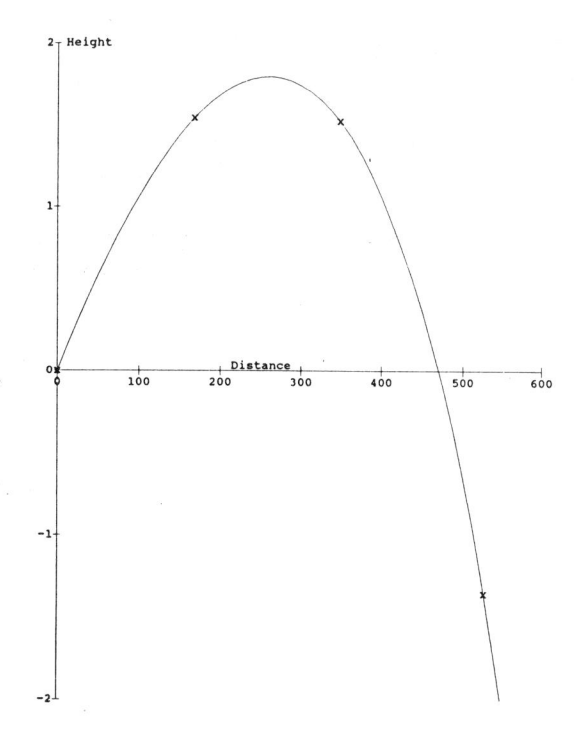

Figure 4.31 Arrow's trajectory together with the theoretical curve

Example 4.12 : The Flight of a Golf Ball with Drag[28]

We also illustrate these results by analyzing the flight of a well-struck golf ball. Experiments have determined that for a golf ball the value of λ is given by $\lambda = k/m \approx 0.25$ sec^{-1}, and a well-struck golf ball has an initial velocity of $w_0 = 200$ ft/sec (≈ 136 mph). With $g = 32$ ft/sec^2 we find $W = 128$ ft/sec (≈ 87 mph.).

Under these circumstances, the trajectory of the golf ball with initial angle α is

$$X(t) = 800 \cos \alpha \left(1 - e^{-t/4}\right),$$

$$Y(t) = -128t + 4 \left(200 \sin \alpha + 128\right) \left(1 - e^{-t/4}\right),$$

and

$$Y = 512 \ln \left(1 - \frac{X}{800 \cos \alpha}\right) + \left(\tan \alpha + \frac{128}{200 \cos \alpha}\right) X.$$

The speed at any time is

$$w(t) = \sqrt{(200 \cos \alpha)^2 e^{-t/2} + \left[128 - (200 \sin \alpha + 128) e^{-t/4}\right]^2}.$$

The hang time satisfies

$$1 - e^{-t_h/4} = \frac{32 \cdot 32}{200 \sin \alpha + 128} t_h,$$

and the range is

$$R = \frac{32 \cdot 128 \cdot 200 \cos \alpha}{200 \sin \alpha + 128} t_h.$$

---

[28] "Maximum projectile range with drag and lift, with particular application to golf" by H. Erichson, *American Journal of Physics*, **51**, 1983, pages 357 through 362.

In Figure 4.32 we show the (X, Y) trajectory for $\alpha = 15°$ to $\alpha = 65°$ in intervals of $10°$. Notice that for these values of α the greatest range occurs when $\alpha = 35°$. In Figure 4.33 we compare the no drag case to the linear drag case for $\alpha = 35°$. As any golfer will tell you, air resistance does play a role! In Table 4.14 we show the numerical computed values for t_h for $\alpha = 30°$ to $\alpha = 35°$ in intervals of $1°$ together with the range. For these values of α the greatest range occurs when $\alpha = 32$ and it is nearly 504 ft. □

Table 4.14 Hang time and range for different initial angles

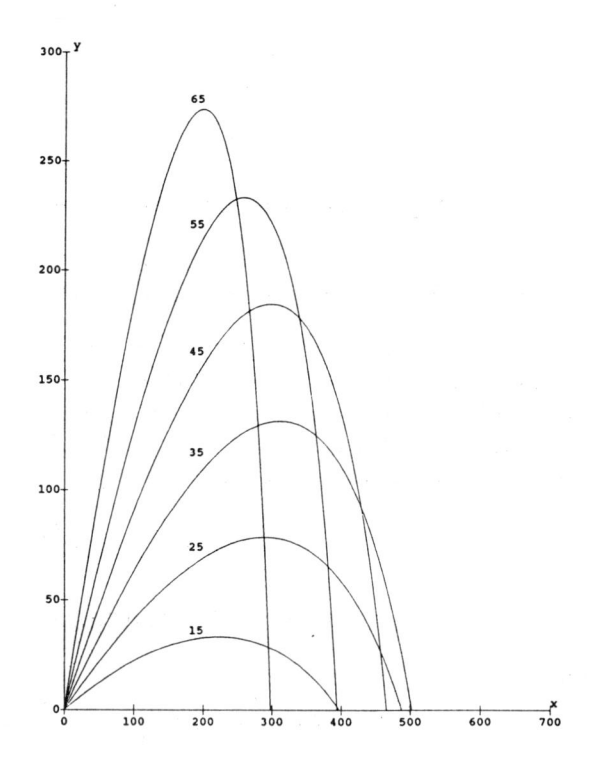

Angle (degrees)	Hang Time (seconds)	Range (feet)
30	5.167	502.43
31	5.300	503.45
32	5.431	503.91
33	5.560	503.84
34	5.686	503.16
35	5.810	501.98

Figure 4.32 Golf ball trajectories with drag for $\alpha = 15°$ to $\alpha = 65°$

Computer Experiment: Plot Y as a function of X for different values of α, and try to find the angle that gives the maximum range.

In Figure 4.34 we show the speed $w(t)$ as a function of time for $\alpha = 15°$ to $\alpha = 65°$ in intervals of $10°$. As t increases all speeds tend to the terminal speed of $W = 128$. But, wait a minute. In every case there are times when $w(t)$ attains a minimum value after which $w(t)$ increases to its terminal speed! Does this occur while the ball is in flight? In Figure 4.35 we show $w(t)$ and $y(t)$ as a function of time for the case $\alpha = 35°$. Notice that the time where

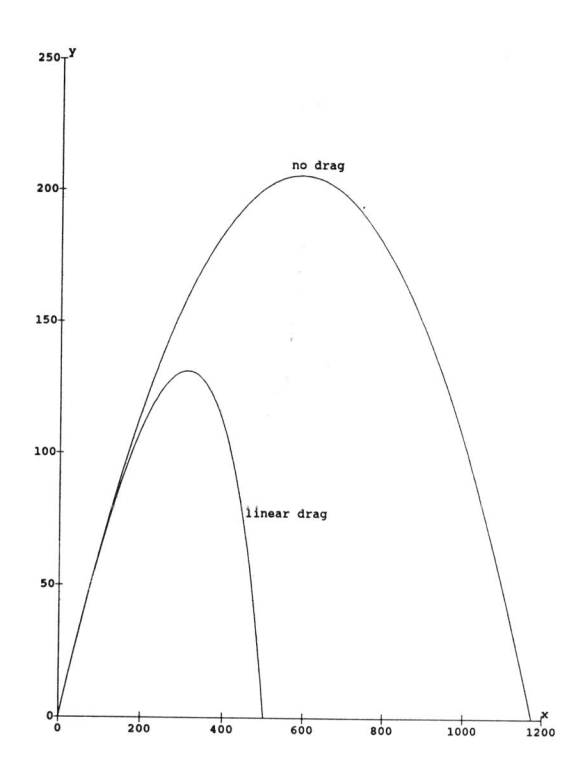

Figure 4.33 Golf ball trajectories with and without drag for $\alpha = 35°$

$w(t)$ attains a maximum is just after the ball starts to fall back to earth. As the ball hits the earth its speed is increasing towards its terminal speed.

Computer Experiment: Find the largest angle α where $w(t)$ does not attain its minimum value while the ball is in the air. Find the angle α where $w(t)$ attains its minimum value at the time when the ball is at its maximum height in the air.

Based on the previous example it is natural to ask under what circumstances $w(t)$ attains a minimum. To answer this we differentiate

$$w^2(t) = u_0^2 e^{-2\lambda t} + \left[W - (v_0 + W) e^{-\lambda t}\right]^2$$

and find

$$2ww' = -2\lambda u_0^2 e^{-2\lambda t} + 2\left[W - (v_0 + W) e^{-\lambda t}\right] \lambda (v_0 + W) e^{-\lambda t}.$$

The right-hand side is zero when

$$-u_0^2 e^{-\lambda t} + \left[W - (v_0 + W) e^{-\lambda t}\right](v_0 + W) = 0,$$

which we can solve for $e^{-\lambda t}$ obtaining

$$e^{-\lambda t} = \frac{W(v_0 + W)}{u_0^2 + (v_0 + W)^2},$$

or

$$t = \frac{1}{\lambda} \ln\left[\frac{u_0^2 + (v_0 + W)^2}{W(v_0 + W)}\right].$$

There will be a time $t > 0$ where $w'(t) = 0$ if

$$\frac{u_0^2 + (v_0 + W)^2}{W(v_0 + W)} > 1,$$

ADDITIONAL MATERIALS

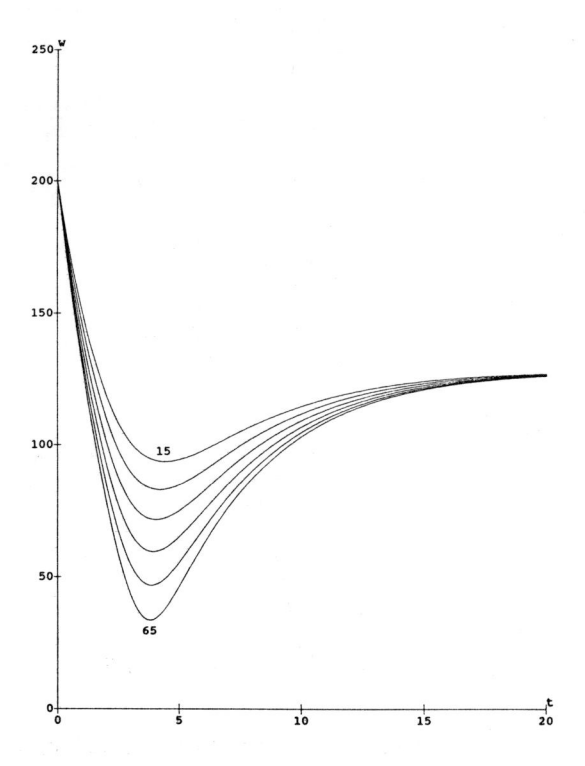

Figure 4.34 The speed $w(t)$ of a golf ball as a function of time for $\alpha = 15°$ to $\alpha = 65°$

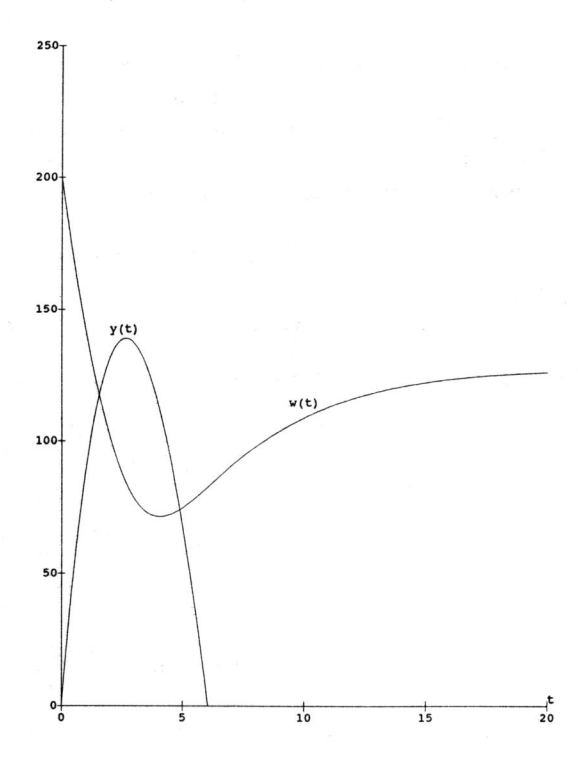

Figure 4.35 The speed $w(t)$ and height $y(t)$ of a golf ball as a function of time for $\alpha = 35°$

that is, if
$$u_0^2 + v_0^2 + v_0 W > 0.$$

At this time the speed will be
$$w = \frac{u_0 W}{\sqrt{u_0^2 + (v_0 + W)^2}}.$$

We can show that
$$\frac{u_0 W}{\sqrt{u_0^2 + (v_0 + W)^2}} < W$$

and
$$\frac{u_0 W}{\sqrt{u_0^2 + (v_0 + W)^2}} < w(0),$$

which shows that there is a time when the particle attains a speed smaller than the terminal speed and smaller than the initial speed.[29] The values of X and Y at which this minimum speed is attained are
$$X(t) = \frac{u_0}{\lambda}\left[1 - \frac{W(v_0 + W)}{u_0^2 + (v_0 + W)^2}\right]$$

and
$$Y(t) = -W\frac{1}{\lambda}\ln\left[\frac{u_0^2 + (v_0 + W)^2}{W(v_0 + W)}\right] + \frac{1}{\lambda}(v_0 + W)\left[1 - \frac{W(v_0 + W)}{u_0^2 + (v_0 + W)^2}\right].$$

---

[29] Based on "Guide to Mathematical Modelling" by D. Edwards and M. Hamsom, Macmillan, 1989, pages 240 through 242.

ADDITIONAL MATERIALS

5. IDEAS FOR CHAPTER 5 — "FIRST ORDER LINEAR DIFFERENTIAL EQUATIONS AND MODELS"

5.1 Page by Page Comments

Chapter 5 requires about 4 class meetings of 50 minutes each.

Section 5.3, *Models That Use Bernoulli's Equation*, is not used in later sections, and so may be selectively omitted.

Page 183. Example 5.1 uses results from Example 2.3 on page 37 of the text.

Pages 186 – 190. We use the $y = uz$ technique to solve linear differential equations. Then we look at the structure of the solution to arrive at the standard integrating factor technique.

Page 192. The differential equation in Example 5.2, namely, $y' = x - y$, is used a number of times. (Example 3.1, page 82; Exercise 1, page 91; and Example 3.5, page 99.)

Page 194. It is worthwhile pointing out the distinction between the uniqueness theorem for linear differential equations and the uniqueness theorem for $y' = g(x, y)$ in general.

Page 198. Note that transient and steady state solutions need not be solutions.

Page 202. Example 5.7 is an example where the forcing term consists of a piecewise defined function. However, it is continuous, so the solution will be smooth.

Page 204. The steps from (5.53), namely,

$$T(t) = -ke^{kt} \int_0^t T_a(u) e^{-ku}\, du + 70 e^{kt} \tag{5.1}$$

where

$$T_a(t) = \begin{cases} 70 + 66t & \text{if } 0 \le t \le 5, \\ 400 & \text{if } 5 < t, \end{cases} \tag{5.2}$$

to (5.54), namely,

$$T(t) = \begin{cases} 70 + 66t + 66(1 - e^{kt})/k & \text{if } 0 \le t \le 5, \\ 400 + [66(1 - e^{5k})/k] e^{k(t-5)} & \text{if } 5 < t, \end{cases} \tag{5.3}$$

are as follows.

If $0 \le t \le 5$ then (5.1) and (5.2) give

$$T(t) = -ke^{kt} \int_0^t (70 + 66u)\, e^{-ku}\, du + 70 e^{kt}. \tag{5.4}$$

Because
$$\int_0^t e^{-ku}\,du = -\frac{1}{k}e^{-ku}\Big|_0^t = \frac{1}{k}\left(1 - e^{-kt}\right)$$
and
$$\int_0^t ue^{-ku}\,du = -e^{-ku}\left(\frac{1}{k}u + \frac{1}{k^2}\right)\Big|_0^t = \frac{1}{k^2} - e^{-kt}\left(\frac{1}{k}t + \frac{1}{k^2}\right)$$
we can write (5.4) as
$$T(t) = -ke^{kt}\left[\frac{70}{k}\left(1 - e^{-kt}\right) + \frac{66}{k^2} - 66e^{-kt}\left(\frac{1}{k}t + \frac{1}{k^2}\right)\right] + 70e^{kt},$$
which simplifies to
$$T(t) = 70 + 66t + \frac{66}{k}\left(1 - e^{kt}\right),$$
the first equation in (5.4).

If $5 < t$ then (5.1) can be written as
$$T(t) = -ke^{kt}\left[\int_0^5 T_a(u)e^{-ku}\,du + \int_5^t T_a(u)e^{-ku}\,du\right] + 70e^{kt}$$
and so (5.2) gives
$$T(t) = -ke^{kt}\left[\int_0^5 (70 + 66u)\,e^{-ku}\,du + \int_5^t 400 e^{-ku}\,du\right] + 70e^{kt}.$$
Doing the indicated integrations gives
$$T(t) = -ke^{kt}\left[\frac{70}{k}\left(1 - e^{-k5}\right) + \frac{66}{k^2} - 66e^{-k5}\left(\frac{1}{k}5 + \frac{1}{k^2}\right) + \frac{400}{k}\left(e^{-k5} - e^{-kt}\right)\right] + 70e^{kt},$$
which simplifies to
$$T(t) = 400 + \frac{66}{k}\left(1 - e^{5k}\right)e^{k(t-5)},$$
the second equation in (5.4).

Pages 202 – 205. An alternative way to solve Example 5.7 is to think of the problem as piecing together differential equations, where $dT/dt = k(70 + 66t)$, $T(0) = 70$, is solved giving the solution valid for $0 \le t \le 5$. Then $dT/dt = 400$ with the initial condition at $t = 5$ being $\lim_{t \to 5^-} T(t)$ [where $T(t)$ is used from the solution valid in $0 \le t \le 5$] is solved. See Section 4.6.

Page 205. Example 5.8 shows how an innocent looking differential equation, namely, $y' = 1 - 2xy$, may have a solution that cannot be expressed in terms of familiar functions, yet can still be graphed using the techniques we have developed. This differential equation occurs throughout the text (Exercise 10, page 92; Exercise 2, page 117; Exercise 27, page 211; and — in its differentiated form, $y'' + 2xy' + 2y = 0$ — in Example 12.3, page 580; and Example 12.5, page 589.)

Page 211. Exercise 28 is worthwhile.

Page 211. There are additional exercises for Section 5.2 on page 163 of this manual.

Page 211. Section 5.3, *Models That Use Bernoulli's Equation*, is not used elsewhere in the text. Bernoulli is pronounced ber-nool'-ee.

Page 212. The differential equation (5.60) goes under various names, perhaps the most common being the VON BERTALANFFY EQUATION.

Page 215. Example 5.10. Note that the inflection point occurs at $(2/3)^3$ of the limiting weight of the fish, and that $(2/3)^3 \approx 0.296$. This distinguishes it from the logistic and Gompertz equations, where the inflection point occurs at 0.5 and $1/e \approx 0.368$ of the limiting value.

Page 217. Equation (5.69) can be expressed in terms of the initial weight w_0 and the limiting weight w_∞ in the following way

$$w(t) = \left[w_\infty^{1/3} - \left(w_\infty^{1/3} - w_0^{1/3} \right) e^{-kt/3} \right]^3.$$

Page 217. There is a very clever way to estimate the parameters w_∞, w_0, and k in (5.69) when written in the form

$$w(t) = \left[w_\infty^{1/3} - \left(w_\infty^{1/3} - w_0^{1/3} \right) e^{-kt/3} \right]^3$$

if $w(t)$ has been measured at equal intervals, say of 1 unit. It is a two step process.[1]

First, note that

$$w^{1/3}(t) = w_\infty^{1/3} - \left(w_\infty^{1/3} - w_0^{1/3} \right) e^{-kt/3},$$

and that

$$w^{1/3}(t+1) = w_\infty^{1/3} - \left(w_\infty^{1/3} - w_0^{1/3} \right) e^{-kt/3} e^{-k/3}.$$

Solving the first equation for $-\left(w_\infty^{1/3} - w_0^{1/3} \right) e^{-kt/3}$ and substituting in the second gives

$$w^{1/3}(t+1) = w_\infty^{1/3} + \left(w^{1/3}(t) - w_\infty^{1/3} \right) e^{-k/3},$$

or

$$w^{1/3}(t+1) = w_\infty^{1/3} \left(1 - e^{-k/3} \right) + w^{1/3}(t) e^{-k/3},$$

This implies that if we plot $w^{1/3}(t+1)$ versus $w^{1/3}(t)$ we should get a straight line. In view of the fact that $w^{1/3}(t+1)$ and $w^{1/3}(t)$ both tend to $w_\infty^{1/3}$ with increasing time, we can estimate $w_\infty^{1/3}$ as the intersection of the straight line just obtained, and the straight line at $45°$ through the origin. We now have an estimate for $w_\infty^{1/3}$.

Second, we rewrite

$$w^{1/3}(t) = w_\infty^{1/3} - \left(w_\infty^{1/3} - w_0^{1/3} \right) e^{-kt/3}$$

in the form

$$w_\infty^{1/3} - w^{1/3}(t) = \left(w_\infty^{1/3} - w_0^{1/3} \right) e^{-kt/3}$$

from which we obtain

$$\ln \left[w_\infty^{1/3} - w^{1/3}(t) \right] = \ln \left(w_\infty^{1/3} - w_0^{1/3} \right) - \frac{k}{3} t.$$

Thus, a plot of $\ln \left[w_\infty^{1/3} - w^{1/3}(t) \right]$ versus t should give a straight line of slope $-k/3$ and intercept $\ln \left(w_\infty^{1/3} - w_0^{1/3} \right)$, from which we can estimate k and w_0. We illustrate this process on the growth of the North Sea sole given in Table 5.1 and shown in Figure 5.1.[2] [Table 5.1 is in TWIDDLE. It is called SOLE.DTA and is in the subdirectory DD-05.]

Table 5.1 Weight of North Sea sole as a function of time

Time (years)	Weight (grams)
1	7.7
2	61.7
3	144.9
4	225.2
5	301.9
6	361.6
7	402.6
8	437.6

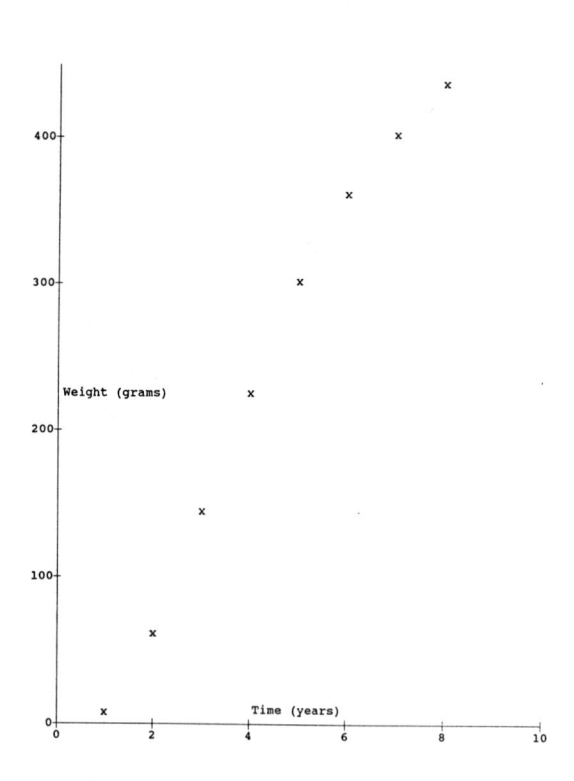

Figure 5.1 Weight of North Sea sole as a function of time

IDEAS FOR CHAPTER 5 — "FIRST ORDER LINEAR DIFFERENTIAL EQUATIONS AND MODELS"

Figure 5.2 shows a plot of $w^{1/3}(t+1)$ versus $w^{1/3}(t)$, the line of best fit ($y = 0.672514x + 2.602687$), and the line $y = x$. Where the two lines intersect ($x = y = 7.9474756$) gives an estimate of $w_\infty^{1/3} \approx 7.947475$, so $w_\infty \approx 502$ grams.

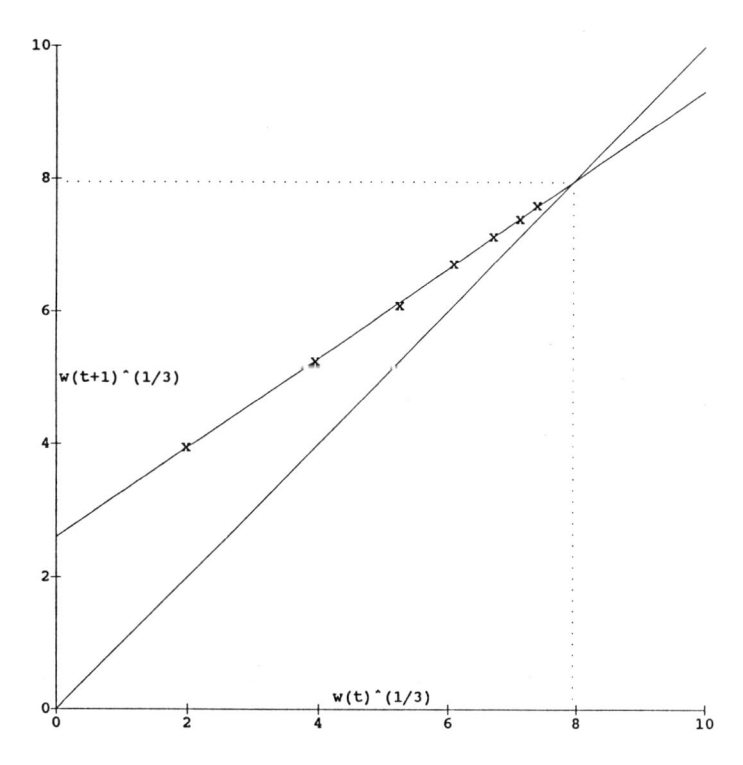

Figure 5.2 A plot of $w^{1/3}(t+1)$ versus $w^{1/3}(t)$, the line $y = 0.672514x + 2.602687$, and the line $y = x$

Figure 5.3 shows a plot of $\ln\left[w_\infty^{1/3} - w^{1/3}(t)\right]$ versus t and the line of best fit ($y = -0.398956t + 2.195061$). From the slope of this line we can estimate $k/3 \approx 0.398956$ and $\ln\left(w_\infty^{1/3} - w_0^{1/3}\right) \approx 2.195061$.

Figure 5.4 shows a plot of Table 5.1 and the function $w(t) = \left(7.9474756 - 8.98055 e^{-0.398956 t}\right)^3$.

Page 218. Table 5.3 is in TWIDDLE. It is called NSPLAICE.DTA and is in the subdirectory DD-05.

Page 220. There is a more illuminating way to analyze the solution

$$P(t) = e^t \left(\int_0^t \frac{e^s}{7 + \sin 6s} ds + C\right)^{-1}$$

than is given in the text. See ANALYZING THE ANIMAL GROWTH EXAMPLE on page 166 of this manual.

Page 222. There are additional exercises for Section 5.3 on page 164 of this manual.

Page 222. A section on Clairaut's Equation can be found on page 170 in this manual. Clairaut is pronounced klay-roh'.

Page 222. A section on Exact Equations can be found on page 183 in this manual.

---

[1] "On the Dynamics of Exploited Fish Populations" by R. J. H. Beverton and S. J. Holt, Fishery Investigations, Series II, **19**, 1957, page 283.

[2] "On the Dynamics of Exploited Fish Populations" by R. J. H. Beverton and S. J. Holt, Fishery Investigations, Series II, **19**, 1957, page 479, Table 16.7..

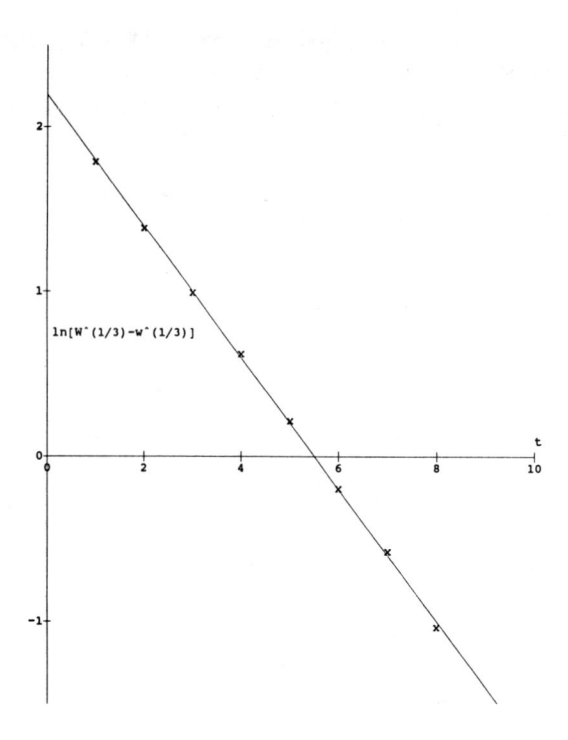

Figure 5.3 A plot of $\ln\left[w_\infty^{1/3} - w^{1/3}(t)\right]$ versus t and the line $y = -0.398956t + 2.195061$

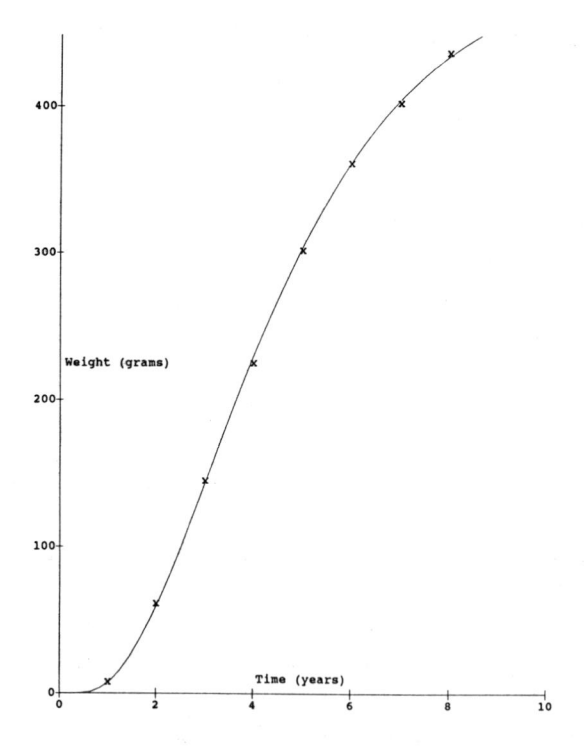

Figure 5.4 Weight of North Sea sole as a function of time and the function $w(t) = \left(7.9474756 - 8.98055e^{-0.398956t}\right)^3$

Page 222. A flowchart showing how to solve first order equations can be found on page 186 in this manual.

5.2 Additional Exercises

Section 5.2

Page 211. Add exercises after the current Exercise 28.

29. Two cars are stationary behind each other at a stoplight at time $t = 0$. The first car's position is characterized by $x(t)$ and the second by $y(t)$, with velocities $u(t) = x'$ and $v(t) = y'$, respectively. A simple model for the motion of these cars[3] (called Pipes' Model) is given by the two equations

$$\begin{align} u' &= a - u \\ v' &= u - v, \end{align}$$

where a is a positive constant. It is said that the second driver adjusts his acceleration to be exactly equal in magnitude to the relative difference in velocity between his vehicle and the vehicle ahead. Explain why this is plausible. Write down a similar statement about the first driver's acceleration. What does a represent? Explain why the initial conditions $x(0) = u(0) = 0$, $y(0) = y_0$, $v(0) = 0$, are reasonable. What does y_0 represent? Solve the first equation for $u(t)$ and, after substituting the result into the second equation, solve the second equation. Now solve $x' = u$, $y' = v$ for $x(t)$ and $y(t)$. Is there any time at which the cars collide?

30. **Mixture Problem.** An open container has 5 pounds of impurities dissolved in 150 liters of water. Pure water is pumped into the container at a rate of 3 liters per minute, and the well-stirred mixture is drained at the rate of 2 liters per minute.

    (a) How many pounds of impurities remain after 10 minutes?

    (b) If the container holds a maximum of 200 liters, what is the concentration of impurities just before the solution overflows?

31. **Mixture Problem.** A conference room contains 3000 cubic meters of air that is free of carbon monoxide. The ventilation system blows in air, free of carbon monoxide, at a rate of 0.3 cubic meters per minute and extracts air at the same rate. If at time $t = 0$, people in the room start smoking and add carbon monoxide to the room at a rate of 0.02 cubic meters per minute, how long before the air in the room contains 0.015% carbon monoxide?

32. **RL Circuit.** Let the circuit of Exercise 11 have a voltage source equal to $e^{-60t} + \sin t$. If $I(0) = 0$, find the explicit solution to this initial value problem. What is the steady state part of the solution in this case? Are there any equilibrium solutions to this problem? Please explain.

33. **Neurobiology.** A simple dynamical model in neurobiology considers a neuron as having three regions: dendrites, a cell body, and an axon. Neurons are connected to other neurons by synapses. The nerve cell body functions as an input-output device,

---

[3] "Car-Following Models" by R.L. Baker, Jr., in "Modules in Applied Mathematics, Volume 1, Differential Equations — Models" edited by M. Braun, C.S. Courteny, and D.A. Drew, Springer-Verlag 1983, Chapter 12

responding to electrical stimuli. One of the differential equations that occurs in this model is
$$C\frac{du}{dt} + \frac{1}{R}(u - u_0) = I(t),$$
where C, R, and u_0 are constants, u is the cell potential (that is, the voltage difference between the inside and outside of this simple neuron), and $I(t)$ is the electrical current injected into the cell. A recent article[4] describing such a model includes the following two statements about this model. (i) "For a typical neuron, $u_0 < u_{thresh}$, so u will decay to u_0 when the injected current vanishes." (ii) "If $I(t)$ is a large constant current I_c, the cell potential will change in an almost linear fashion between u_0 and u_{thresh}."

(a) Explain these two statements by just analyzing the differential equation. (You may wish to use some specific values of the constants and look at a typical slope field.)

(b) Find the explicit solution of the differential equation when $I(t)$ equals a constant, and explain these two statements using your solution.

(c) Find an explicit formula for u_{thresh} in terms of the other constants.

34. **Projectile Motion with Arbitrary Resistance.** Read the article "Projectile Motion with Arbitrary Resistance" by T. A. de Alwis, College Mathematics Journal, **26**, 1995, pages 361 through 366. Rewrite the proofs of Theorems 1 through 3 of that article in a way that a calculus student could understand by including all the calculations and theorems required.

Section 5.3

Page 222. Add exercises after the current Exercise 10.

11. The logistic, Gompertz, and von Bertalanffy equations have their inflection points at $0.5b$, $0.368b$, and $0.296b$ respectively, where $y = b$ is the carrying capacity. Sketch these three curves with initial condition $y(0) = 0.16$ assuming that all three inflection points occur when $x = a$.

12. **Richards' Growth Model.**[5] The Richards' Growth Model is governed by
$$\frac{dy}{dx} = \frac{k}{1-m} y \left[\left(\frac{b}{y}\right)^{1-m} - 1 \right]$$
where k, m, and b are constants. Identify the special cases $m = 0$, $m = 2/3$, $m \to 1$, and $m = 2$ of this differential equation, in terms of previously investigated differential equations. Show that the solution of Richards' equation is
$$y(x) = b\left(1 - Ce^{-kx}\right)^{1/(1-m)}.$$
Compare the Richards' Growth Model to the Generalized Logistic Equation on page 82 of this manual.

13. **The Population of the USA.** The discussion on page 131 of this manual suggests that the carrying capacity of the USA has changed with time. Explain why that might

---
[4] J. Hopfield "Neurons, Dynamics, and Computation", Physics Today **47** No. 2, February, 1994, pages 40–46.

[5] "A flexible growth function for fitting equations to growth curves" by F. J. Richards, Journal of Experimental Botany, **10**, 1959, pages 290 through 300.

have happened. This suggests that the USA population may be modeled by a logistic differential equation in which the carrying capacity is not constant, so that

$$\frac{dP}{dt} = aP\left[1 - \frac{P}{b}\right] \tag{5.5}$$

where a is constant but $b = b(t)$. One way in which the carrying capacity might change is by satisfying its own logistic equation

$$\frac{db}{dt} = cb\left[1 - \frac{b}{B}\right] \tag{5.6}$$

where c and B are constants, and B is the carrying capacity of the carrying capacity. Solve (5.6) subject to $b(0) = b_0$. Then substitute your solution for $b(t)$ into (5.5) and solve the Bernoulli equation obtained in this way. Seeing how well the subsequent model fits the actual data for the USA population would be a good project.

14. **Riccati Differential Equation.** A differential equation of the form $y' = a_0(x) + a_1(x)y + a_2(x)y^2$, with $a_0(x)$ and $a_2(x)$ not identically zero, is called a RICCATI DIFFERENTIAL EQUATION.

   (a) Show that if we know a solution $y_1(x)$ of a Riccati differential equation, then the change of variable $y = y_1(x) - 1/u$ results in the following linear differential equation for u:
   $$u' + [a_1(x) + 2a_2(x)y_1(x)]u = a_2(x).$$

   (b) To find this first solution of a Riccati differential equation, we often try simple functions like ax^b or ae^{bx}. Find values of a and b such that ax^b is a solution of $y' = 1 + x + 2x^2\cos x - (1 + 4x\cos x)y + 2(\cos x)y^2$.

   (c) Make the change of variable given in part (a) to the differential equation in part (b) and obtain the differential equation $u' - u = 2\cos x$.

   (d) Solve the differential equation in part (c), and show that the general solution of our original differential equation in part (b) is $y = x - (\sin x - \cos x + Ce^x)^{-1}$, where C is an arbitrary constant.

15. Solve the following Riccati differential equations.

   (a) $y' = 5 + \frac{4}{x}y + \frac{1}{4x^2}y^2$
   (b) $y' = (1-y)\left(\frac{1}{x} - \frac{1}{10} + \frac{y}{10}\right)$
   (c) $y' = \frac{6}{x^2} - 2y^2$
   (d) $y' = -1 + 2y - y^2$
   (e) $y' = 3e^{4x} + 2y - 12y^2$
   (f) $y' = e^{2x} + \left(1 + \frac{5}{2}e^x\right)y + y^2$
   (g) $y' = \frac{4}{x^2} - \frac{y}{x} - y^2$

16. Show that if y is a solution of the Riccati differential equation (in the form from Exercise 5.2), then $u = 1/y$ is a solution of the differential equation $u' = -a_2(x) - a_1(x)u - a_0(x)u^2$.

17. Show that if we know one solution, $y_1(x)$, of a Riccati equation, the substitution $y = y_1(x) + v$ converts the general Riccati equation to the Bernoulli equation $v' - [a_1(x) + 2a_2(x)y_1(x)]v = a_2(x)v^2$. Explain the relationship between this result and that of Exercise 5.2.

18. The logistic differential equation $y' = ay(b-y)$ can be classified either as a Bernoulli or as a Riccati differential equation.

   (a) Use the method in Exercise 5.2 to solve it as a Riccati equation.
   (b) Use the appropriate change of dependent variable to solve it as a Bernoulli equation.

5.3 Additional Materials

Analyzing the Animal Growth Example

This deals with the solution

$$P(t) = e^t \left(\int_0^t \frac{e^s}{7 + \sin 6s} ds + C \right)^{-1},$$

where $C = 1/P(0)$, of the differential equation

$$\frac{dP}{dt} = P \left(1 - \frac{P}{7 + \sin 6t} \right).$$

(Example 5.11 on animal population growth, page 219.) If we introduce the abbreviation

$$I(t) = \int_0^t \frac{e^s}{7 + \sin 6s} ds, \tag{5.7}$$

then the solution can be written

$$P(t) = \frac{e^t}{I(t) + C}, \tag{5.8}$$

The slope field and the numerical solutions in Figure 5.22 of the text suggest that might be a particular solution that is periodic. If we look at the solution $P(t)$ the only hint of a periodic function is $\sin 6s$ in the integrand, and, because $\sin[6(s + \pi/3)] = \sin 6s$ the function $\sin 6s$ is periodic of period $p = \pi/3$. Is there a particular solution $P(t)$ that is periodic of period $p = \pi/3$? In other words, can we select C so that $P(t + p) = P(t)$ for every t? Thus, we try to find a C for which

$$\frac{e^{t+p}}{I(t+p) + C} = \frac{e^t}{I(t) + C}, \tag{5.9}$$

for every t.

In order to do this, we first relate $I(t + p)$ to $I(t)$. From (5.7) we have

$$I(t + p) = \int_0^{t+p} \frac{e^s}{7 + \sin 6s} ds.$$

We split this integral from 0 to $t + p$ into an integral from 0 to p and an integral from p to $t + p$

$$I(t + p) = \int_0^p \frac{e^s}{7 + \sin 6s} ds + \int_p^{t+p} \frac{e^s}{7 + \sin 6s} ds.$$

We recognize that the first integral is $I(p)$, while in the second integral we change variables from s to u where $s = u + p$, to find

$$I(t + p) = I(p) + \int_0^t \frac{e^{u+p}}{7 + \sin 6(u + p)} du,$$

or, because $\sin[6(u + p)] = \sin 6u$,

$$I(t + p) = I(p) + e^p \int_0^t \frac{e^u}{7 + \sin 6u} du.$$

We recognize the last integral is just $I(t)$ so we have that $I(t + p)$ and $I(t)$ are related by

$$I(t + p) = I(p) + e^p I(t). \tag{5.10}$$

If we substitute (5.10) into (5.9) we find

$$\frac{e^{t+p}}{I(p) + e^p I(t) + C} = \frac{e^t}{I(t) + C}.$$

If we cancel e^t from both sides, and then cross multiply, we have

$$e^p [I(t) + C] = I(p) + e^p I(t) + C,$$

or

$$e^p C = I(p) + C,$$

which when solved for C yields

$$C = \frac{1}{e^p - 1} I(p).$$

Because this equation is independent of t then there is a value of C which gives a periodic solution of period $p = \pi/3$. (What would we have concluded if C had depended on t?) Because $P(0) = 1/C$, then the solution that starts at $(e^p - 1)/I(p)$ will be periodic of period p, and, from (5.8), this solution is given by

$$P_p(t) = \frac{e^t}{I(t) + I(p)/(e^p - 1)}.$$

In order to make this equation appear simpler, we introduce the constant κ, (the Greek letter kappa) by

$$\kappa = \frac{1}{e^p - 1} I(p) = \frac{1}{e^{\pi/3} - 1} \int_0^{\pi/3} \frac{e^s}{7 + \sin 6s} ds,$$

so that

$$P_p(t) = \frac{e^t}{I(t) + \kappa},$$

and the initial value of this curve is $P_p(0) = 1/\kappa$.

Because we are not able to integrate

$$I(t) = \int_0^t \frac{e^s}{7 + \sin 6s} ds$$

in terms of familiar functions, we use numerical integration to obtain Table 5.2, a table of values for $P_p(t)$ where $0 \leq t < p = \pi/3 \approx 1.047$. (Why don't we consider larger values of t?) From this table we see that $P_p(0) \approx 6.771$ so this is the initial condition for the periodic solution. The curve with this initial condition, namely the curve $P_p(t)$ is sketched in Figure 5.5 along with the values given in Table 5.2.

The slope field and the numerical solutions in Figure 5.15 also suggest that two things eventually happen: The effect of the initial condition is lost, and that all solutions tend to the same oscillating solution, presumably $P_p(t)$. Can we confirm these suggestions?

Both of these suggestions are concerned with the behavior of $P(t)$ as $t \to \infty$, and possible comparison to the periodic function $P_p(t)$. We can think of t consisting of two parts: the number of periods p it has gone through, and the remainder. Thus, we can think of t in the form

$$t = np + \ell$$

where

$$0 \leq \ell < p$$

and n is a positive integer. This means that we are considering t that satisfies

$$np \leq t < (n+1)p,$$

Table 5.2 The function $P_p(t)$

t	$P_p(t)$
0.0	6.771
0.1	6.818
0.2	6.898
0.3	6.989
0.4	7.062
0.5	7.094
0.6	7.069
0.7	6.989
0.8	6.886
0.9	6.801
1.0	6.767

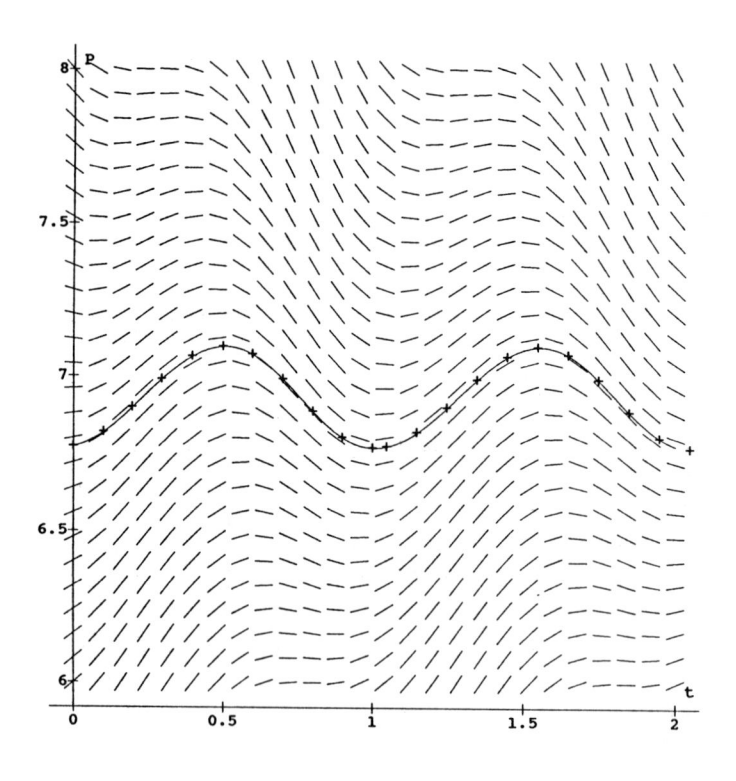

Figure 5.5 The numerical solution of $dP/dt = P[1 - P/(7 + \sin 6t)]$ through the point $P_p(0) \approx 6.771$ and the numerical values of $P_p(t)$

and $t \to \infty$ is equivalent to $n \to \infty$. Thus, from (5.8), we have

$$P(t) = P(np + \ell) = \frac{e^{np+\ell}}{I(np+\ell) + C}. \tag{5.11}$$

We will concentrate on the term $I(np + \ell)$. Writing $np + \ell = [\ell + (n-1)p] + p$ and using (5.10) once leads to

$$I(np+\ell) = I(p) + e^p I(\ell + (n-1)p),$$

while using it again, this time on $I(\ell + (n-1)p)$, gives

$$I(np+\ell) = I(p) + e^p I(\ell + (n-1)p) = I(p) + e^p \left[I(p) + e^p I(\ell + (n-2)p) \right].$$

This equation can be rewritten as

$$I(np+\ell) = I(p)(1 + e^p) + e^{2p} I(\ell + (n-2)p).$$

We repeat this process $n - 2$ more times until we finally find

$$I(np+\ell) = I(p)\left[1 + e^p + e^{2p} + \cdots + e^{(n-1)p}\right] + e^{np} I(\ell).$$

We recognize the quantity in square brackets as a geometric series, which we can sum as follows

$$1 + e^p + e^{2p} + \cdots + e^{(n-1)p} = \frac{1 - e^{np}}{1 - e^p},$$

so we find

$$I(np+\ell) = I(p)\frac{1 - e^{np}}{1 - e^p} + e^{np} I(\ell).$$

If we rewrite this in the form

$$I(np+\ell) = (e^{np} - 1)\frac{1}{e^p - 1} I(p) + e^{np} I(\ell),$$

and recall that

$$\kappa = \frac{1}{e^p - 1} I(p),$$

we have

$$I(np+\ell) = (e^{np} - 1)\kappa + e^{np} I(\ell).$$

We now substitute this equation into (5.11) to find

$$P(t) = P(np + \ell) = \frac{e^{np+\ell}}{(e^{np} - 1)\kappa + e^{np} I(\ell) + C}.$$

Dividing by e^{np} and rearranging gives

$$P(t) = P(np + \ell) = \frac{e^{\ell}}{I(\ell) + \kappa + e^{-np}(C - \kappa)}.$$

In this form we can see that

$$\lim_{t \to \infty} P(t) = \lim_{n \to \infty} P(np + \ell) = \frac{e^{\ell}}{I(\ell) + \kappa} = P_p(\ell) = P_p(np + \ell) = P_p(t).$$

[How do we know that $P_p(\ell) = P_p(np + \ell)$?] Thus, as $t \to \infty$ all solutions tend to $P_p(t)$ and the dependence on C — the initial condition — is lost. The solution $P_p(t)$ is stable. Figure 5.6 is the text's Figure 5.15 with $P_p(t)$ superimposed for $8\pi/3 \le t < 3\pi$.

ADDITIONAL MATERIALS

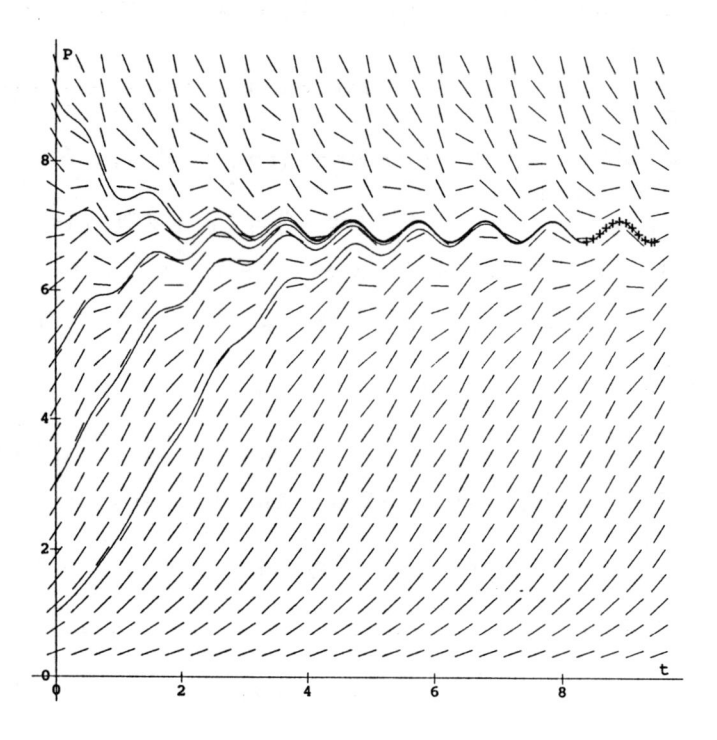

Figure 5.6 Numerical solution curves, the curve $P_p(t)$, and the slope field for $dP/dt = P[1 - P/(7 + \sin 6t)]$

Clairaut's Equation

In this section we look at an example from string art. (In place of String Art, you could do THE GARAGE DOOR on page 179 of this manual, which is the same as Exercise 5.) The differential equation that arises in this case is not one we have encountered previously. We develop a technique for solving it and show how it is used in other examples.

Example 5.1 : String Art

Imagine we put nails along the x- and y-axes at the points $0, 1, 2, \cdots, 10$. We tie one end of a piece of string to the nail at the point $(0,0)$ and the other end to the nail at the point $(0, 10)$. Now we join the nail at $(1,0)$ to the one at $(9,0)$, the nail at $(2,0)$ to the one at $(0,8)$, and so on until we have joined $(10,0)$ to $(0,0)$. Figure 5.7 illustrates the end result. If we look at this figure we see that the straight lines have created the illusion of a curve. The question we want to answer is, what is the equation of this curve?[6]

Let's denote the function we are looking for by $y = f(x)$. A typical string will be tangent to $y = f(x)$ at a point (x_0, y_0), say. This string is tied to a nail on the x-axis, at, say, $(u, 0)$, and its other end is tied to the point $(0, 10 - u)$ on the y-axis, as shown in Figure 5.8.

The relationship between the string and the curve is that the slope of the string is the slope of the curve—namely, $f'(x_0)$. The equation for a straight line with slope $m = f'(x_0)$ is
$$y - y_0 = m(x - x_0).$$

---

[6]Based on an idea in "Calculus in a Real and Complex World" by F. Wattenberg, PWS 1995, page 232.

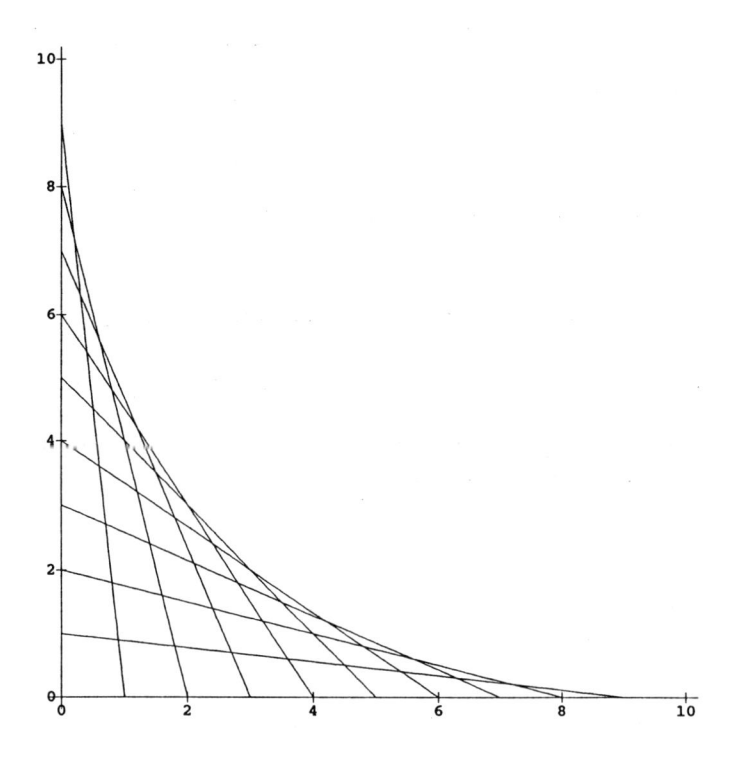

Figure 5.7 String art

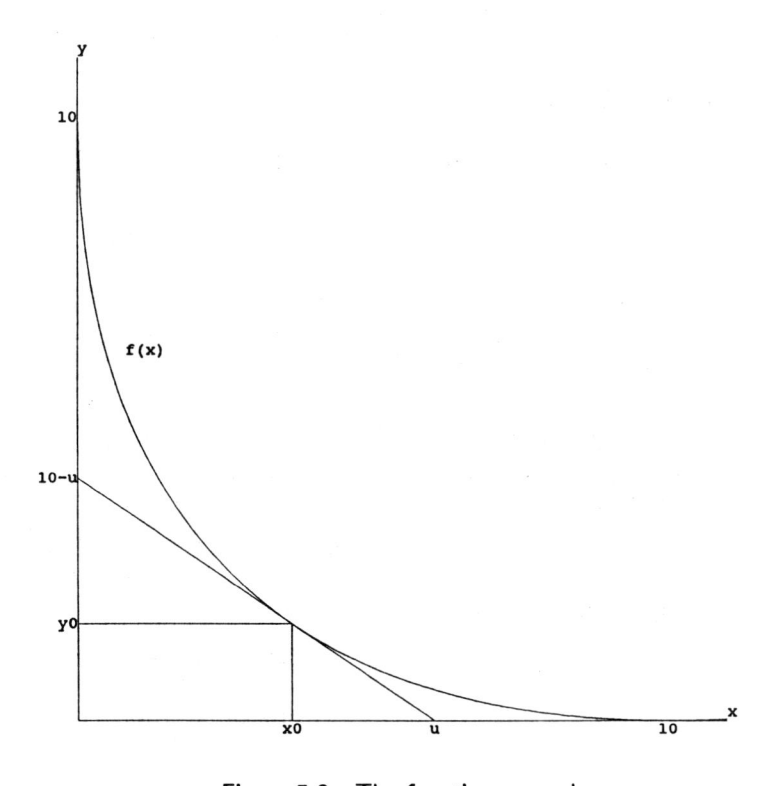

Figure 5.8 The function we seek

ADDITIONAL MATERIALS

Because this straight line passes through the point $(u, 0)$, we have

$$-y_0 = m(u - x_0). \tag{5.12}$$

However, from Figure 5.8 we clearly have

$$m = \frac{u - 10}{u},$$

which, when solved for u, gives

$$u = \frac{10}{1 - m}. \tag{5.13}$$

If we substitute (5.13) into (5.12) and rearrange, we find

$$y_0 - x_0 m = \frac{10m}{m - 1}, \tag{5.14}$$

where $m = f'(x_0)$. However, we want (5.14) to be valid at all points (x, y) on the curve, so (5.14) gives the differential equation

$$y - xy' = \frac{10 y'}{y' - 1}. \tag{5.15}$$

Furthermore, the curve we seek must pass through the point $(10, 0)$, so (5.15) is subject to the initial condition

$$y(10) = 0. \tag{5.16}$$

If we multiply equation (5.15) by $y' - 1$ and simplify, we find

$$x(y')^2 + (10 - x - y)y' + y = 0.$$

This is a quadratic equation for y', which can be solved to find

$$y' = \frac{1}{2x}\left[-(10 - x - y) \pm \sqrt{(10 - x - y)^2 - 4xy}\right]. \tag{5.17}$$

This differential equation is not one we have seen before, and so far we have no way to solve it.

However, if we write (5.15) in the form

$$y = xy' + \frac{10 y'}{y' - 1}, \tag{5.18}$$

we see that if we could find y' as a function of x, then we could substitute it into the right-hand side of (5.18) and read off $y = y(x)$ immediately. So the problem becomes finding y' as a function of x. If we define

$$p = y',$$

then we want to find $p(x)$. One way to do this is to find a differential equation that p must satisfy and solve it for $p(x)$. This suggests that we should differentiate (5.15) with respect to x, which gives

$$y' - xy'' - y' = 10 \frac{y''(y' - 1) - y' y''}{(y' - 1)^2},$$

or, upon simplification,

$$-xy'' = 10 \frac{-y''}{(y' - 1)^2}.$$

We can rewrite this last equation in the form

$$y''\left(x - \frac{10}{(y' - 1)^2}\right) = 0,$$

from which we conclude that either
$$y'' = 0 \tag{5.19}$$
or
$$x - \frac{10}{(y'-1)^2} = 0. \tag{5.20}$$

We treat each of these equations in turn.

In the case of (5.19), we see that an integration yields $y' = C$. Substituting this into (5.18) gives
$$y = Cx + \frac{10C}{C-1}, \tag{5.21}$$
which is a family of straight lines.

In the case of (5.20), we see that we can write it in the form
$$(y'-1)^2 = \frac{10}{x},$$
or
$$y' = 1 \pm \sqrt{\frac{10}{x}}.$$

When the latter is substituted into (5.18), we find the solution
$$y = 10 + x \pm 2\sqrt{10x}. \tag{5.22}$$

These are singular solutions[7], because they are not contained in (5.21). (Why?)

Thus, all the solutions of (5.15) have the form of (5.21) or (5.22).

The solution (5.21) satisfies the initial condition (5.16) only if $C = 0$, which leads to the solution $y(x) = 0$, which is of no interest. In fact, the solution (5.21) is exactly the family of straight lines that was used to construct the string art. Part of the singular solution (5.22), the part with the minus sign, does satisfy our initial condition (5.16). Thus, the shape generated by this string art is given by
$$y = 10 + x - 2\sqrt{10x}.$$

This is a parabola with axis $y = x$ and vertex at $(5/2, 5/2)$—see Exercise 3 on page 176. What shape is generated by the part of (5.22) with the plus sign? □

A differential equation of the form
$$y - xy' = f(y') \tag{5.23}$$
is known as CLAIRAUT'S EQUATION. It arises in a number of different problems and is easy to solve. In order to simplify the notation, we introduce the variable p where
$$p = \frac{dy}{dx}. \tag{5.24}$$

If we differentiate (5.23) with respect to x we find
$$-xp' = \frac{df}{dp} p',$$

---

[7] Recall that a singular solution of
$$\frac{dy}{dx} = g(x, y)$$
is a particular solution that cannot be obtained from the family of solutions (containing the arbitrary constant C) by selecting a finite value for C.

or equivalently
$$\left(\frac{df}{dp} + x\right)p' = 0,$$
from which we have either
$$p' = 0 \tag{5.25}$$
or
$$\frac{df}{dp} = -x. \tag{5.26}$$
Equations (5.25) and (5.24) imply that
$$y' = p = C. \tag{5.27}$$
If we substitute (5.27) into (5.23), we find
$$y = Cx + f(C),$$
which is a family of straight lines. If we solve (5.26) subject to (5.23), we obtain the singular solution of (5.23).

How to Solve Clairaut's Equation

Purpose: To find the explicit and singular solutions of Clairaut's equation
$$y - xy' = f(y'). \tag{5.28}$$

Technique:

1. To make the notation less confusing, set $y' = p$ to get
$$y - xp = f(p).$$

2. Differentiate this last equation with respect to x to obtain
$$-xp' = \frac{df}{dp}p',$$
or
$$\left(\frac{df}{dp} + x\right)p' = 0.$$

3. There are two cases to consider.

    (a) $p' = 0$. This implies that $y' = p = C$. Substitute this into (5.28) to find
    $$y = xC + f(C).$$
    This gives the family of explicit solutions of (5.28).

    (b) $df/dp = -x$. We try to solve this for $p = p(x)$ and then use the fact that $y' = p$ to obtain
    $$y = xp(x) + f(p(x)).$$
    This is the singular solution of (5.28).

Comments:

- It is sometimes quite difficult to decide whether an equation is of the Clairaut type. Solving for y and seeing whether the result may be expressed in the form

$$y = xy' + f(y')$$

is usually very successful.

- The family of explicit solutions can be written down immediately from Clairaut's equation by replacing y' with C. The family of explicit solutions is always a family of straight lines.

- Usually the singular solution cannot be obtained from the family of explicit solutions.

- If we are unable to solve $df/dp = -x$ for $p = p(x)$, we can treat p as a parameter, and the singular solution will be given parametrically by

$$\begin{aligned} x &= -df/dp, \\ y &= -pdf/dp + f(p). \end{aligned}$$

Example 5.2 :

Solve the differential equation

$$y - xy' = \frac{1}{4}(y')^2. \qquad (5.29)$$

We introduce the variable $p = y'$ and differentiate (5.29) to find

$$-xp' = \frac{1}{2}pp',$$

from which we have either

$$p' = 0$$

or

$$p = -2x.$$

If $p' = 0$, then $y' = p = C$, which, when substituted into (5.29), yields the family of explicit solutions

$$y = xC + \frac{1}{4}C^2.$$

If $y' = p = -2x$, then when we substitute it into (5.29) we find the singular solution

$$y = -2x^2 + \frac{1}{4}(2x)^2 = -x^2.$$

Notice that this singular solution cannot be obtained from the family of explicit solutions for any choice of the constant C. □

Exercises

1. Solve the following differential equations.

    (a)
    $$y - xy' = (y')^2$$

    (b)
    $$y - xy' = -(y')^3$$

(c)
$$y - xy' = \sqrt{1 + (y')^2}$$

(d)
$$y(y')^2 - x(y')^3 - 1 = 0$$

(e)
$$x(y')^2 - (1+y)y' + 1 = 0$$

Use technology to plot the slope field for (5.17). Notice that the curve we seek is the boundary of the region in the xy plane where the derivative exists. From this information, find the equation of this curve.

3. Confirm that the change of variables from (x, y) to (X, Y), where

$$\begin{aligned} X &= (x - y)/\sqrt{2}, \\ Y &= (x + y)/\sqrt{2}, \end{aligned}$$

is a clockwise rotation through $\pi/4$ of the xy-plane. Show that under this change of variables, the equation

$$y = 10 + x - 2\sqrt{10x}$$

becomes

$$Y = \frac{1}{10\sqrt{2}} X^2 + \frac{5}{\sqrt{2}},$$

which is a parabola symmetric about the Y-axis and vertex at $X = 0$, $Y = 5/\sqrt{2}$ —that is, a parabola with axis of symmetry $y = x$ and vertex at $x = 5/2$, $y = 5/2$.

4. Consider the string art that is constructed by joining nails on the positive x- and y-axes, which are placed so that the area under the string is always 1. Show that the differential equation that characterizes the curve created by these straight lines is

$$y - xy' = \sqrt{-2y'}.$$

   (a) What is the equation of this curve?

   (b) Use the idea suggested in Exercise 2 (that is, find an explicit formula for y' and see where it is undefined) to confirm the equation you found in part (a) for the curve.

5. Consider the string art that is constructed by joining nails on the positive x- and y-axes which are placed so that the length of the string is always 1. Show that the differential equation that characterizes the curve created by these straight lines is

$$y - xy' = \frac{-y'}{\sqrt{1 + (y')^2}}.$$

What is the equation of this curve?

6. By changing from the unknown variable y to a new unknown variable v where $v = y^3$, solve

$$y - 3xy' = 6y^2 (y')^2.$$

7. By changing from the variables x and y to the new variables u and v, where $u = x^2$ and $v = y^2$, show that the differential equation

$$xy(y')^2 + (x^2 - a^2 - y^2)y' - xy = 0$$

reduces to
$$u\left(\frac{dv}{du}\right)^2 + (u - a^2 - v)\frac{dv}{du} - v = 0.$$

Solve this equation for v in terms of u and dv/dx and then show that it is a Clairaut equation. Now solve the Clairaut equation for $v = v(u)$, and then find the solution of the original equation.

8. The ellipse has the property that all light rays emitted from one focus pass through the other focus after reflection off the ellipse. This exercise is designed to show that the ellipse is the only curve with this reflective property. If you choose the coordinate system so the light that is emitted from the point $(a, 0)$ is received at the point $(-a, 0)$, show that the differential equation that the curve must satisfy is the one in Exercise 7.

9. Identify the following differential equations as separable equations, those with homogeneous coefficients, linear equations, Bernoulli equations, Clairaut equations, or none of these. (Some may fall into more than one category; see Table 5.3.) Do not attempt to solve any of these equations.

(a)
$$(2 - y)\frac{dy}{dx} = x^2 + 1$$

(b)
$$\tan\frac{dy}{dx} - x\frac{dy}{dx} + y = 0$$

(c)
$$(x + y)\frac{dy}{dx} + y - x = 0$$

(d)
$$2xy + (x^2 + \cos y)\frac{dy}{dx} = 0$$

(e)
$$y\sin x + \frac{1}{y} - \frac{dy}{dx} = 0$$

(f)
$$3x(y^2 + 1) + y(x^2 + 1)\frac{dy}{dx} = 0$$

(g)
$$x\frac{dy}{dx} + 3y - x^2 = 0$$

(h)
$$x^3 + y^3 - xy^2\frac{dy}{dx} = 0$$

(i)
$$\frac{dy}{dx} = \frac{xy^2 - 1}{1 - x^2 y}$$

(j)
$$y\sin x + e^x - \frac{dy}{dx} = 0$$

(k)
$$y + \left(\frac{dy}{dx}\right)^2 - x\frac{dy}{dx} = 0$$

(1)
$$xy^2 - y + \frac{dy}{dx} = 0$$

Table 5.3 Solving First Order Differential Equations

Differential Equation	Response		Type
$y' = f(y)g(x)$?	Yes	$\longrightarrow$	Separable
$y' = g(y/x)$?	Yes	$\longrightarrow$	Homogeneous coefficients
$y' + p(x)y = q(x)$?	Yes	$\longrightarrow$	Linear
$y' + p(x)y = q(x)y^n$?	Yes	$\longrightarrow$	Bernoulli
$y' = a_0(x) + a_1(x)y + a_2(x)y^2$?	Yes	$\longrightarrow$	Riccati
$y - xy' = f(y')$?	Yes	$\longrightarrow$	Clairaut

The Garage Door

A colleague of ours[8] was going to replace his garage door. The least expensive one he could find was a solid plank 10 feet tall that worked in a particular way. It had rollers at the top and bottom that fit into runners. The bottom rollers moved in vertical runners on either side the garage entrance, while the top rollers moved in a horizontal runner attached to the ceiling 10 feet above the ground. Figure 5.9 shows the door moving from closed (vertical) to open (horizontal). The problem with this mechanism is that in the process of opening the door, it intrudes into the garage and consumes space. This can be seen in Figure 5.9. Thus, before buying this door our colleague wants to know whether his car will fit in his garage with this door fitted. If we look at Figure 5.9 we see that the lines have created the illusion of a curve. If we can find the equation of this curve we could work out whether a particular car will fit in the garage so the door can be closed.

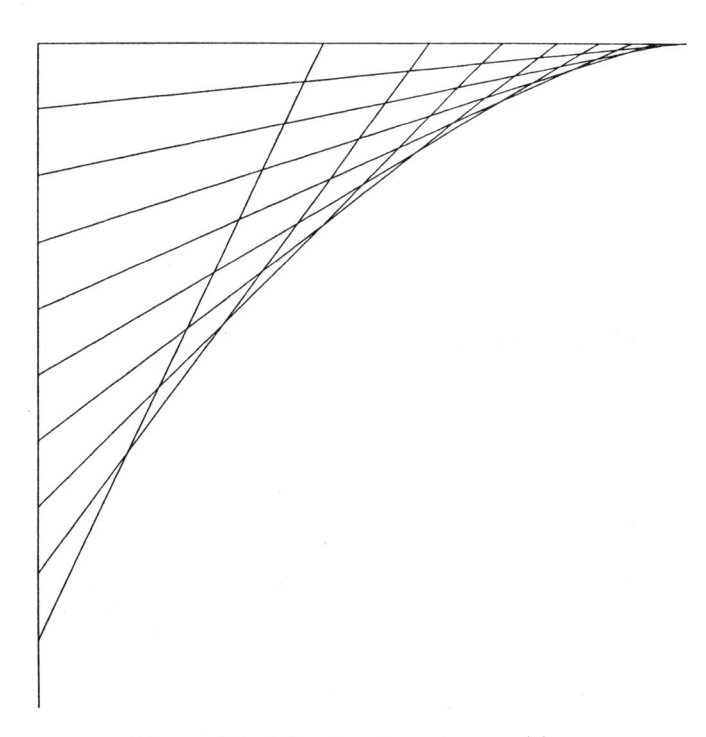

Figure 5.9 The door in various positions

Let's denote the function we are looking for by $y = f(x)$. A door position will be tangent to $y = f(x)$ at a point (x_0, y_0), say. Let's look at a typical door position starting on the y-axis, at, say, $(0, u)$, and its other end is at the point $(v, 10)$ on the ceiling, as shown in Figure 5.10.

The relationship between the door and the curve is that the slope of the door is the slope of the curve — namely, $f'(x_0)$. The equation for a straight line with slope $m = f'(x_0)$ is

$$y - y_0 = m(x - x_0).$$

Because this straight line passes through the point $(0, u)$, we have

$$u - y_0 = -mx_0. \qquad (5.30)$$

---

[8] Dr. J. Yasom.

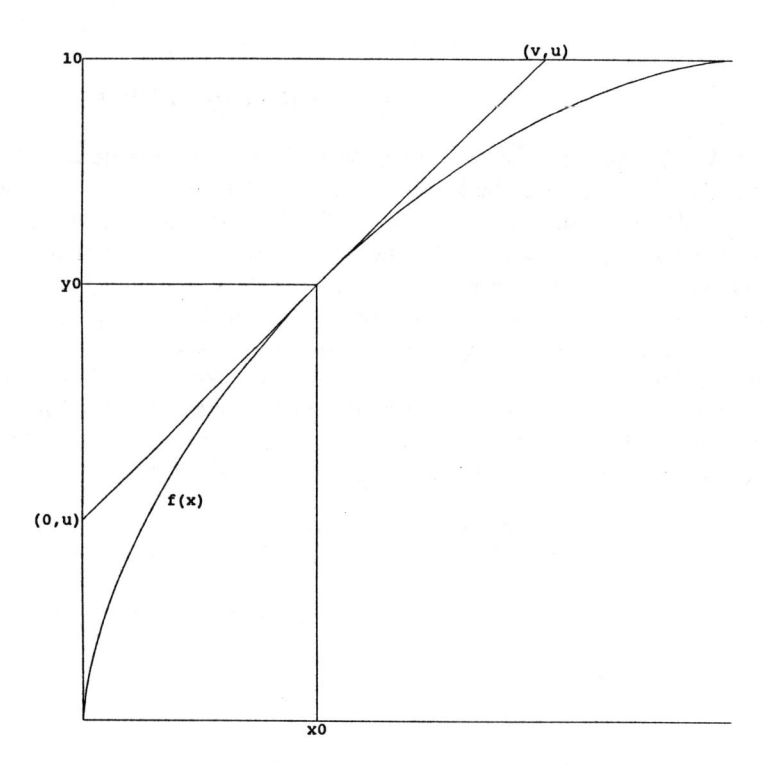

Figure 5.10 The function we seek

However, because the door has a constant height of 10 feet from Figure 5.10 we have
$$v^2 + (10 - u)^2 = 10^2,$$
or
$$v = \sqrt{10^2 - (10 - u)^2}.$$
[Why didn't we include the minus sign.] The slope of the door is given by
$$m = \frac{10 - u}{v} = \frac{10 - u}{\sqrt{10^2 - (10 - u)^2}},$$
which, when solved for u, gives
$$u = 10 - \frac{10m}{\sqrt{m^2 + 1}}. \tag{5.31}$$
If we substitute (5.31) into (5.30) and rearrange, we find
$$y_0 - x_0 m = 10 - \frac{10m}{\sqrt{m^2 + 1}}, \tag{5.32}$$
where $m = f'(x_0)$. However, we want (5.32) to be valid at all points (x, y) on the curve, so (5.32) gives the differential equation
$$y - xy' = 10 - \frac{10y'}{\sqrt{(y')^2 + 1}}. \tag{5.33}$$
Furthermore, the curve we seek must pass through the point $(10, 10)$, so (5.33) is subject to the initial condition
$$y(10) = 10. \tag{5.34}$$

If we subtract -10 from both sides of (5.33), multiply by $\sqrt{(y')^2 + 1}$, square, and then collect like powers of y', we find

$$x^2(y')^4 - 2x(y-10)(y')^3 + \left[(y-10)^2 + x^2 - 10^2\right](y')^2 - 2x(y-10)y' + (y-10)^2 = 0.$$

This is a quartic equation for y', which we cannot solve. All of our previous techniques for solving differential equations required that we start from $y' = g(x, y)$, but here we are unable to write our equation in this form, so those techniques will not help us now.

However, if we write (5.33) in the form

$$y = xy' + 10 - \frac{10y'}{\sqrt{(y')^2 + 1}}, \qquad (5.35)$$

we see that if we could find y' as a function of x, then we could substitute it into the right-hand side of (5.35) and read off $y = y(x)$ immediately. So the problem becomes finding y' as a function of x. If we define

$$p = y',$$

then we want to find $p(x)$. One way to do this is to find a differential equation that p must satisfy and solve it for $p(x)$. This suggests that we should differentiate (5.33) with respect to x, which gives

$$y' - xy'' - y' = -\frac{10}{(y')^2 + 1}\left[y''\sqrt{(y')^2 + 1} - \frac{(y')^2 y''}{\sqrt{(y')^2 + 1}}\right],$$

or, upon simplification,

$$-xy'' = -\frac{10}{\left[(y')^2 + 1\right]^{3/2}} y''.$$

We can rewrite this last equation in the form

$$y''\left\{x - \frac{10}{\left[(y')^2 + 1\right]^{3/2}}\right\} = 0,$$

from which we conclude that either

$$y'' = 0 \qquad (5.36)$$

or

$$x - \frac{10}{\left[(y')^2 + 1\right]^{3/2}} = 0. \qquad (5.37)$$

We treat each of these equations in turn.

In the case of (5.36), we see that an integration yields $y' = C$. Substituting this into (5.35) gives

$$y = xC + 10 - \frac{10C}{\sqrt{C^2 + 1}}, \qquad (5.38)$$

which is a family of straight lines.

In the case of (5.37), we see that we can write it in the form

$$\left[(y')^2 + 1\right]^{3/2} = \frac{10}{x},$$

or
$$y' = \sqrt{\left(\frac{10}{x}\right)^{2/3} - 1} = \sqrt{\frac{10^{2/3} - x^{2/3}}{x^{2/3}}} = \frac{1}{x^{1/3}}\sqrt{10^{2/3} - x^{2/3}}.$$

When the latter is substituted into (5.35) and simplified, we find the solution

$$y = 10 - \left(10^{2/3} - x^{2/3}\right)^{3/2}. \tag{5.39}$$

These are singular solutions[9], because they are not contained in (5.38). (Why?)

Thus, all the solutions of (5.33) have the form of (5.38) or (5.39).

The solution (5.38) satisfies the initial condition (5.34) only if $C = 0$. This leads to the solution $y(x) = 0$, which is of no interest, because it means the door is permanently shut!. In fact, the solution (5.38) is exactly the family of lines that represent different positions of the door. The solution (5.39) satisfies the initial condition (5.34) automatically. Thus, the shape generated by the door is given by (5.39).[10]

In order to decide whether a particular car will fit in the garage so the door can be closed we construct Table 5.4. For example, this table shows that a car which is parked so that its rear end is 1 foot from the door will allow the door to close if its rear height is less than 3.05 feet. □

Table 5.4 The function $10 - \left(10^{2/3} - x^{2/3}\right)^{3/2}$ evaluated at various values of x

x	$10 - \left(10^{2/3} - x^{2/3}\right)^{3/2}$
0	0.000
1	3.051
2	4.662
3	5.900
4	6.909
5	7.749
6	8.449
7	9.026
8	9.486
9	9.823
10	10.000

---

[9]Recall that a singular solution of
$$\frac{dy}{dx} = g(x, y)$$
is a particular solution that cannot be obtained from the family of solutions (containing the arbitrary constant C) by selecting a finite value for C.

[10]If we write (5.39) in the form
$$x^{2/3} + (y - 10)^{2/3} = 10^{2/3}$$
we can identify this curve as a special case of the hypocycloid centered at $(0, 10)$ — namely, an astroid. This is the curve that is traced by a point on the circumference of a circle of radius $10^{2/3}/4$ when rolling inside a fixed circle of radius $10^{2/3}$.

Exact Equations

If we are given an equation $u(x,y) = c$, we can think of it as implicitly defining a function $y = y(x)$, although we may not be able to solve it for $y(x)$ explicitly. However, we can always obtain the differential equation satisfied by $y(x)$ by differentiating $u(x, y(x)) = c$ implicitly with respect to x. Doing this gives

$$\frac{\partial u}{\partial x} + \frac{\partial u}{\partial y}\frac{dy}{dx} = 0,$$

or

$$\frac{dy}{dx} = -\frac{\frac{\partial u}{\partial x}}{\frac{\partial u}{\partial y}}.$$

Consequently, given any differential equation of this form, its solution will be $u(x,y) = c$.

This leads to the following definition. The differential equation

$$\frac{dy}{dx} = -\frac{M(x,y)}{N(x,y)},$$

which can also be written as

$$M(x,y) + N(x,y)\frac{dy}{dx} = 0,$$

is said to be EXACT if there exists a function $u(x,y)$ for which

$$\frac{\partial u}{\partial x} = M(x,y) \qquad \text{and} \qquad \frac{\partial u}{\partial y} = N(x,y).$$

In this case, the differential equation can be written

$$\frac{\partial u}{\partial x} + \frac{\partial u}{\partial y}\frac{dy}{dx} = 0,$$

or

$$\frac{d}{dx}[u(x,y)] = 0,$$

which, when integrated with respect to x, yields

$$u(x,y) = c.$$

How do we decide whether or not a differential equation is exact? The answer lies in the fact that, under fairly general conditions,

$$\frac{\partial^2 u}{\partial x \partial y} = \frac{\partial^2 u}{\partial y \partial x},$$

that, is

$$\frac{\partial}{\partial x}\left(\frac{\partial u}{\partial y}\right) = \frac{\partial}{\partial y}\left(\frac{\partial u}{\partial x}\right).$$

So, if the equation is exact then

$$\frac{\partial N}{\partial x} = \frac{\partial M}{\partial y}.$$

However, it can be shown that if

$$\frac{\partial N}{\partial x} = \frac{\partial M}{\partial y}$$

and M and N have continuous first partial derivatives, then the equation is exact.

ADDITIONAL MATERIALS

Example 5.3 : *Consider the differential equation*

$$(y + 3x^3y^2)\frac{dy}{dx} + 3x^2y^3 - 5x^4 = 0.$$

Here $M(x,y) = 3x^2y^3 - 5x^4$ and $N(x,y) = y + 3x^3y^2$, so that $\frac{\partial M}{\partial y} = 9x^2y^2$ and $\frac{\partial N}{\partial x} = 9x^2y^2$, so the equation is exact.

The question is, how do we use this knowledge to find $u(x,y)$ so as to solve the differential equation? The answer lies in the fact that $u(x,y)$ must satisfy

$$\frac{\partial u}{\partial x} = M(x,y) \quad \text{and} \quad \frac{\partial u}{\partial y} = N(x,y),$$

at the same time.

In this case

$$\frac{\partial u}{\partial x} = 3x^2y^3 - 5x^4 \quad \text{and} \quad \frac{\partial u}{\partial y} = y + 3x^3y^2.$$

We can proceed in two different ways: either integrate the first equation, remembering that the "constant" of integration could contain an arbitrary function of y, and then substitute the result in the second to find $u(x,y)$, or, integrate the second equation, remembering that the "constant" of integration could contain an arbitrary function of x, and then substitute the result in the first to find $u(x,y)$. The decision as to which of these two methods to use, is usually based on selecting the easier function to integrate.

In this example there is nothing to chose between the two methods, so we will use the first method. Integrating

$$\frac{\partial u}{\partial x} = 3x^2y^3 - 5x^4$$

with respect to x, gives

$$u(x,y) = x^3y^3 - x^5 + f(y),$$

where $f(y)$ is an arbitrary function of y. From this we find

$$\frac{\partial u}{\partial y} = 3x^3y^2 + \frac{df}{dy}.$$

But we know that

$$\frac{\partial u}{\partial y} = y + 3x^3y^2,$$

so

$$\frac{df}{dy} = y.$$

Integrating this with respect to y gives $f(y) = \frac{1}{2}y^2 + c_1$, so $u(x,y) = x^3y^3 - x^5 + \frac{1}{2}y^2 + c_1$. Thus the solution of the differential equation is $x^3y^3 - x^5 + \frac{1}{2}y^2 + c_1 = c$, or

$$x^3y^3 - x^5 + \frac{1}{2}y^2 = c,$$

where we have absorbed c_1 into c.

Exercises

1. Check the following equations for exactness and solve those that are exact.

   (a) $x^2 - y^2 - 2xyy' = 0$.

   (b) $x\cos y - y + (x\sin y + x)y' = 0$

   (c) $e^x + y - 1 + (3e^y + x - 7)y' = 0$

(d) $y\cos xy + 3y - 1 + (x\cos xy + 3x)y' = 0$

(e) $4x^3y^3 + x^2/4 + (3x^4y^2 - 16)y' = 0$

(f) $3x^2y + e^y + (x^3 + xe^y + \sin y)y' = 0$

(g) $x\cos(x+y) + \sin(x+y) + [x\cos(x+y) + y]y' = 0$

(h) $xy^3 - 1/(xy) + (3x^2y^2 - 1/y + x^2)y' = 0$

(i) $1 + y\cos xy + (x\cos xy + 2)y' = 0$

(j) $2xy + x^3 + (y + x^2)y' = 0$

(k) $3x^2 \ln y + x^3 y^{-1} y' = 0$

2. Find values for the constant r which make the following equations exact.

(a) $2y + (y^2 - rx)y' = 0$

(b) $\sin y + (x^r \cos y + y^3)y' = 0$

(c) $y^4 + 2rxy + (4xy^3 + rx^2)y' = 0$

(d) $3x^2 - 3y^r + rx + (3xy^{-2} - 3r + 7)y' = 0$

3. Show that the solution of the differential equation

$$\frac{dy}{dx} = \frac{cx - ay + f}{ax + by + e}$$

is the conic $-\frac{1}{2}cx^2 + axy + \frac{1}{2}by^2 + ey - fx - k = 0$. (If $cb + a^2 < 0$ the curve is an ellipse, circle, point, or no curve; if $cb + a^2 = 0$ the curve is a parabola, two parallel lines, one line, or no curve; and if $cb + a^2 > 0$ the curve is a hyperbola or two intersecting lines.)

4. Show that the linear differential equation $p(x)y - q(x) + y' = 0$, is not exact, but that, after multiplying by the usual integrating factor, $\mu(x)$, it is exact.

Flowchart

To find an analytical solution of $\frac{dy}{dx} = g(x, y)$

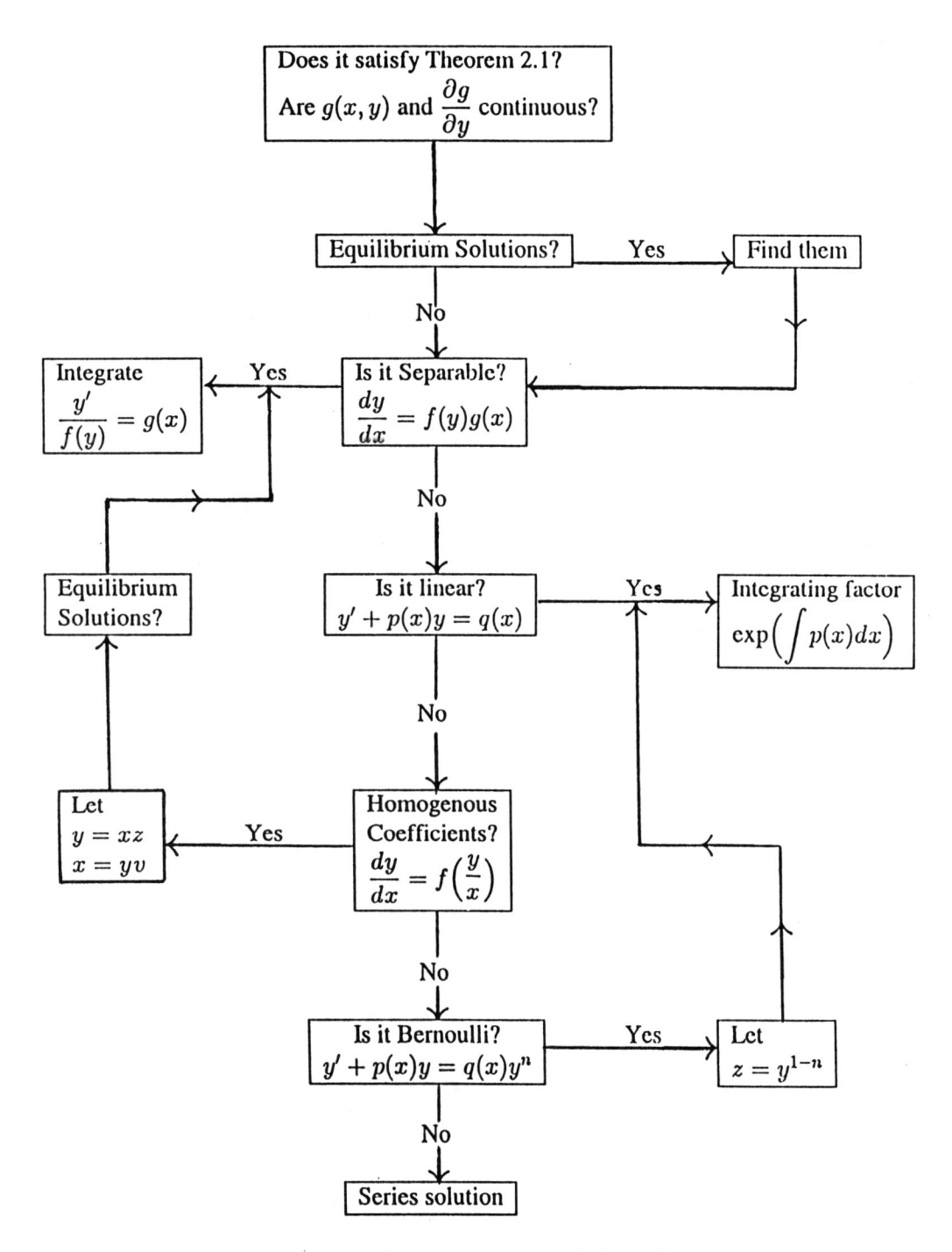

Does it satisfy Theorem 2.1?
Are $g(x,y)$ and $\frac{\partial g}{\partial y}$ continuous?

↓

Equilibrium Solutions? —Yes→ **Find them**

↓ No

Is it Separable? $\frac{dy}{dx} = f(y)g(x)$ —Yes→ **Integrate** $\frac{y'}{f(y)} = g(x)$ → **Equilibrium Solutions?**

↓ No

Is it linear? $y' + p(x)y = q(x)$ —Yes→ **Integrating factor** $\exp\left(\int p(x)dx\right)$

↓ No

Homogenous Coefficients? $\frac{dy}{dx} = f\left(\frac{y}{x}\right)$ —Yes→ **Let** $y = xz$, $x = yv$

↓ No

Is it Bernoulli? $y' + p(x)y = q(x)y^n$ —Yes→ **Let** $z = y^{1-n}$

↓ No

Series solution

<u>Be sure to check your analytical solution with a slope field, calculus, and a numerical solution.</u>

Figure 5.11   Flowchart

ADDITIONAL MATERIALS

6. IDEAS FOR CHAPTER 6 — "INTERPLAY BETWEEN FIRST ORDER SYSTEMS AND SECOND ORDER EQUATIONS"

6.1 Page by Page Comments

Chapter 6 requires 4 to 6 class meetings of 50 minutes each.

In this chapter we extend the work and ideas from Chapters 1 through 5 to linear second order differential equations with constant coefficients by relating them to first order differential equations.

Page 224. Example 6.1 could be replaced by SEA BATTLES on page 207 of this manual. This is Exercise 14 on page 249.

Page 224. Example 6.1 is based on the relationship between one of the author's daughters, Denise Lovelock, and her boyfriend when she was 18, Chad. For reasons of privacy we refrain from supplying Chad's last name. Other examples concerning relationship moods may be found in "Nonlinear Dynamics and Chaos" by S.H. Strogatz, Addison-Wesley, 1994, Section 5.3, and in "The Lighter Side of Differential Equations" by J.M. McDill and B. Felsager, College Mathematics Journal, **25**, 1994, pages 448 through 452.

Page 226. The fact that Figure 6.3 has orbits that have vertical tangent lines and are not functions does not contradict our earlier work. Here the orbits are actually parametrically defined solutions of the system of equations (6.1) rather than the solution curves of (6.2).

Page 230. Students enjoy Exercise 2 of Section 6.1. They will have an easier time answering this question if they are aware of the fact that t_1, t_3, and t_5 are zeros of $y(t)$; t_2, t_4, and t_6 are zeros of $x(t)$.

Page 231. There are additional exercises for Section 6.1 on page 192 of this manual.

Page 232. The solution $x(t)$ and $y(t)$ of the system of differential equations $dx/dt = P(t, x, y)$, $dy/dt = Q(t, x, y)$ is sometimes called the **time series** of the system.

Page 233. Students sometimes think that if any of the terms a, b, c, or d, in a linear system are zero, then the system is not linear.

Page 236. In $x'' + p(t)x' + q(t)x = f(t)$, the function $f(t)$ is called the FORCING FUNCTION.

Page 236. Students sometimes think that if any of the terms p or q in a linear differential equation are zero, then the equation is not linear.

Page 238. Theorem 6.4. It is worth pointing out that, in the first order case, we exploited the fact that solutions do not intersect by using the uniqueness property. Here a unique solution is specified by two conditions, the initial position and the initial slope, so this theorem implies that a solution that starts from a specified position and with a specified slope is unique. Thus, we should not be surprised to find that solutions that start from the same point, but with different slopes, intersect. This is what happens in Figures 6.10 — 6.12, for example.

Page 238. Theorem 6.4. The function $x(t) = 0$ satisfies $x'' + p(t)x' + q(t)x = 0$ — that is (6.23) with $f(t) = 0$ — and the initial conditions $x(t_0) = 0$ and $x'(t_0) = 0$. Consequently, $x(t) = 0$ is the only solution of $x'' + p(t)x' + q(t)x = 0$ that passes through the origin with zero slope. No other solution can have the property that $x(t_0) = 0$ and $x'(t_0) = 0$. This solution, $x(t) = 0$, is called the **trivial solution**.

Page 238. Some students have considerable trouble with Exercises 1 and 2 of Section 6.2. If they do, then more can be assigned. See Additional Exercises for Section 6.2 on page 193 of this manual.

Page 240. The system of differential equations in Exercise 16 of Section 6.2 is van der Pol's equation in disguise. See Exercises 2(g) and 7 on pages 238 and 239.

Page 240. The differential equation in Exercise 18 of Section 6.2 is solved in Exercise 8 of Section 8.5.

Page 241. There are additional exercises for Section 6.2 on page 193 of this manual.

Page 241. Theorem 6.5 is used repeatedly in future sections without reference. The proof is optional.

Page 245. Some students may have little or no knowledge of the elementary properties of complex numbers, so you may need to spend extra time covering Appendix A.4 of the text.

Page 245. This is also a place where you can point out the usefulness of the hyperbolic functions for the case of the real roots having the form $r_1 = -r_2$.

Page 246. Some students appreciate the simplicity of finding the characteristic equation when the differential equation is given in operational form. Because many students have never seen operator notation, in class it is worth drawing attention to the last bullet under Comments About Second Order Linear Differential Equations with Constant Coefficients.

Page 249. Exercises 11 and 13 of Section 6.3 are worth assigning.

Page 249. Exercise 14 of Section 6.3. For other ideas on the use of differential equations to model battles see "Combat Models" by C. S. Coleman in "Modules in Applied Mathematics: Volume 1 — Differential Equation Models" edited by M. Braun, C. S. Coleman, and D. A. Drew, Springer-Verlag, 1983, Chapter 8, pages 109 through 131.

Page 250. There are additional exercises for Section 6.3 on page 193 of this manual.

Page 250. There are applications of second order differential equations to SYSTEMS OF FUNCTIONAL EQUATIONS. See page 214 of this manual.

Page 250. We can obtain (6.37) as follows. Suspend a spring of length L vertically from one end. If a mass m is suspended from the other end of the spring — and the spring obeys Hooke's law — then the differential equation governing the motion of the spring-mass system is
$$m\frac{d^2X}{dt^2} = mg - k(X - L),$$
where X is the distance measured vertically from the point of suspension and k is the spring constant. We can rewrite this as
$$m\frac{d^2X}{dt^2} = -k\left[X - \left(L + \frac{gm}{k}\right)\right].$$
If we make the change of variable $x = X - (L + gm/k)$ then the differential equation becomes $mx'' = -kx$. One solution of this is $x(t) = 0$ which corresponds to the mass being at rest at $x = 0$ for all time. This is the equilibrium position, and corresponds to a distance $X = L + gm/k$ from the top of the spring. Here L is the unstretched length of the spring and gm/k is the distance that the spring stretches to achieve equilibrium due to the mass m.

Page 250. It is generally believed that assuming the damping force is proportional to the velocity v — rather than to v^2 — is a reasonable model for the spring-mass system (and for the simple pendulum) because the velocities are relatively small.

Page 253. There is a discussion of KIRCHHOFF'S LAWS on page 217 of this manual. Kirchhoff is pronounced kirkh'huf.

Page 254. Exercise 6 of Section 6.4 is interesting.

Page 254. Table 6.1 is in TWIDDLE. It is called HOOKE1.DTA and is in the subdirectory DD-06.

Page 254. Table 6.2 is in TWIDDLE. It is called HOOKE2.DTA and is in the subdirectory DD-06.

Page 257. Table 6.3 is in TWIDDLE. It is called UNDER.DTA and is in the subdirectory DD-06.

Page 257. Table 6.4 is in TWIDDLE. It is called OVERDAMP.DTA and is in the subdirectory DD-06.

Page 259. There are additional exercises for Section 6.4 on page 193 of this manual.

Page 260. Interpreting the effect of the resistance term a in terms of
$$\left(\frac{dy}{dx}\right)_{a\neq 0} = \left(\frac{dy}{dx}\right)_{a=0} - a$$
is very useful.

Page 264. Undamped motion of a linearized pendulum predicts that no matter how large the initial velocity v_0, the motion is always bounded and oscillatory. This is clearly unrealistic, because if we have a sufficiently large initial velocity we would expect the pendulum to go over the top, and, because there is no friction, continue going over the top forever. In this case the motion would be unbounded. This reinforces our earlier statement that for the model to be realistic, x must be restricted to small values. It is instructive to ask the students to interpret the elliptical orbit in the phase plane that starts at $x = 5$, $x' = 0$, in terms of the motion of a simple, frictionless, linear pendulum. Remember, x is measured in radians, so $x = 5$ is an angle of approximately 286 degrees.

Page 266. Exercise 12 of Section 6.5 is interesting, and the result surprises some students. It is solved in a different way in Exercise 11 of Section 7.3.

Page 266. There are additional exercises for Section 6.5 on page 202 of this manual.

Page 266. The technique described in Section 6.6 of the text is really a technique for plotting y versus x from their parametric representations $x = x(t)$, $y = y(t)$.

Page 271. There are additional exercises for Section 6.6 on page 203 of this manual.

Page 272. For a discussion of the period of a pendulum — both linear and nonlinear, see THE PERIOD OF A PENDULUM on page 219 of this manual.

Page 277. There are additional exercises for Section 6.7 on page 204 of this manual.

6.2 Additional Exercises

Section 6.1

Page 231. Add exercises after the current Exercise 5.

6. **Diana and Charles I.** The affection between Diana x and Charles y is modeled by different differential equations from those of Denise and Chad because a third party is involved with one of them. Their affection is modeled by $x' = y$, $y' = x$.

   (a) Draw the direction field and various orbits in the phase plane.

   (b) What happens to the long-term Diana-Charles relationship under the following initial impressions?

   (i) $x(0) = -1$ $y(0) = 5$ (iv) $x(0) = -5$ $y(0) = 1$
   (ii) $x(0) = 1$ $y(0) = -5$ (v) $x(0) = 5$ $y(0) = 5$
   (iii) $x(0) = 5$ $y(0) = -1$ (vi) $x(0) = -5$ $y(0) = 5$

   (c) In what sense is the Diana-Charles relationship less stable than the Denise-Chad relationship?

7. **Diana and Charles II.** The affection between Diana x and Charles y is modeled by different differential equations from those of Denise and Chad because a third party is involved with one of them. However, this time their affection is modeled by $x' = -y$, $y' = -x$.

   (a) Draw the direction field and various orbits in the phase plane.

   (b) What happens to this long-term Diana-Charles relationship under the following initial impression?

   (i) $x(0) = -1$ $y(0) = 5$ (iv) $x(0) = -5$ $y(0) = 1$
   (ii) $x(0) = 1$ $y(0) = -5$ (v) $x(0) = 5$ $y(0) = 5$
   (iii) $x(0) = 5$ $y(0) = -1$ (vi) $x(0) = -5$ $y(0) = 5$

(c) In what sense is this Diana-Charles relationship less stable than the Denise-Chad relationship?

(d) Which of the three relationships — Denise-Chad, Diana-Charles I, Diana-Charles II — is most likely to produce a stable long term relationship, independent of initial impression?

Section 6.2

Page 241. Add exercises after the current Exercise 20.

21. Classify the following systems of equations using the terms autonomous, nonautonomous, linear, nonlinear, homogeneous, nonhomogeneous, and constant coefficients. Do not attempt to solve any of these equations.

(a) $\begin{aligned} x' &= y + t^2 \\ y' &= tx \end{aligned}$
(b) $\begin{aligned} x' &= yt^2 \\ y' &= tx \end{aligned}$
(c) $\begin{aligned} x' &= y^2 + t \\ y' &= tx \end{aligned}$
(d) $\begin{aligned} x' &= y + t^2 \\ y' &= x \end{aligned}$
(e) $\begin{aligned} x' &= 2x - y \\ y' &= x + 3y \end{aligned}$
(f) $\begin{aligned} x' &= -y \\ y' &= x \end{aligned}$

22. Classify the following second order differential equations using the terms autonomous, nonautonomous, linear, nonlinear, homogeneous, nonhomogeneous, and constant coefficients. Do not attempt to solve any of these equations.

(a) $x'' + 2x' + x = \cos t$
(b) $x'' + 2x' + x^2 = \cos t$
(c) $x'' + 2x' + x = 0$
(d) $x'' + 2x' + x = \cos t^2$
(e) $x'' + 2x' + tx = \cos t$
(f) $x'' + 2x' + t^2 x = \cos t$

Section 6.3

Page 250. Add exercise after the current Exercise 14.

15. Two students have modeled a particular situation. One says the situation is governed by the differential equation $y'' - a^2 y = 0$, where a is a positive constant, subject to $y(x) \to 0$ as $x \to \infty$. The other says the situation is governed by the differential equation $y' + ay = 0$. Will any experiment be able to distinguish between the two models?

Section 6.4

Page 259. Add exercises after the current Exercise 26.

27. Perform the following experiment. Take a piece of string and fix a weight on one end. Anchor the other end, and measure the length of the pendulum. Displace the pendulum through a small angle, release it from rest, and measure the time it takes to return to the initial angle.

(a) Repeat this a few times with different initial angles. Is the resulting period independent of the small initial angle?

(b) Repeat this a few times with different masses. Is the resulting period independent of the mass?

(c) Repeat the experiment for pendulums of different lengths. Do your results agree with Exercise 12?

28. Repeat the experiment described in Exercise 27 for larger initial angles. Show that the model is less reliable as the initial angle increases.

29. Consider a spring-mass system that is governed by $x'' + 8x = 0$, where motion is started from the equilibrium position by giving the mass an initial velocity y_0.

(a) Write the corresponding differential equation for the orbits in the phase plane (in terms of x and $y = dx/dt$).

(b) What are the appropriate initial conditions for the statement of this problem in terms of both the original second order differential equation and the one for the orbits.

(c) Find the equation of the orbit associated with this initial value problem.

(d) Find an explicit solution for the displacement as a function of time.

30. A Nonlinear Spring.[1] A mass is attached to one end of a spring whose properties are such that the motion of the mass is governed by the differential equation

$$x'' = g - \frac{x-1}{x(2-x)}, \qquad (6.1)$$

where x is measured in meters from the anchored end of the spring and $g \approx 9.8$ m/sec^2.

(a) For what values of x is this equation valid?

(b) What is the natural length of the spring?

(c) What is the equilibrium length x_e of the spring? Give both a formula and a numerical value for x_e.

(d) The differential equation (6.1) can be replaced by the system $x' = y$, $y' = g - \frac{x-1}{x(2-x)}$. Show that the orbits in the phase plane satisfy

$$\frac{dy}{dx} = \frac{gx(2-x) - (x-1)}{yx(2-x)}. \qquad (6.2)$$

The direction field for (6.2) is shown in Figure 6.1 for the window $0 \leq x \leq 2$, $0 \leq y \leq 2$. Use this information and symmetry to add the direction field for the window $0 \leq x \leq 2$, $-2 \leq y \leq 0$. Sketch various orbits for different values of $x(0) = a$, some near x_e. Find some values of a for which the orbits are approximately ellipses.

(e) Show that the orbits of (6.2) which satisfy $x = a$ when $y = 0$ are characterized by

$$y^2 = 2g(x-a) + \ln\left[\frac{x(2-x)}{a(2-a)}\right].$$

Plot these orbits for different values of a, some near x_e. Find some values of a for which the orbits are approximately ellipses. Reconcile these orbits with those you sketched in part (d).

---

[1] This exercise is based on an idea of a colleague, Bruce Bayly.

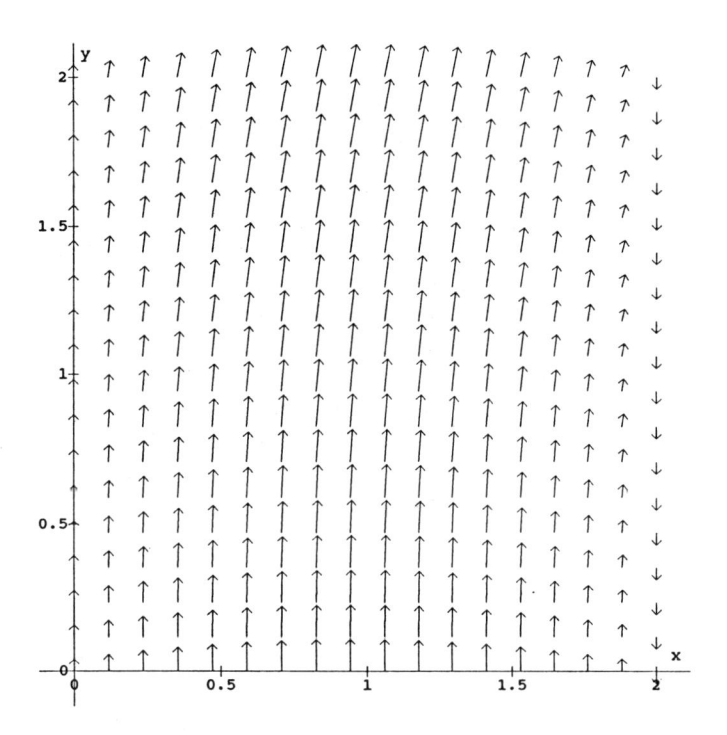

Figure 6.1 The direction field for the nonlinear spring

(f) Suppose that the spring oscillates with small amplitude about its equilibrium position x_e. Show how the nonlinear differential equation (6.1) can be approximated by the linear differential equation

$$x'' = -k(x - x_e), \tag{6.3}$$

where k is some constant. What is k? Give both a formula and a numerical value for x_e.

(g) Introduce the new variable $X(t) = x(t) - x_e$ and show that the differential equation (6.3) can be written $X'' = -kX$ with solution $X(t) = C_1 \cos \omega t + C_2 \sin \omega t$, so that

$$x(t) = x_e + C_1 \cos \omega t + C_2 \sin \omega t \tag{6.4}$$

is the solution of (6.3). What is the value of the constant ω? Give both a formula and a numerical value for ω. What is the general shape of the solutions (6.3)?

(h) Find specific values of C_1 and C_2 in (6.4) that satisfy $x(0) = 1.5$, $x'(0) = 0$. Do you expect the mass to stay close to the equilibrium point? Is this a reasonable approximation to the solution of (6.1)?

(i) Find specific values of C_1 and C_2 in (6.4) that satisfy $x(0) = 1.95$, $x'(0) = 0$. Do you expect the mass to stay close to the equilibrium point? Is this a reasonable approximation to the solution of (6.1)?

(j) Reconcile your answers to parts (g) and (h) with those you gave for parts (d) and (e). Estimate the values of a for which (6.4) is a reasonable approximation to the solution of (6.1).

31. Find the solution of the linearized equation for the simple pendulum (6.33) if the pendulum is released from rest at an angle of $1/10$ radians. The relative difference between x and $\sin x$ is $(x - \sin x)/x$. Show that it satisfies

$$\frac{x - \sin x}{x} < \frac{1}{3!}x^2.$$

ADDITIONAL EXERCISES

What is the maximum relative difference between x and $\sin x$ for this motion?

32. **The Bobbing Buoy — A Simple Experiment.**[2] Show that the period of oscillation T for the floating object described in Exercise 10 which is displaced a distance d under the water when in equilibrium — so that $a = g/d$ — is $T = (2/\pi)\sqrt{d/g}$. Either conduct the following simple experiment to confirm this periodic behavior, or use Table 6.1 which is the result of such an experiment. Find a cigar-tube — or any lightweight cylindrical container that is closed on one end — which easily accepts dimes. Measure its length. Add enough dimes so that the tube floats upright when placed in a container of water. Measure the distance from the water level to the top of the tube. Determine d, the distance from the water level to the bottom of the tube. Displace the tube slightly, and measure the time it takes to complete three oscillations. From this time, estimate T for this d. Continue adding dimes, one at a time, and repeating the experiment until the conditions of the model are violated — that is, until the tube sinks!

Table 6.1 Period T of a bobbing buoy as a function of distance d

d (mm)	T (sec)
71.0	0.50
77.5	0.54
85.0	0.58
92.5	0.60
101.0	0.63
107.5	0.67
116.0	0.70
123.0	0.71
130.5	0.74

33. **Coupled Pendulums I — A Simple Experiment.**[3] Construct the following piece of apparatus, which is a coupled pendulum. Suspend two identical pendulums (same length wire, same mass) and join the vertical wires with a horizontal light rod. Make sure that the points of suspension of the pendulums are the same distance apart as the length of the rod so that the wires hang vertically.[4] We are first going to perform two experiments, and then we are going to explain what we have found. All motion is assumed to be in the plane of the wires.

   (a) Perform the following experiment. Displace the two pendulums through the same small initial angle and simultaneously release both from rest. Describe what happens.

   (b) Perform the following experiment. Displace the two pendulums through the same small initial angle but in opposite directions, and simultaneously release both from rest. Describe what happens.

   (c) Perform the following experiment. Displace one of the pendulums through a small initial angle holding the other pendulum vertical. Now simultaneously release both from rest. Describe what happens.

   (d) It can be shown that if $x(t)$ and $y(t)$ are the horizontal displacements of the two pendulums from vertical, then the linearized differential equations governing the subsequent motion, neglecting air resistance, are $x'' + \lambda^2 x = \lambda^2 \omega$, $y'' + \lambda^2 y = \lambda^2 \omega$,

---

[2] Based on "The Bobbing Buoy" by J. Casey, The Physics Teacher, January 1988, page 33.

[3] "Teaching physics with coupled pendulums" by J. Priest and J. Poth, The Physics Teacher, February 1982, pages 80 through 85.

[4] An interesting in-class version of this pendulum can be constructed from two bowling balls suspended with two steel wires about 7 feet long. For the horizontal rod a 4 foot dowel rod attached to the wires by rubber bands can be used.

where $\lambda^2 = g/H$ and $\omega = \frac{1}{2}(x+y)(1-H/L)$. Here L is the length of the pendulums from the pivot and H is the length of the pendulums from the rod.

  i. Does the experiment in part (a) suggest that there is a solution of these equations for which $x(t) = y(t)$? If so, then this suggests we should consider $x(t) - y(t)$ as a new variable, say $X(t) = x(t) - y(t)$, which we would obtain by subtracting the two differential equations.

  ii. Does the experiment in part (b) suggest that there is a solution of these equations for which $x(t) = -y(t)$? If so, then this suggests we should consider $x(t) + y(t)$ as a new variable, say $Y(t) = x(t) + y(t)$, which we would obtain by adding the two differential equations.

  iii. Show that this leads to the differential equations $X'' + \lambda^2 X = 0$, $Y'' + \mu^2 Y = 0$, where $\mu^2 = g/L$. Solve these two equations for $X(t)$ and $Y(t)$, and then find $x(t)$ and $y(t)$.

(e) What are the solutions of the differential equations in part (d) subject to the initial conditions $x(0) = y(0) = x_0$ and $x'(0) = y'(0) = 0$? How does this agree with part (a)?

(f) What are the solutions of the differential equations in part (d) subject to the initial conditions $x(0) = x_0$, $y(0) = -x_0$, and $x'(0) = y'(0) = 0$? How does this agree with part (b)?

(g) What are the solutions of the differential equations in part (d) subject to the initial conditions $x(0) = x_0$, $y(0) = 0$, and $x'(0) = y'(0) = 0$? How does this agree with part (c)? Show that according to this model the period T of the beats — that is, the time it takes for the mass at rest to return to rest — is $T = 2\pi/(\lambda - \mu)$ which can be written as $1/T = \sqrt{g}\left(1/\sqrt{H} - 1/\sqrt{L}\right)/(2\pi)$. Thus, a plot of $1/T$ (the frequency) versus $1/\sqrt{H}$ should yield a straight line of slope $\sqrt{g}/(2\pi)$ and intercept $-\sqrt{g}/\left(\sqrt{L}2\pi\right)$. Conduct an experiment where the beat period T is measured for various values of H. Then plot $1/T$ versus $1/\sqrt{H}$. Do you get a straight line? Table 6.2 and Figure 6.2 shows the results of such an experiment.[5]

Table 6.2 Frequency $1/T$ of coupled pendulums as a function of $H^{-1/2}$

$1/\sqrt{H}$ (cm$^{-1/2}$)	$1/T$ (sec^{-1})
0.247	0.044
0.256	0.097
0.266	0.141
0.280	0.209
0.295	0.276
0.315	0.375
0.340	0.502
0.364	0.633

34. **Energy Ripples.**[6] Consider the underdamped solution of $mx'' + bx' + kx = 0$, subject to $x(0) = x_0$, $x'(0) = v_0$.

---

[5] "Teaching physics with coupled pendulums" by J. Priest and J. Poth, The Physics Teacher, February 1982, pages 80 through 85, Figure 6.

[6] "Ripples in the energy of a damped harmonic oscillator" by E. A. Karlow, The American Journal of Physics, **62**, 1994, pages 634 through 636.

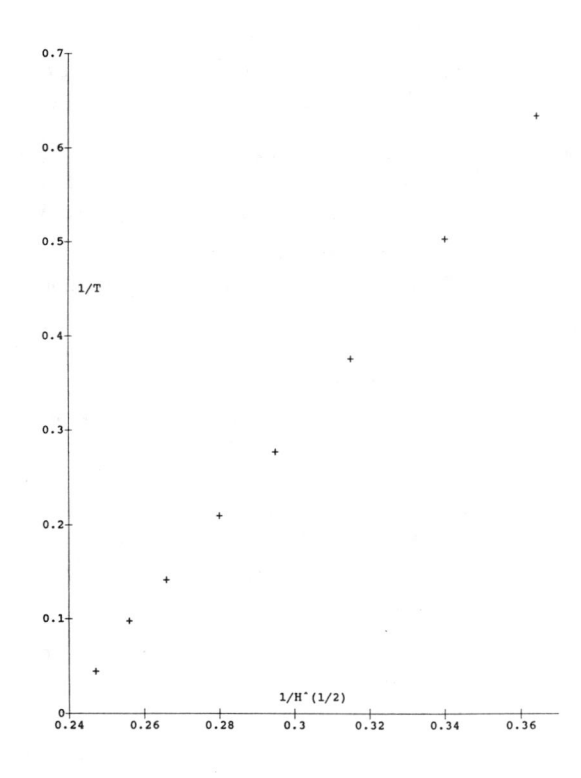

Figure 6.2 Frequency $1/T$ of coupled pendulums as a function of $H^{-1/2}$

(a) Use the results from Exercise 6 of Section 6.3 to write the velocity in the form

$$x'(t) = -A\sqrt{\frac{k}{m}} e^{\alpha t} \sin(\beta t + \phi - \psi),$$

where $\sin\psi = \alpha/\sqrt{\alpha^2 + \beta^2}$, $\cos\psi = \beta/\sqrt{\alpha^2 + \beta^2}$, and $\sqrt{\alpha^2 + \beta^2} = \sqrt{\frac{k}{m}}$.

(b) Show that if $v_0 = 0$ then $\phi = \psi$.

(c) Show that the energy $E(t)$ at time t, namely, $E(t) = \frac{1}{2}m(x')^2 + \frac{1}{2}kx^2$, can be expressed as $E(t) = \frac{1}{2}kA^2 e^{2\alpha t}\left[\sin^2(\beta t + \phi - \psi) + \cos^2(\beta t + \phi)\right]$. Use the identities $\cos^2(\beta t + \phi) = 1 - \sin^2(\beta t + \phi)$ and $\sin^2 A - \sin^2 B = \sin(A+B)\sin(A-B)$ to find $E(t) = \frac{1}{2}kA^2 e^{2\alpha t}\left[1 - \sin\psi \sin(2\beta t + 2\phi - \psi)\right]$. In this form we can see that the energy has an exponentially decaying amplitude $E_0(t) = kA^2 e^{2\alpha t}/2$, but it is modulated by a ripple — $\sin(2\beta t + 2\phi - \psi)$ — with amplitude $\sin\psi$. Figure 6.3 shows $E(t)$ and $E_0(t)$ graphed for the case $m = 8$, $b = 2$, $k = 32$, $x_0 = 0.5$, $v_0 = 0$.

(d) Figure 6.4 shows $x'(t)$ and $E(t)$ graphed for the case $m = 8$, $b = 2$, $k = 32$, $x_0 = 0.5$, $v_0 = 0$. Explain how these graphs are consistent with the result from Exercise 9 — namely, $E' = -b(x')^2$.

35. **Spring-Mass-Spring System.** The spring-mass-spring system consists of two springs with a mass attached between them, the other ends of the springs being secured to two walls, as shown in Figure 6.5. The mass is displaced horizontally from its equilibrium position and allowed to oscillate. If the springs obey Hooke's law with spring constants k_1 and k_2, and if x denotes the distance from the equilibrium position, then, in the absence of friction, the motion is governed by $mx'' + kx = 0$, where m is the mass and $k = k_1 + k_2$.

(a) If the mass is released from rest from $x = x_0$ at time $t = 0$, show that the subsequent motion is given by $x(t) = x_0 \cos\beta t$ where $\beta = \sqrt{k/m}$.

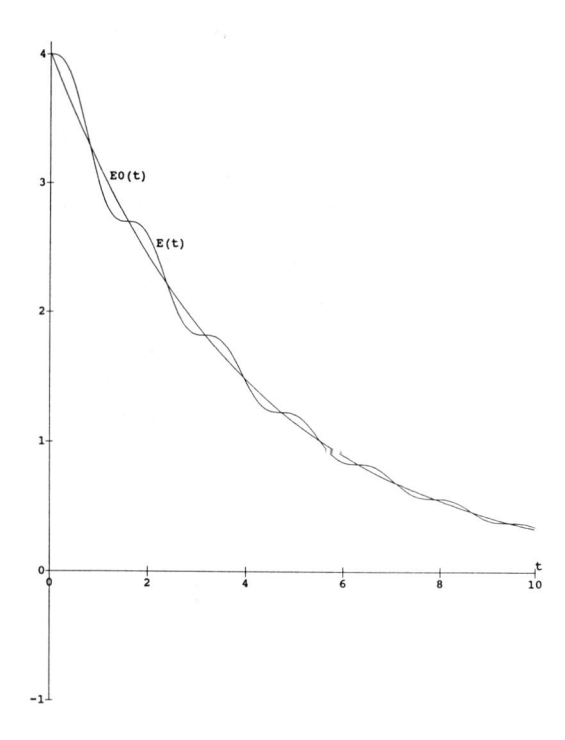

Figure 6.3 The energy $E(t)$ and $E_0(t) = kA^2 e^{2\alpha t}/2$ for underdamped motion

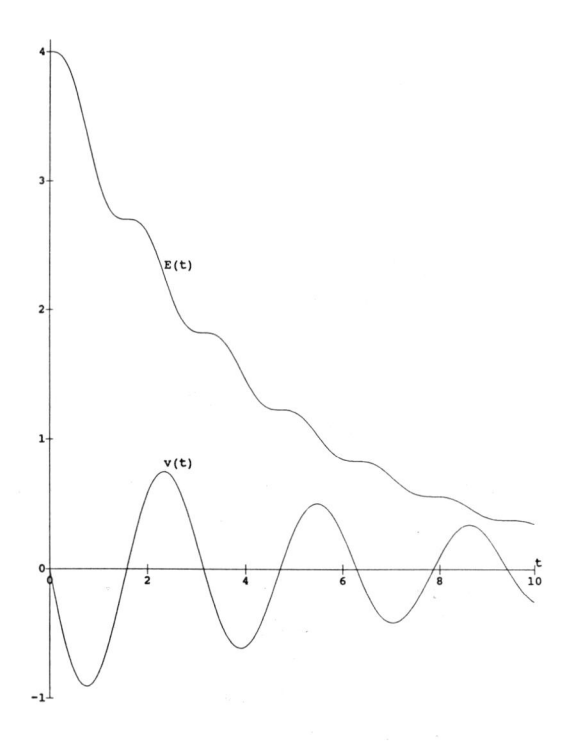

Figure 6.4 The energy $E(t)$ and velocity $v(t)$ for underdamped motion

ADDITIONAL EXERCISES

(b) Show that the mass returns to its starting position at times $t = 2n\pi/\beta$, $n = 1, 2, 3, \cdots$. Show the period T of the oscillation is $T = 2\pi\sqrt{m/k}$.

(c) To test whether the preceding is confirmed by experiment, we could measure the period for different masses and see whether the period is proportional to $\sqrt{m}$. Table 6.3 and Figure 6.6 show the results of such an experiment where friction was effectively eliminated by using an air track.[7] How well does the model fit this data set? What is a reasonable estimate for k?

Table 6.3   The period as a function of mass

Mass (kg)	Period (sec)
0.1256	0.9920
0.1758	1.1686
0.2549	1.4068
0.3051	1.5370
0.3549	1.6556
0.3805	1.7182
0.4307	1.8256
0.4805	1.9292
0.5105	1.9924
0.5607	2.0846

(d) It is suggested that the springs used in this experiment are the same ones used in the experiment to establish Hooke's law in Exercise 11. Could this be true?

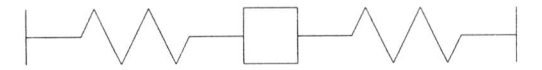

Figure 6.5   The spring-mass-spring system

36. **Coupled Pendulums II — A Simple Experiment.** Construct the following piece of apparatus, which is a coupled pendulum. Suspend two identical pendulums (same length wire, same mass) and join the masses with a light spring. Make sure that the points of suspension of the pendulums are the same distance apart as the unstretched spring. We are first going to perform two experiments, and then we are going to explain what we have found.

   (a) Perform the following experiment. Displace the two pendulums through the same small initial angle and simultaneously release both from rest. Describe what happens.

   (b) Perform the following experiment. Displace the two pendulums through the same small initial angle, but in opposite directions, and simultaneously release both from rest. Describe what happens.

   (c) Perform the following experiment. Displace one of the pendulums through a small initial angle holding the other pendulum vertical. Now simultaneously release both from rest. Describe what happens.

---

[7] These experiments were conducted by David Harman and Mark Zerella while students at the University of Arizona.

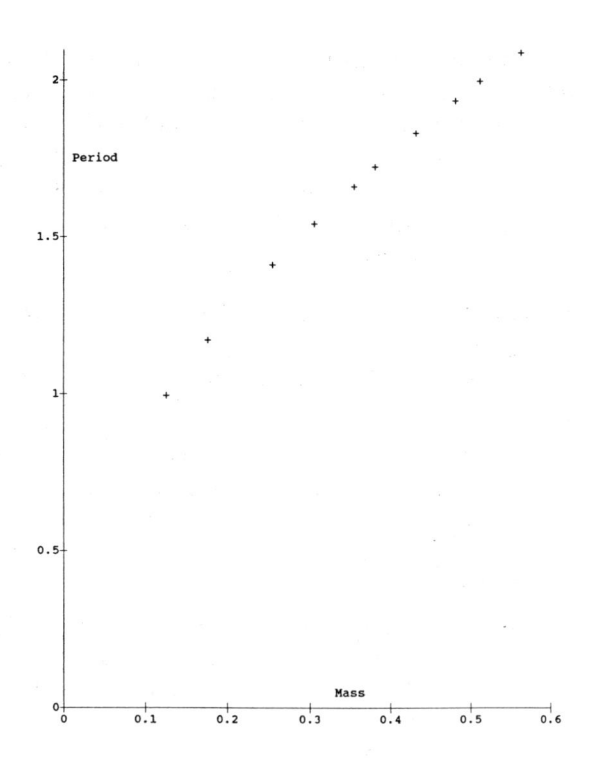

Figure 6.6 The period as a function of mass

(d) Show that if $x(t)$ and $y(t)$ are the angles displaced by the two pendulums from vertical, and $k > 0$ is the spring constant, then the linearized differential equations governing the subsequent motion, neglecting air resistance, are $x'' + \lambda^2 x = k(y-x)$, $y'' + \lambda^2 y = k(x-y)$.

(e) Does the experiment in part (a) suggest that there is a solution of these equations for which $x(t) = y(t)$? If so, then this suggests we should consider $x(t) - y(t)$ as a new variable, say $X(t) = x(t) - y(t)$, which we would obtain by subtracting the two differential equations.

(f) Does the experiment in part (b) suggest that there is a solution of these equations for which $x(t) = -y(t)$? If so, then this suggests we should consider $x(t) + y(t)$ as a new variable, say $Y(t) = x(t) + y(t)$, which we would obtain by adding the two differential equations.

(g) Show that this leads to the differential equations $X'' + \lambda^2 X = -2kX$, $Y'' + \lambda^2 Y = 0$. Solve these two equations for $X(t)$ and $Y(t)$, and then find $x(t)$ and $y(t)$.

(h) What are the solutions of the differential equations in part (d) subject to the initial conditions $x(0) = y(0) = x_0$ and $x'(0) = y'(0) = 0$? How does this agree with part (a)?

(i) What are the solutions of the differential equations in part (d) subject to the initial conditions $x(0) = x_0$, $y(0) = -x_0$ and $x'(0) = y'(0) = 0$? How does this agree with part (b)?

(j) What are the solutions of the differential equations in part (d) subject to the initial conditions $x(0) = x_0$, $y(0) = 0$ and $x'(0) = y'(0) = 0$? How does this agree with part (c)?

37. **The Spring Pendulum.**[8] A spring pendulum is a pendulum where the mass is

---

[8] "On the recurrence phenomenon of a resonant spring pendulum" by H. M. Lai, The American Journal of Physics, **52**, 1984, pages 219 through 223.

suspended by a spring in place of a rigid rod. We are concerned with two dimensional motion where the motion is in a single vertical plane. Let (x, y) denote the coordinates of the mass measured from the equilibrium point of the spring pendulum, that is, measured from where the mass would be if it were just to hang at rest. For small x and y the motion of the mass m can be approximated by $x'' + \omega_p^2 x = 0$, $y'' + \omega_s^2 y = 0$, where $\omega_p = \sqrt{g/h}$, $\omega_s = \sqrt{k/m}$, h is the natural length of the unstretched spring, and k is the spring constant.

(a) Show that the solutions of these equations subject to the initial conditions $x(0) = x_0$, $x'(0) = u_0$, $y(0) = y_0$, $y'(0) = v0$, can be written in the form $x(t) = x_0 \cos \omega_p t + \frac{u_0}{\omega_p} \sin \omega_p t$, $y(t) = y_0 \cos \omega_s t + \frac{v_0}{\omega_s} \sin \omega_s t$.

(b) By defining the new variable $T = \omega_p t$, so that $U_0 = u_0/\omega_p$ and $V_0 = v_0/\omega_p$, show that these solutions can be rewritten as $x(T) = x_0 \cos T + U_0 \sin T$, $y(t) = y_0 \cos \omega T + \frac{V_0}{\omega} \sin \omega T$, where $\omega = \omega_s/\omega_p$. Notice that in this form the motion depends on ω, the ratio ω_s/ω_p, rather than w_p and w_s independently. Show that ω can be changed by varying the natural length h, the spring constant k, or the mass m.

(c) For the initial conditions $x_0 = y_0 = 1$, $U_0 = V_0 = 0$, the first row of Figure 6.7 shows the motion of the mass with $\omega = 2, 3$, and 4. From these it is possible immediately to show the motion for $\omega = 1/2, 1/3$, and $1/4$. How? [Hint: Rewrite the equations in terms of $\tilde{T} = \omega T$ and interchange the roles of x and y.] What do these motions look like?

(d) For the initial conditions $x_0 = y_0 = 0$, $U_0 > 0$, $V_0 > 0$, the second row of Figure 6.7 shows the motion of the mass with $\omega = 2, 3$, and 4. From these it is possible immediately to show the motions for $\omega = 1/2, 1/3$, and $1/4$. What do these motions look like?

(e) For the initial conditions $x_0 = y_0 = 1$, $U_0 = V_0 = 0$, the first row of Figure 6.7 shows the motion of the mass with $\omega = 3/2, 4/3$, and $5/4$. From these it is possible immediately to show the motions for $\omega = 2/3, 3/4$, and $4/5$. What do these motions look like?

(f) The graphs in Figure 6.7 are special cases of Lissajous curves, which are also known as Bowditch curves.[9] It is claimed that Lissajous figures result from two simple harmonic motions interacting at right angles. Comment on this claim.

Section 6.5

Page 266. Add exercise after the current Exercise 12.

13. Two cars are stationary behind each other at a stoplight at time $t = 0$. The first car's position is characterized by $x(t)$ and the second by $y(t)$, with velocities $u(t) = x'$ and $v(t) = y'$, respectively. A simple model for the motion of these cars[10] (called Pipes' Model) is given by the system of equations

$$\begin{aligned} u' &= a - u \\ v' &= u - v, \end{aligned}$$

where a is a positive constant. It is said that the second driver adjusts her acceleration to be exactly equal in magnitude to the relative difference in velocity between her vehicle and the vehicle ahead. Explain why this is plausible. Write down a similar

---

[9] "A Catalog of Special Plane Curves" by J. D. Lawrence, Dover, 1972, page 178.

[10] "Car-Following Models" by R.L. Baker, Jr., in "Modules in Applied Mathematics, Volume 1, Differential Equations — Models" edited by M. Braun, C.S. Courteny, and D.A. Drew, Springer-Verlag 1983, Chapter 12

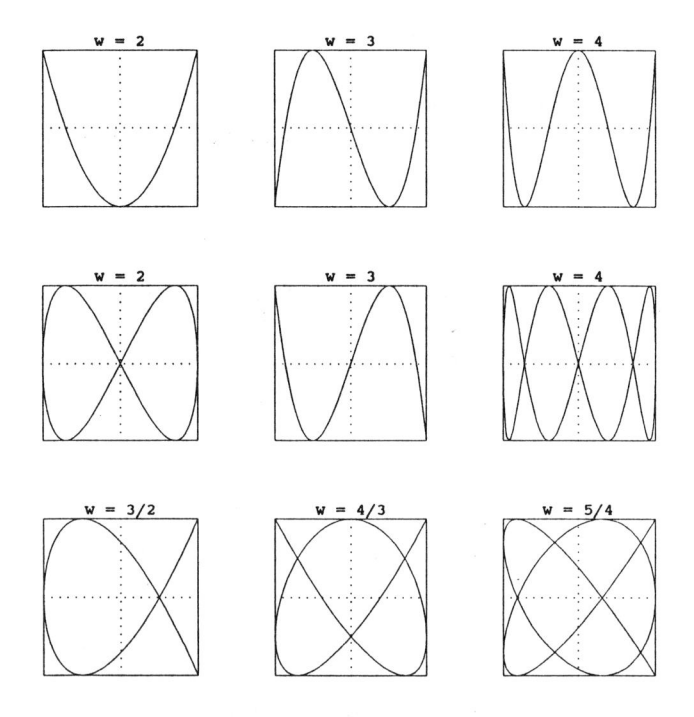

Figure 6.7 The motion of the mass of different spring pendulums

statement about the first driver's acceleration. What does a represent? Explain why the initial conditions $x(0) = u(0) = 0$, $y(0) = y_0$, $v(0) = 0$, are reasonable. What does y_0 represent? Show that the trajectories of this system in the (u, v) phase plane satisfy

$$\frac{dv}{du} = \frac{u-v}{a-u}.$$

Construct the direction field for this system and explain why trajectories that start at $u = v = 0$ have the property that $v(t) < u(t)$. Is there any time at which the cars collide?

Section 6.6

Page 271. Add exercise after the current Exercise 6.

7. **The Cubic Oscillator.** The cubic oscillator is governed by the differential equation $x'' + \lambda^2 x^3 = 0$ where λ is a constant. This can occur in at least two different ways. First, it arises when a mass m is forced to move in a straight horizontal channel under the action of a spring of force constant k and natural length ℓ — in which case $\lambda^2 = k/(2m\ell^2)$.[11] Second, it occurs when two springs are attached on opposite sides of a mass, the other ends of the unstretched springs being fixed to a horizontal surface — in which case $\lambda^2 = k/(m\ell^2)$.[12] If x is the distance from the mass to the unstretched position of the springs and if x is small then x satisfies $x'' + \lambda^2 x^3 = 0$ where $\lambda^2 = k/(2M\ell^2)$. Here $M = m$ in the first case and $M = m/2$ in the second. We will concentrate on the initial conditions $x(0) = x_0$, $x'(0) = 0$.

---

[11] "The x^3 oscillator" by A. Cromer, The Physics Teacher, **30**, 1992, pages 249 through 250. "A simple nonlinear oscillator: analytical and numerical solutions" by V. Detcheva and V. Spassov, Physics Education, **20**, 1993, pages 39 through 42.

[12] "A cube-law air track oscillator" by S. Whinery, European Journal of Physics, **12**, 1991, pages 90 through 95.

ADDITIONAL EXERCISES

(a) Investigate the phase plane of this equation and predict the types of motion possible.

(b) Multiply $x'' + \lambda^2 x^3 = 0$ by x' and integrate with respect to t to find $(x')^2 = \frac{1}{2}\lambda^2 \left(x_0^4 - x^4\right)$. Integrate once more to find

$$t = \frac{\varepsilon\sqrt{2}}{\lambda} \int_{x_0}^{x} \frac{du}{\sqrt{x_0^4 - u^4}},$$

where $\varepsilon = \pm 1$ and $\varepsilon = 1$ is used when we are moving to the right and $\varepsilon = -1$ to the left. Thus, initially we use $\varepsilon = 1$. Explain why this integral is an improper integral.

(c) Make the change of variable $\xi = u/x_0$ to write t as

$$t = 2\varepsilon\sqrt{\frac{M}{k}\frac{\ell}{x_0}} \int_{1}^{x/x_0} \frac{d\xi}{\sqrt{1-\xi^4}}.$$

This integral cannot be evaluated in terms of familiar functions.

(d) Explain why

$$-2\sqrt{\frac{M}{k}\frac{\ell}{x_0}} \int_{1}^{0} \frac{d\xi}{\sqrt{1-\xi^4}}$$

is the time taken to complete one-quarter of a period, so the period T is

$$T = 8\sqrt{\frac{M}{k}\frac{\ell}{x_0}} \int_{0}^{1} \frac{d\xi}{\sqrt{1-\xi^4}}.$$

(e) Use a numerical integration package to evaluate

$$\int_0^1 \frac{d\xi}{\sqrt{1-\xi^4}}.$$

[Because this is an improper integral, the usual numerical integration techniques may be unreliable. If this is the case, make the change of variable $\xi = \sin z$ and simplify the result to show

$$\int_0^1 \frac{d\xi}{\sqrt{1-\xi^4}} = \int_0^{\pi/2} \frac{dz}{\sqrt{1+\sin^2 z}}.$$

This is no longer an improper integral and should cause no problems when evaluated numerically.] Now show that

$$T \approx 10.48 \sqrt{\frac{M}{k}\frac{\ell}{x_0}}.$$

(f) As x_0 decreases what happens to the period T? Figure 6.8 shows three solutions of the same differential equation. Explain why the differential equation is not $x'' + \lambda^2 x^3 = 0$ for fixed λ and three different initial conditions.

(g) Table 6.4 and Figure 6.9 show the results of an experiment where the period of a cubic oscillator was measured for different initial amplitudes. How well does the model fit these data?

Section 6.7

Page 277. Add exercises after the current Exercise 7.

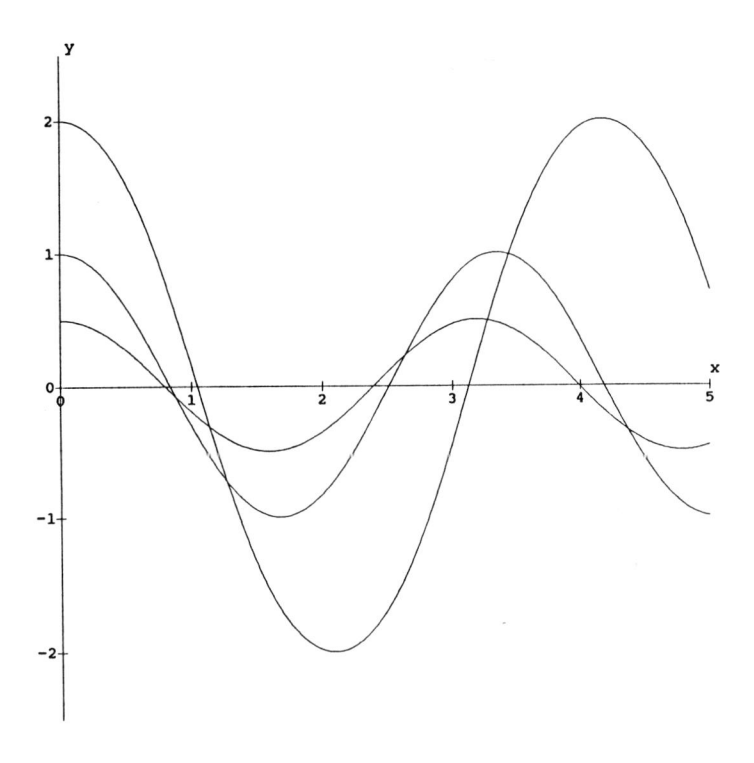

Figure 6.8 Three solutions of the same differential equation

Table 6.4 The period of a cubic oscillator for different initial amplitudes

Amplitude (cm)	Period (sec)
101.64	7.84
119.34	6.93
137.05	5.86
160.66	4.95
178.36	4.50
196.07	3.98
213.11	3.55
229.51	3.41
245.90	3.07
269.51	2.86
301.64	2.50
321.97	2.41
337.70	2.20
350.82	2.14
367.21	2.11
390.16	1.91

ADDITIONAL EXERCISES

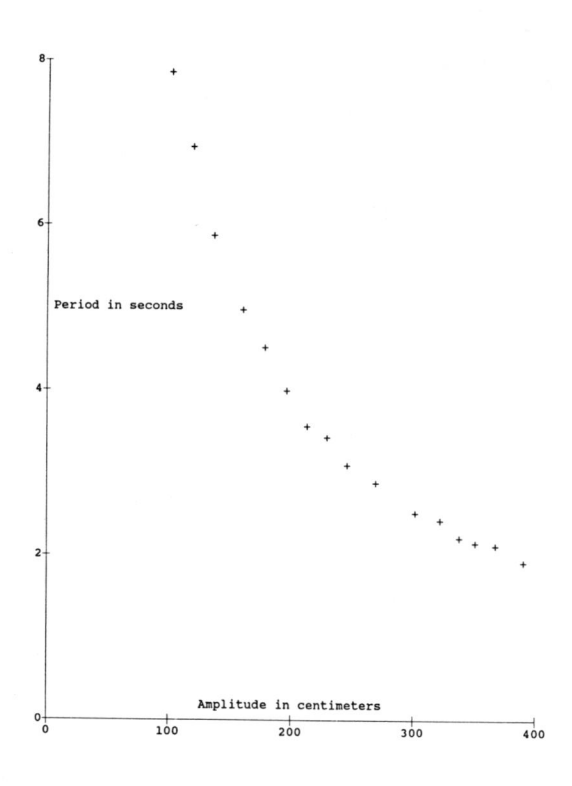

Figure 6.9 The period of a cubic oscillator for different initial amplitudes

8. A frictionless pendulum is released from rest one radian from its equilibrium position. Two differential equations used to model this situation is the linear model, $x''+\lambda^2 x = 0$, and the nonlinear model, $x''+\lambda^2 \sin x = 0$. When the pendulum reaches the equilibrium position, which model predicts the higher velocity? [Hint: Look at the phase plane.] What does this tell you about the period of the linearized pendulum compared to the nonlinear pendulum?

9. Show that the frictionless nonlinear pendulum, $x'' + 9\sin x = 0$, when released from $x = 0$ with the velocity $x' = 6$, has the orbit $y^2 = 18(\cos x + 1)$, where $y = x'$. Show that the point $x = \pi$, $y = 0$, lies on the curve $y^2 = 18(\cos x + 1)$. Explain why the pendulum takes an infinite amount of time to move from $x = 0$ to $x = \pi$. [Hint: Use the Existence-Uniqueness Theorem and note that $x = \pi$, $y = 0$ is an equilibrium point.] Explain why the orbit $y^2 = 18(\cos x + 1)$ remains in the first quadrant of the phase plane. Now use a computer/calculator to solve the initial value problem $x''+9\sin x = 0$, $x = 0$, $x' = 6$, numerically using different step-sizes. Does the numerical solution end at $x = \pi$? Explain what you see.

10. A frictionless nonlinear pendulum of length h is released from $x = 0$ with an initial **horizontal** velocity v. It swings to a maximum height a above horizontal, where $a \leq h$. Show that $v = \sqrt{2ga}$. Thus, the initial horizontal velocity can be determined from the height a. (This arrangement has been used to estimate the velocity of a bullet of mass m, by firing it into the mass M of a pendulum at rest. The velocity of the bullet is then $(m + M)v/m$.)[13]

---

[13] "Mathematics for Exterior Ballistics" by G.A. Bliss, Wiley, 1944, page 25.

6.3 Additional Materials

Sea Battles

Example 6.1 : Sea Battles

It is the year 1805. You are the commander of a fleet of warships about to do battle with an enemy. Your ships and those of your enemy are equal in quality. The only difference is in numbers: You have 37 ships, and your opponent, 12. You engage the enemy in battle. Who will win the battle? How many of the winner's ships will survive? How long will the battle last? The first question does not require much intuition. You should win the battle, because you have a larger force and the ships are evenly matched. The second question could be answered with the guess that 25 ships will survive, because $37 - 12 = 25$. The third question might receive the response, "Who knows?"

We will try to answer these questions by looking at a simple model, in which we assume that the rate of change of the number of ships is proportional to the number of enemy ships. If x is the number of ships in the first fleet and y the number of ships in the second fleet — so x and y are nonnegative — then the appropriate differential equations are

$$\frac{dx}{dt} = -ay, \qquad \frac{dy}{dt} = -bx, \tag{6.5}$$

where a and b measure the effectiveness of the two fleets and are both positive. [Why is there a negative sign on the right-hand sides of (6.5)?] If the fleets are equally matched in quality, then $a = b$, so (6.5) becomes

$$\frac{dx}{dt} = -ay, \qquad \frac{dy}{dt} = -ax. \tag{6.6}$$

This is a coupled set of first order differential equations, in which a solution consists of x and y given as functions of the time t. From (6.6) we observe that although it is apparent that x and y will be decreasing functions of time (as is to be expected in a battle without reinforcements), we currently have no means of solving (6.6). We also have one more variable than we usually have when we draw a slope field. However, because x and y are both functions of the time t, we could (in principle) solve for t in terms of one of the dependent variables — say x — and substitute the result into the expression for y. This would eliminate the variable t and give an equation relating x and y. For such situations the graph of y versus x is called a TRAJECTORY (or ORBIT), and the xy-plane is called the PHASE PLANE associated with (6.6). We proceed to analyze (6.6) in this light.

Because the right-hand sides of the two equations (6.6) do not involve the variable t explicitly, we may use the chain rule to combine these two equations into one. If we treat y as a function of x, and x as a function of t, using the chain rule gives

$$\frac{dy}{dx} = \left(\frac{dy}{dt}\right)\left(\frac{dt}{dx}\right) = \frac{dy/dt}{dx/dt} = \frac{-ax}{-ay},$$

or

$$\frac{dy}{dx} = \frac{x}{y}. \tag{6.7}$$

Notice that because x and y are both nonnegative, the slope field for (6.7) (shown in Figure 6.10) has positive slopes. This means that both x and y must increase or decrease together in time, but the slope field does not tell us which of these happens. However, the original differential equations (6.6) show that x and y must both decrease with time at rate a. We could indicate this on the slope field by adding arrows to the slopes to indicate the direction the orbits follow as time increases, creating a field of vectors. Such a slope field is known as a DIRECTION FIELD and is shown in Figure 6.11, where the arrows clearly indicate the direction of travel. The length of each vector has been scaled so that those with small values of $(dx/dt, dy/dt)$ are short, and those with large values are long.

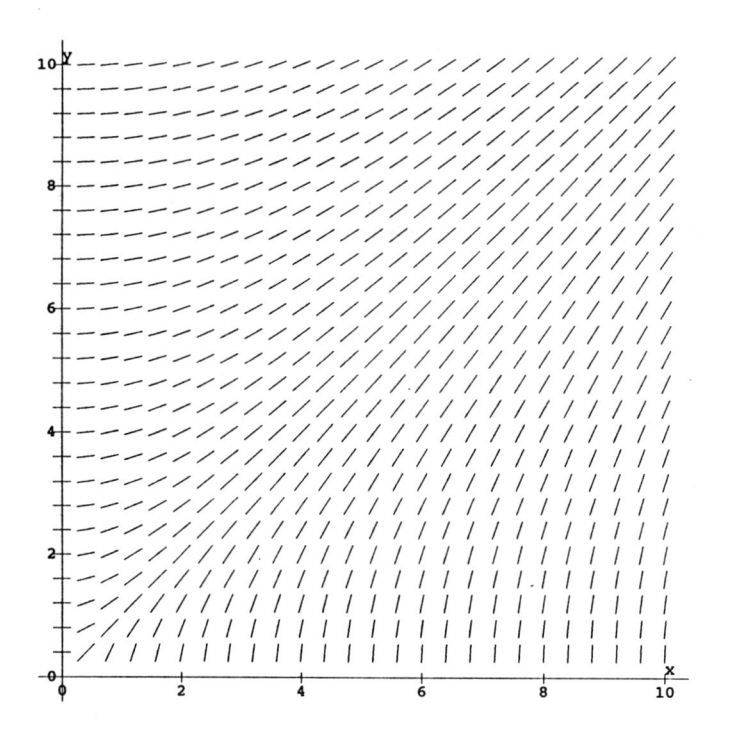

Figure 6.10 Slope field for $dy/dx = x/y$

It easy to see that $y = x$ is a solution of (6.7). So if initially x and y have equal values, they will continue to do so, and as time increases this solution approaches the origin — namely, $x = y = 0$, which is when the two fleets are eliminated simultaneously.

The direction field also suggests that, for the case $a = b$, the side that starts with the greater number of ships will always have ships surviving when the other fleet is destroyed completely. That is, if the initial value of x is greater than the initial value of y, the initial point on the direction field picture will be placed below the line $y = x$, and because (by the uniqueness theorem) other solution curves cannot cross the solution curve $y = x$, the x-intercept will be positive. That means that x is positive when y is zero, so x wins the battle. Figure 6.12 shows a few orbits for this situation, including $y = x$. Each orbit corresponds to a battle that started with different numbers of ships — that is, different initial conditions. Similarly, if the initial value of y is greater than the initial value of x, the initial point on the direction field picture will be placed above the line $y = x$, and the y-intercept will be positive. That means that y is positive when x is zero, so y wins the battle (see Figure 6.13).

The regions where solutions are concave up or concave down are clearly evident in both of these figures — namely, above or below the line $y = x$. We confirm this by differentiating (6.7) to find

$$\frac{d^2y}{dx^2} = \left(\frac{y - x\,(dy/dx)}{y^2}\right) = \left(\frac{y - x\,(x/y)}{y^2}\right),$$

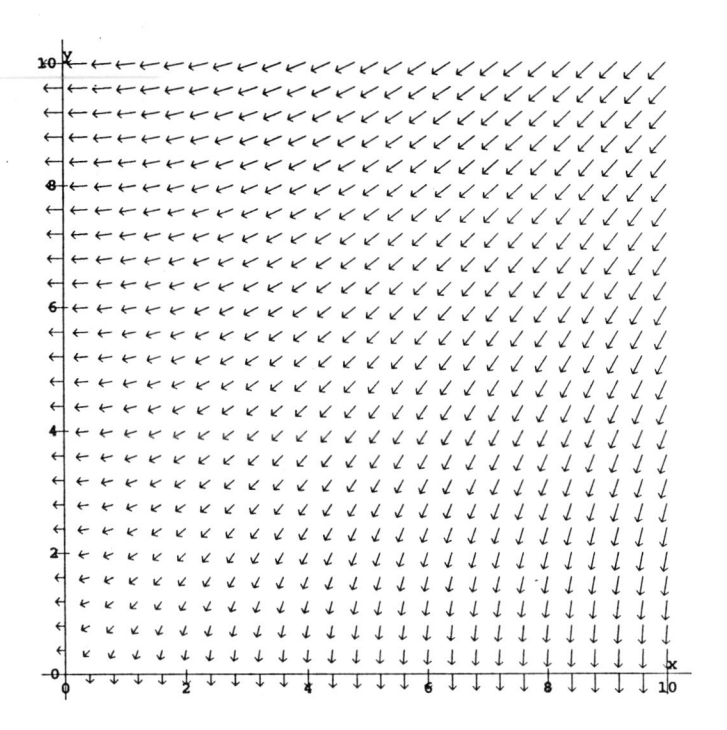

Figure 6.11 Direction field for $dy/dx = x/y$

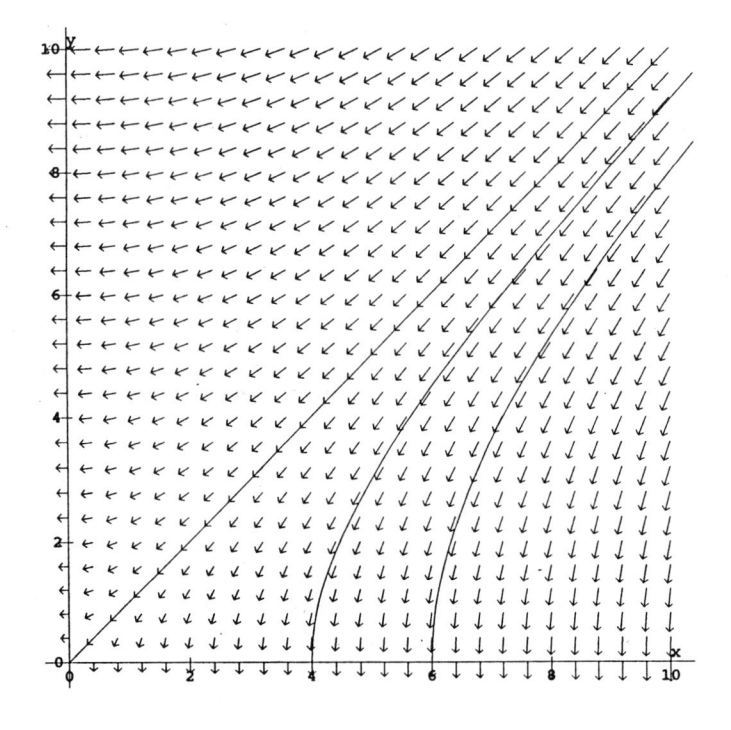

Figure 6.12 Trajectories and direction field for $dy/dx = x/y$, where initially $x \geq y$

ADDITIONAL MATERIALS

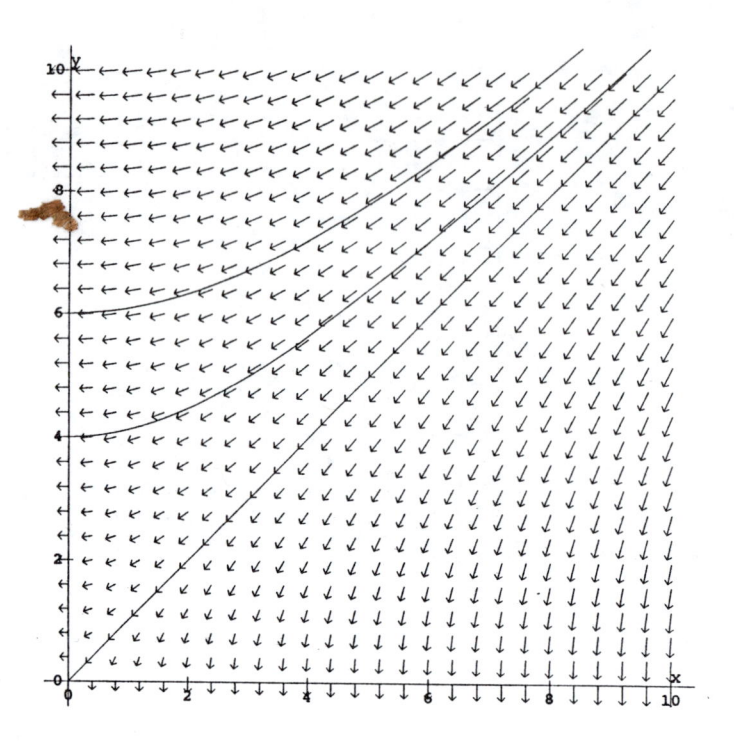

Figure 6.13 Trajectories and direction field for $dy/dx = x/y$, where initially $x \leq y$

or

$$\frac{d^2y}{dx^2} = \frac{1}{y^3}(y-x)(y+x).$$

This result means that $y = x$ is the dividing line between orbits that are concave up and those that are concave down. Thus, initial values of x and y that place the initial point below this dividing line will result in y going to zero before x, whereas if the initial point is above this line, the opposite will happen.

We can discover how many ships survive if we can find the orbits in the phase plane. This we can do by integrating the separable equation (6.7). We find

$$x^2(t) - y^2(t) = C, \tag{6.8}$$

where C is the constant of integration. These curves are hyperbolas. They open up if $C < 0$ and open sideways if $C > 0$. Because the expression in (6.8) is a constant for all values of t, if we start at $t = 0$ with x_0 and y_0, respectively, that is,

$$\begin{aligned} x(0) &= x_0, \\ y(0) &= y_0, \end{aligned}$$

then (6.8) implies that

$$x^2(t) - y^2(t) = x_0^2 - y_0^2 \tag{6.9}$$

for all values of t. Thus, for the case where $x_0 > y_0$, y will die out first. If we denote the value of x when $y = 0$ with x_r and call it the residual value of x, then from (6.9) we have that

$$x_r^2 = x_0^2 - y_0^2. \tag{6.10}$$

That is, we can predict who wins and that the number of survivors will be

$$x_r = \sqrt{x_0^2 - y_0^2}.$$

In the case of 37 ships versus 12, the number of surviving ships is 35, a long way from our guess of 25.

Surprisingly, we have answered the first two questions without knowing the exact time dependence of each variable.

We now turn to the third question. When will the battle end? This requires an analysis in which the time is explicitly involved, so we proceed to determine the time behavior of x and y. If we solve the second equation in (6.6) for x we have

$$x = -\frac{1}{a}\frac{dy}{dt},$$

which, when differentiated, gives

$$\frac{dx}{dt} = -\frac{1}{a}\frac{d^2y}{dt^2}.$$

If we substitute this result into the first equation in (6.6), we obtain

$$\frac{d^2y}{dt^2} = a^2 y,$$

or

$$\frac{d^2y}{dt^2} - a^2 y = 0. \tag{6.11}$$

This is a SECOND ORDER LINEAR DIFFERENTIAL EQUATION. It is SECOND ORDER because it contains the second derivative as its highest derivative. It is LINEAR because it is a linear combination of d^2y/dt^2, dy/dt, and y with coefficients 1, 0, and a^2. A precise definition will be given in the next section, and Section 6.3 is devoted to solving such equations.

In Section 6.3 we will show that the solution of (6.11) is

$$y(t) = C_1 e^{at} + C_2 e^{-at}, \tag{6.12}$$

where C_1 and C_2 are arbitrary constants. [The fact that this is a solution may easily be seen by substituting (6.12) into (6.11).]

To find the explicit solution for $x(t)$, we substitute (6.12) into the second equation in (6.6) to obtain

$$x(t) = -\frac{1}{a}\frac{dy}{dt} = -\frac{1}{a}\left(aC_1 e^{at} - aC_2 e^{-at}\right) = -C_1 e^{at} + C_2 e^{-at}. \tag{6.13}$$

Let us now use these explicit solutions to determine their time history, starting with the initial values $y(0) = y_0$ and $x(0) = x_0$. Setting $t = 0$ in (6.12) and (6.13) leads to the algebraic equations

$$y_0 = C_1 + C_2 \text{ and } x_0 = -C_1 + C_2, \tag{6.14}$$

with solution $C_1 = -(x_0 - y_0)/2$ and $C_2 = (x_0 + y_0)/2$. This gives our solution of the system of equations (6.6) as

$$\begin{array}{l} x(t) = \frac{1}{2}(x_0 - y_0)e^{at} + \frac{1}{2}(x_0 + y_0)e^{-at}, \\ y(t) = -\frac{1}{2}(x_0 - y_0)e^{at} + \frac{1}{2}(x_0 + y_0)e^{-at}. \end{array} \tag{6.15}$$

From these expressions it is clear that if $x_0 > y_0$, then x will always be positive, and there will be a time $t = T$ when y vanishes, that is — $y(T) = 0$. At that time the value of x will be $x(T)$, and this represents the number of surviving ships. From (6.15) we have that $y(T) = 0$ when

$$e^{2aT} = \frac{x_0 + y_0}{x_0 - y_0}.$$

Thus, the battle ends at time

$$T = \frac{1}{2a}\ln\frac{x_0 + y_0}{x_0 - y_0}.$$

ADDITIONAL MATERIALS

This time T depends not only on the initial values x_0 and y_0 but also on a, the efficiency of the ships. (The value of a could be determined experimentally. How? See Exercise 1 on page 212.)

Using this expression for T in (6.15) gives the number of surviving ships as

$$x_r = x(T) = e^{-aT}\left(\frac{1}{2}(x_0 - y_0)e^{2aT} + \frac{1}{2}(x_0 + y_0)\right)$$

or

$$x_r = x(T) = \sqrt{\frac{x_0 - y_0}{x_0 + y_0}}(x_0 + y_0) = \sqrt{x_0^2 - y_0^2}, \tag{6.16}$$

which is what we discovered earlier in (6.10). Figure 6.14 shows the solutions (6.15) for the case $x_0 = 37$, $y_0 = 12$, and $a = (1/40)\ln(7/5)$.. It is clear from this figure that both x and y decrease steadily, and when the number y is zero the residual value of x (the number of survivors) is 35, which can be confirmed by using (6.16). □

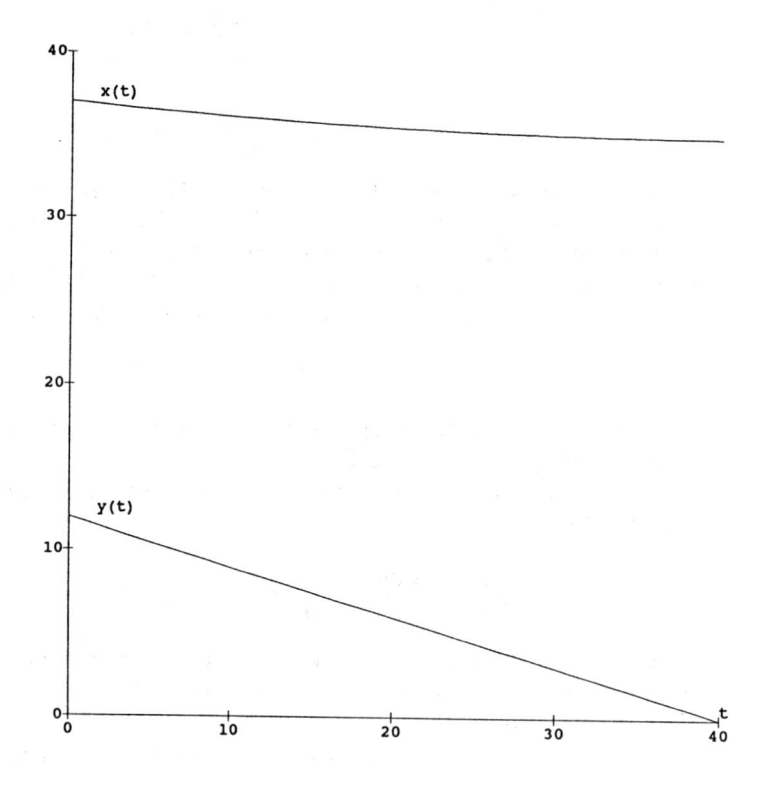

Figure 6.14 The solutions $x(t) = 25e^{at}/2 + 49e^{-at}/2$, $y(t) = -25e^{at}/2 + 49e^{-at}/2$

During the First World War, these ideas were used to design a successful strategy for airplane dogfights. There is also evidence that in the Napoleonic Wars, the British naval commander used a strategy similar to this at the Battle of Trafalgar. A discussion of the use of this model at the Battle of Trafalgar can be found in "Mathematics in Warfare" by F. W. Lanchester, in Volume 4 of "The World of Mathematics" by J. R. Newman, Tempus, 1988, pages 2113 through 2131, in particular pages 2125 through 2131.

Exercises

1. If the sea battle characterized by (6.6) began with $x_0 = 37$ and $y_0 = 12$ and lasted for 40 hours, what is the value of a?

2. Find the time-dependent behavior of $x(t)$ and $y(t)$ for Example 6.1 when $a \neq b$. That is, first eliminate the variable x from the system of equations in (6.5) to obtain $y'' - aby = 0$, and then solve for $y(t)$ and $x(t)$, assuming that $y(0) = y_0$ and $x(0) = x_0$. If $b > a$, and $y_0 = x_0$, who will win the battle? How many ships will remain after the battle is over?

3. **Computer Experiment.** Imagine we wanted to simulate a sea battle of the early 1800s where all ships are equally equipped. We could do this as follows. To start, imagine that we had an $x = 1$ versus $y = 1$ sea battle. We pick up a fair die with ones and twos on it, and throw it. If a 1 appears then one of y's ships is removed. If a 2 appears then one of x's ships is removed. In the first case we have $x = 1$, $y = 0$, and in the second $x = 0$, $y = 1$. In either case the battle is over. Now imagine that we had an $x = 2$ versus $y = 1$ sea battle. We pick up a fair die with ones, twos, and threes on it, and throw it. If a 1 or a 2 appears then one of y's ships is removed. If a 3 appears then one of x's ships is removed. In the first case we have $x = 2$, $y = 0$, and the battle is over. In the second case we have $x = 1$, $y = 1$. But this is just the first battle we fought, so we put down the fair die with ones, twos, and threes on it, and pick up the fair die with ones and twos on it, and continue until the battle is over. If we had an $x = 2$ versus $y = 2$ sea battle we pick up a fair die with ones, twos, threes, and fours on it, and throw it. If a 1 or a 2 appears then one of y's ships is removed. If a 3 or a 4 appears then one of x's ships is removed. In either case we have a 2 versus 1 battle which we simulate as before.

While this process is straightforward it has some disadvantages. First, in the case of, say, $x = 37$ versus $y = 12$ we would need a fair die with 1 through 49, and after that die is used we would need a fair die with 1 through 48, and after that die is used we would need a fair die with 1 through 47, and so on. Second, we would want to repeat this simulation a large number of times, perhaps 1000 to acquire a feeling for the general behavior. However, these disadvantages can be overcome using a computer program designed specifically for this task. The program SEA BATTLE-MANY does this. Run the program and record the number of ships remaining after 1000 simulations for each of the following battles: 37 versus 12, 25 versus 7, 17 versus 8, and 13 versus 5. Can you see a pattern?[14]

---

[14] It is useful to assign this exercise and have the students bring their results to class the next day. They are usually amazed to see they all have very similar results. In a class of 30 students usually one student spots that these are Pythagorean triples, as predicted by (7.7).

ADDITIONAL MATERIALS

Systems of Functional Equations

Example 6.2 : A System of Functional Equations

The functions $\sin t$ and $\cos t$ are differentiable functions that satisfy the identities

$$\sin(t_1 + t_2) = \sin t_1 \cos t_2 + \cos t_1 \sin t_2$$

and

$$\cos(t_1 + t_2) = \cos t_1 \cos t_2 - \sin t_1 \sin t_2,$$

for all t_1, t_2. This leads to the problem of finding all differentiable functions $x(t)$, and $y(t)$, that satisfy the identities

$$x(t_1 + t_2) = x(t_1)y(t_2) + y(t_1)x(t_2) \tag{6.17}$$

and

$$y(t_1 + t_2) = y(t_1)y(t_2) - x(t_1)x(t_2) \tag{6.18}$$

for all t_1, t_2. We see that $x(t) = 0$ and $y(t) = 0$ satisfy (6.17) and (6.18), so we seek functions that are not identically zero.

At first sight, finding $x(t)$ and $y(t)$ that satisfy (6.17) and (6.18) does not appear to be related to differential equations. However, we have already seen a number of functional equation problems that can be converted to differential equations that can be solved.

We construct the derivatives of x and y with respect to t,

$$x'(t) = \lim_{h \to 0} \frac{x(t+h) - x(t)}{h} \tag{6.19}$$

and

$$y'(t) = \lim_{h \to 0} \frac{y(t+h) - y(t)}{h}. \tag{6.20}$$

If we use (6.17) in (6.19), we find

$$x'(t) = \lim_{h \to 0} \frac{x(t)y(h) + y(t)x(h) - x(t)}{h} = \lim_{h \to 0} \left[x(t)\frac{y(h) - 1}{h} + y(t)\frac{x(h)}{h} \right]. \tag{6.21}$$

The first term on the right-hand side would contain $y'(0)$ if $y(0) = 1$, and the second term would contain $x'(0)$ if $x(0) = 0$. Let's return to (6.17) and (6.18) to see what they tell us about $x(0)$ and $y(0)$.

Putting $t_1 = t$ and $t_2 = 0$ in (6.17) and (6.18), we find

$$x(t) = x(t)y(0) + y(t)x(0) \tag{6.22}$$

and

$$y(t) = y(t)y(0) - x(t)x(0). \tag{6.23}$$

Now we put $t = 0$ in (6.22) and (6.23) and find

$$x(0) = 2x(0)y(0) \tag{6.24}$$

and

$$y(0) = y^2(0) - x^2(0). \tag{6.25}$$

Equation (6.24) gives rise to two cases: $y(0) = 1/2$ or $x(0) = 0$. If we substitute $y(0) = 1/2$ in (6.25), we find $x^2(0) = -1/4$, which is not possible, so we are forced to conclude that $x(0) = 0$. If we substitute $x(0) = 0$ in (6.23), we find $y(t) = 0$, or $y(0) = 1$. From (6.17) we see that $y(t) = 0$ implies that $x(t) = 0$, which is not what we are looking for. So the only possibilities are $x(0) = 0$ and $y(0) = 1$.

Using these facts we can rewrite (6.21) as

$$x'(t) = \lim_{h \to 0} \left[x(t) \frac{y(h) - 1}{h} + y(t) \frac{x(h)}{h} \right] = \lim_{h \to 0} \left[x(t) \frac{y(h) - y(0)}{h} + y(t) \frac{x(h) - x(0)}{h} \right],$$

or

$$x'(t) = ax(t) + by(t), \tag{6.26}$$

where $a = y'(0)$ and $b = x'(0)$. In the same way, we can use (6.20) to find

$$y'(t) = -bx(t) + ay(t). \tag{6.27}$$

Thus, our problem is to solve the system of first order linear differential equations (6.26) and (6.27), subject to the initial conditions

$$x(0) = 0, \qquad y(0) = 1. \tag{6.28}$$

Notice that we can obtain no more information about the values of a and b from (6.26) or (6.27), because putting $t = 0$ in either (6.26) or (6.27) leads to identities. However, we also notice that if $b = 0$, then (6.26) and (6.27) subject to (6.28), can be solved immediately to yield

$$x(t) = 0 \text{ and } y(t) = e^{at}, \tag{6.29}$$

which do satisfy (6.17) and (6.18). But, when we posed the original question, we were thinking of functions that were not identically zero, unlike $x(t) = 0$. Thus, from now on, we concentrate on $b \neq 0$. Nevertheless, it did come as a surprise that (6.29) satisfied (6.17) and (6.18) — we were expecting the trigonometric functions.

If we differentiate (6.27) with respect to t, and then substitute from (6.26) for $x'(t)$, we find

$$y'' = -bx' + ay' = -b(ax + by) + ay'.$$

If we now solve (6.27) for $x = (ay - y')/b$ (remember $b \neq 0$) and then substitute this in the last equation, we get

$$y'' = -b(ax + by) + ay' = -b\left(\frac{a}{b}(ay - y') + by\right) + ay',$$

or

$$y'' - 2ay' + (a^2 + b^2)y = 0.$$

This is a second order linear differential equation with constant coefficients, with solution

$$y(t) = e^{at}(C_1 \cos bt + C_2 \sin bt).$$

The condition $y(0) = 1$ implies that $C_1 = 1$, so that the last equation becomes

$$y(t) = e^{at}(\cos bt + C_2 \sin bt). \tag{6.30}$$

We substitute (6.30) into (6.27) and solve for $x(t)$ to find

$$x(t) = e^{at}(-C_2 \cos bt + \sin bt). \tag{6.31}$$

The condition $x(0) = 0$ implies that $C_2 = 0$, so that (6.30) and (6.31) become

$$x(t) = e^{at} \sin bt, \qquad y(t) = e^{at} \cos bt. \tag{6.32}$$

We see that the functions in (6.32) satisfy (6.17) and (6.18), and so, contrary to expectation, $\sin t$ and $\cos t$ are not the unique solutions of (6.17) and (6.18). Rather $e^{at} \sin bt$ and $e^{at} \cos bt$ are. □

Exercises

1. The functions $\sinh t$ and $\cosh t$ are defined by

$$\sinh t = \frac{1}{2}\left(e^t - e^{-t}\right) \quad \text{and} \quad \cosh t = \frac{1}{2}\left(e^t + e^{-t}\right)$$

and satisfy the identities

$$\sinh(t_1 + t_2) = \sinh t_1 \cosh t_2 + \cosh t_1 \sinh t_2$$

and

$$\cosh(t_1 + t_2) = \cosh t_1 \cosh t_2 + \sinh t_1 \sinh t_2$$

for all t_1, t_2. Find the most general differentiable functions $x(t)$ and $y(t)$ that satisfy the identities

$$x(t_1 + t_2) = x(t_1)y(t_2) + y(t_1)x(t_2)$$

and

$$y(t_1 + t_2) = y(t_1)y(t_2) + x(t_1)x(t_2)$$

for all t_1, t_2.

2. Find the most general differentiable functions $x(t)$ and $y(t)$ that satisfy the identities

$$x(t_1 + t_2) = x(t_1)y(t_2) + y(t_1)x(t_2)$$

and

$$y(t_1 + t_2) = y(t_1)y(t_2)$$

for all t_1, t_2.

Kirchhoff's Laws

If R is the resistance of a resistor (measured in ohms), L is the coefficient of inductance of an inductor (measured in henries), C is the capacitance of a capacitor (measured in farads), $q(t)$ is the charge (measured in coulombs), and $I(t)$ is the current (defined by $I = dq/dt$ and measured in amperes) then the current in one of these components is related to the voltage drop V (measured in volts) across it as follows:

- For the resistor is $V = RI$
- For the inductor is $V = LdI/dt$
- For the capacitor is $V = q/C$, so that $I = CdV/dt$.

KIRCHHOFF'S LAWS are:

1. At any junction of two or more components of a circuit the total current entering the junction is equal to the total current leaving the junction.

2. In a branch — that is, between any two junctions — the current passing through each of the components is the same.

3. In a loop the sum of the voltage drops across each component of the loop is equal to the applied voltage $E(t)$ in that loop.

Exercises

1. Show that if the current $I(t)$ and the applied voltage $E(t)$ are identical for the two circuits shown in Figures 6.15 and 6.16 then by Kirchhoff's laws
$$R = R_1 + R_2.$$

2. If the circuit in Figure 6.16 had n resistors, $R_1, R_2, R_3, \cdots, R_n$, in series, how would they be related to R?

3. Show that if the current $I(t)$ and the applied voltage $E(t)$ are identical for the two circuits shown in Figures 6.15 and 6.17 then by Kirchhoff's laws
$$\frac{1}{R} = \frac{1}{R_1} + \frac{1}{R_2}.$$

4. If the circuit in Figure 6.17 had n resistors, $R_1, R_2, R_3, \cdots, R_n$, in parallel, how would they be related to R?

5. Show that if there are n inductances $L_1, L_2, L_3, \cdots, L_n$ in series then they are equivalent to the single inductance L where
$$L = L_1 + L_2 + L_3 + \cdots + L_n.$$

6. Show that if there are n inductances $L_1, L_2, L_3, \cdots, L_n$ in parallel then they are equivalent to the single inductance L where
$$\frac{1}{L} = \frac{1}{L_1} + \frac{1}{L_2} + \frac{1}{L_3} + \cdots + \frac{1}{L_n}.$$

7. Show that if there are n capacitances $C_1, C_2, C_3, \cdots, C_n$ in series then they are equivalent to the single capacitance C where
$$\frac{1}{C} = \frac{1}{C_1} + \frac{1}{C_2} + \frac{1}{C_3} + \cdots + \frac{1}{C_n}.$$

8. Show that if there are n capacitances $C_1, C_2, C_3, \cdots, C_n$ in parallel then they are equivalent to the single capacitance C where
$$C = C_1 + C_2 + C_3 + \cdots + C_n.$$

ADDITIONAL MATERIALS

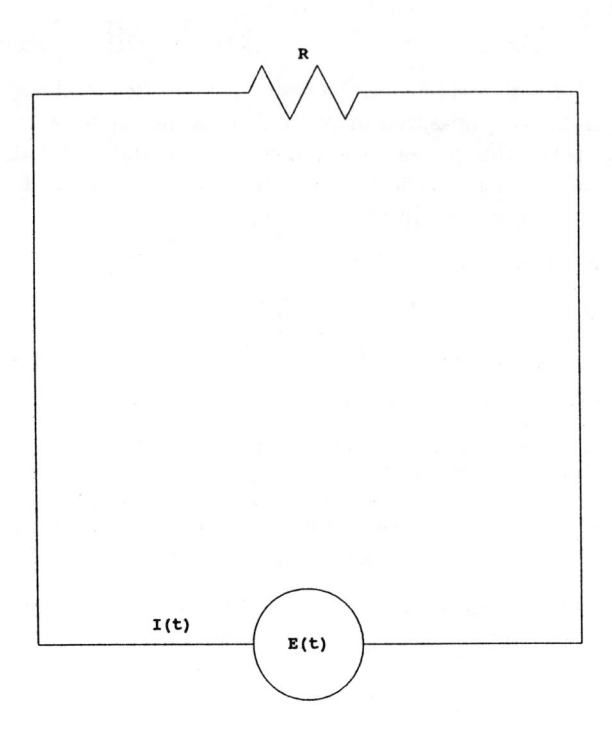

Figure 6.15 The circuit with one resistor

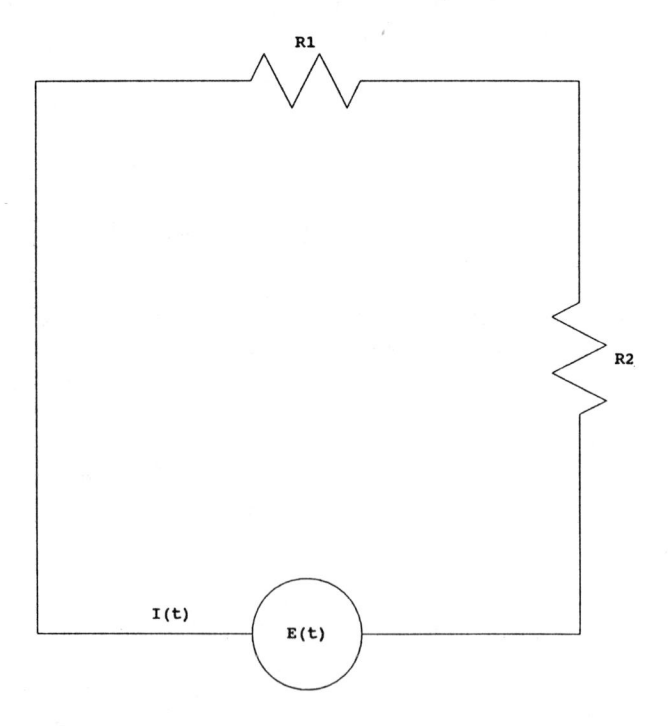

Figure 6.16 The circuit with two resistors in series

218 IDEAS FOR CHAPTER 6 — "INTERPLAY BETWEEN FIRST ORDER SYSTEMS AND SECOND ORDER EQUATIONS"

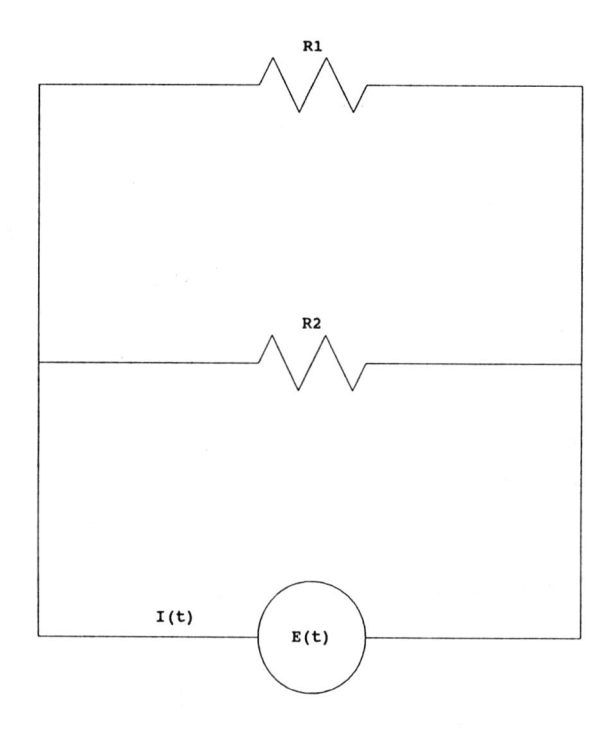

Figure 6.17 The circuit with two resistors in parallel

The Period of a Pendulum

Example 6.3 : Simple Frictionless Pendulum, Linearized

The motion of linearized pendulum is modeled

$$\frac{d^2x}{dt^2} + \lambda^2 x = 0. \tag{6.33}$$

if the angle x is near the rest position We have seen this differential equation many times, with its general solution as

$$x(t) = C_1 \cos \lambda t + C_2 \sin \lambda t, \tag{6.34}$$

where C_1 and C_2 are arbitrary constants. If the pendulum is released from rest, so that

$$x(0) = x_0, \qquad \frac{dx}{dt}(0) = 0,$$

its subsequent motion is

$$x(t) = x_0 \cos \lambda t.$$

The period T of the pendulum is the smallest time it takes to return to its maximum displacement of x_0. Thus, the period is the smallest T for which $x(T) = x_0$, so T must satisfy $x_0 = x_0 \cos \lambda T$. Solving for the smallest $T > 0$, we find

$$T = \frac{2\pi}{\lambda} = 2\pi \sqrt{\frac{h}{g}}. \tag{6.35}$$

Notice that the period is independent of the mass and the initial angle and that it varies as the square root of the length of the pendulum. The validity of (6.35) can be checked by simple experiments (see Exercise 1 on page 223) of this manual.

ADDITIONAL MATERIALS

We can obtain the period T in a different way, without solving (6.33), as follows. Multiplying (6.33) by dx/dt, gives

$$\frac{dx}{dt}\frac{d^2x}{dt^2} + \lambda^2 x \frac{dx}{dt} = 0,$$

which can be written

$$\frac{1}{2}\left[\left(\frac{dx}{dt}\right)^2 + \lambda^2 x^2\right]' = 0.$$

Integration gives

$$\left(\frac{dx}{dt}\right)^2 + \lambda^2 x^2 = C, \tag{6.36}$$

where C is an arbitrary constant. If the pendulum is released from rest, then this equation can be written

$$\left(\frac{dx}{dt}\right)^2 = \lambda^2 \left(x_0^2 - x^2\right). \tag{6.37}$$

This results in

$$\frac{dx}{dt} = \pm\lambda\sqrt{x_0^2 - x^2},$$

or

$$\frac{dt}{dx} = \frac{1}{\pm\lambda\sqrt{x_0^2 - x^2}}, \tag{6.38}$$

where $+$ applies to the case where x is increasing (moving to the right) and $-$ applies to the case where x is decreasing (moving to the left). Thus, one period would correspond to four different motions, from the initial (maximum) value to $x = 0$, from $x = 0$ to the minimum value of x, and then back again. The extreme values of x will occur when $dx/dt = 0$ — namely, $x(t) = \pm x_0$. Thus, the pendulum will swing an equal distance in both directions and will return to its original angle in one period. This period will be the sum of the time the pendulum takes to move from x_0 to 0, then from 0 to $-x_0$, then from $-x_0$ to 0, and finally back to x_0. The time the pendulum takes to move from x_0 to 0 is

$$\int_{x=x_0}^{x=0} dt,$$

with similar integrals for the other three time intervals. Thus the period T will be

$$T = \int_{x=x_0}^{x=0} dt + \int_{x=0}^{x=-x_0} dt + \int_{x=-x_0}^{x=0} dt + \int_{x=0}^{x=x_0} dt,$$

corresponding to the four successive motions, or

$$T = \int_{x=x_0}^{x=0} \frac{dt}{dx} dx + \int_{x=0}^{x=-x_0} \frac{dt}{dx} dx + \int_{x=-x_0}^{x=0} \frac{dt}{dx} dx + \int_{x=0}^{x=x_0} \frac{dt}{dx} dx.$$

However, we must make sure we substitute the correct dt/dx from (6.38) into the integrals in this last equation. Remember that $-$ corresponds to the angle decreasing (the first and second integrals), whereas $+$ corresponds to the angle increasing (the third and fourth integrals). We thus have

$$T = \int_{x_0}^{0} \frac{1}{-\lambda\sqrt{x_0^2 - x^2}} dx + \int_{0}^{-x_0} \frac{1}{-\lambda\sqrt{x_0^2 - x^2}} dx + \int_{-x_0}^{0} \frac{1}{\lambda\sqrt{x_0^2 - x^2}} dx + \int_{0}^{x_0} \frac{1}{\lambda\sqrt{x_0^2 - x^2}} dx. \tag{6.39}$$

If we make use of $\int_a^b f(x)\,dx = -\int_{-a}^{-b} f(-x)\,dx$ and $\int_a^b f(x)\,dx = -\int_b^a f(x)\,dx$, we see that each of the four integrals in (6.39) is equivalent to the first one, so the full period is four times the time it takes for the pendulum to swing from $x = x_0$ to $x = 0$ (in complete agreement with our intuition). That is,

$$T = -4 \int_{x_0}^{0} \frac{1}{\lambda\sqrt{x_0^2 - x^2}} \, dx,$$

or

$$T = 4\sqrt{\frac{h}{g}} \int_{0}^{x_0} \frac{1}{\sqrt{x_0^2 - x^2}} \, dx. \qquad (6.40)$$

This last integral has two unusual features: it appears to depend on x_0 (the initial angle), and it is an improper integral (why?). We could check to see whether this integral actually depends on x_0 by computing T for different values of x_0 by a standard numerical integration technique, such as Simpson's rule. Unfortunately, such numerical techniques are not reliable when applied to improper integrals. One way to get around this problem is to use a change of variable that converts this improper integral into a proper integral. A typical change of variables for this type of integral would be

$$x = x_0 \sin u, \qquad (6.41)$$

in which case

$$T = 4\sqrt{\frac{h}{g}} \int_0^{x_0} \frac{1}{\sqrt{x_0^2 - x^2}} \, dx = 4\sqrt{\frac{h}{g}} \int_0^{\pi/2} \frac{1}{\sqrt{x_0^2 - x_0^2 \sin^2 u}} x_0 \cos u \, du = 4\sqrt{\frac{h}{g}} \int_0^{\pi/2} du. \qquad (6.42)$$

In this case the integral reduces to a proper integral, which we can evaluate directly without recourse to numerical integration. Also notice that the dependence on x_0 vanished, so in fact we find

$$T = 2\pi\sqrt{\frac{h}{g}},$$

in agreement with (6.35). □

Example 6.4 : Simple Frictionless Pendulum

We are unable to solve the nonlinear differential equation for the frictionless pendulum

$$\frac{d^2x}{dt^2} + \lambda^2 \sin x = 0,$$

where $\lambda = \sqrt{g/h}$, in terms of familiar functions. However, we can use it to obtain useful information by following steps similar to the linearized case. We multiply the differential equation by dx/dt to find

$$\frac{dx}{dt}\frac{d^2x}{dt^2} + \lambda^2 \sin x \frac{dx}{dt} = 0,$$

or,

$$\frac{1}{2}\left[\left(\frac{dx}{dt}\right)^2 + 2\lambda^2 \cos x\right]' = 0.$$

This may be integrated to yield

$$v^2(t) - 2\lambda^2 \cos x(t) = c, \qquad (6.43)$$

where $v = dx/dt$ and c is an arbitrary constant. This is the counterpart of our earlier model (6.36). Note that we found (6.36) by approximating $\sin x$ with x, for small x. Thus, for small values of x, (6.43) should be equivalent to (6.36). As stated (6.43) is not too revealing. To explore this further, we use the trigonometric identity

$$\cos x = 1 - 2\sin^2 \frac{x}{2}$$

to rewrite (6.43) in the form

$$v^2(t) + 4\lambda^2 \sin^2 \frac{x(t)}{2} = C,$$

where $C = c + 2\lambda^2$. With the initial conditions $x(0) = x_0$, $v(0) = 0$, this conservation of energy equation becomes

$$\left(\frac{dx}{dt}\right)^2 = 4\lambda^2 \left(\sin^2 \frac{x_0}{2} - \sin^2 \frac{x(t)}{2}\right). \tag{6.44}$$

In this form, the approximation $\sin x/2 \approx x/2$ reduces (6.44) to (6.37) as we expected.

To compute the period, we want the time the pendulum takes to return to the original angle x_0. Looking at the various orbits in the phase plane (Figure 6.41 of the text), we see that the only orbits for which the notion of period makes sense would be closed, ellipse-like orbits. (Why?) Thus, we require $-\pi < x_0 < \pi$.

From (6.44) we see that the velocity will be zero whenever $\sin^2(x_0/2) - \sin^2(x(t)/2) = 0$ — that is, when $x(t) = \pm x_0$. The counterpart of (6.38) is

$$\frac{dt}{dx} = \frac{1}{\pm 2\lambda\sqrt{\sin^2(x_0/2) - \sin^2(x/2)}},$$

where $+$ applies to the case in which x is increasing (moving to the right) and $-$ applies when x is decreasing (moving to the left). Thus, one period would correspond to four different motions, from the initial (maximum) value of x_0 to $x = 0$, from $x = 0$ to the minimum value of x, $-x_0$, and then back again. The pendulum will therefore swing an equal distance in both directions and will return to its original angle in one period. Using arguments similar to those used in the previous example, we see that this period will also be four times the time it takes for the pendulum to swing from $x = x_0$ to $x = 0$, that is,

$$T = 4\int_{x=x_0}^{x=0} dt = 4\int_{x_0}^{0} \frac{dt}{dx} dx = -4\int_{x_0}^{0} \frac{1}{2\lambda\sqrt{\sin^2(x_0/2) - \sin^2(x/2)}} dx.$$

In view of $\lambda = \sqrt{g/h}$, this last integral can be rewritten as

$$T = 2\sqrt{\frac{h}{g}} \int_0^{x_0} \frac{1}{\sqrt{k^2 - \sin^2(x/2)}} dx, \tag{6.45}$$

where

$$k = \sin \frac{x_0}{2}.$$

This integral is the counterpart of (6.40). Again the period appears to depend on x_0 and again we have an improper integral. We follow the pattern suggested by the linearized case and make a substitution similar to (6.41) to try to convert the improper integral into a proper one, namely,

$$\sin \frac{x(t)}{2} = k \sin u.$$

From this we see that

$$\frac{1}{2} \cos \frac{x(t)}{2} dx = k \cos u\, du,$$

or
$$dx = \frac{2k\cos u}{\cos x/2} du = \frac{2k\cos u}{\sqrt{1-\sin^2 x/2}} du = \frac{2k\cos u}{\sqrt{1-k^2\sin^2 u}} du.$$

Because $u = 0$ when $x = 0$, and $u = \pi/2$ when $x = x_0$, the integral in (6.45) can be written

$$\int_0^{x_0} \frac{1}{\sqrt{k^2 - \sin^2(x/2)}} dx = \int_0^{\pi/2} \frac{1}{\sqrt{k^2 - k^2\sin^2 u}} \frac{2k\cos u}{\sqrt{1-k^2\sin^2 u}} du = 2\int_0^{\pi/2} \frac{du}{\sqrt{1-k^2\sin^2 u}}.$$

Thus, (6.45) becomes

$$T = 4\sqrt{\frac{h}{g}} \int_0^{\pi/2} \frac{du}{\sqrt{1-k^2\sin^2 u}}, \qquad (6.46)$$

where

$$k = \sin\frac{x_0}{2}. \qquad (6.47)$$

This is a proper integral provided $k^2 < 1$, which is always satisfied for the orbits we are considering ($-\pi < x_0 < \pi$). There are two major differences between (6.46) and its counterpart (6.42) in the linearized case. First, T depends on the initial angle x_0 (through k), and second, the integral in (6.46) cannot be evaluated in terms of familiar functions.

We can always evaluate the integral numerically for given x_0 (and hence given k) by any of the standard numerical integration techniques. For example, using Simpson's rule, we can construct Table 6.5, which shows that the period depends on the initial angle x_0.

Table 6.5 Simpson's rule for period

x_0	T
$\pi/180$ (1°)	$6.2833\sqrt{h/g}$
$\pi/36$ (5°)	$6.2862\sqrt{h/g}$
$\pi/6$ (30°)	$6.3926\sqrt{h/g}$
$\pi/4$ (45°)	$6.5343\sqrt{h/g}$
$\pi/3$ (60°)	$6.7430\sqrt{h/g}$

The integral in (6.46) occurs in a number of different applications, and it is known as the COMPLETE ELLIPTIC INTEGRAL OF THE FIRST KIND. □

Exercises

1. Perform the following experiment. Take a piece of string and fix a weight on one end. Anchor the other end, and measure the length of the pendulum. Displace the pendulum through a small angle, release it from rest, and measure the time it takes to return to the initial angle.

    (a) Repeat this a few times with different initial angles. Is the resulting period independent of the small initial angle?

    (b) Repeat this a few times with different masses. Is the resulting period independent of the mass?

    (c) Repeat the experiment for pendulums of different lengths. Do your results agree with (6.35)?

2. Repeat the experiment described in Exercise 1 for larger initial angles. Show that the model is less reliable as the initial angle increases.

3. Show that if $-\pi < x_0 < \pi$, then

$$\int_0^{x_0} \frac{1}{\sqrt{\sin^2(x_0/2) - \sin^2(x(t)/2)}}\, dx > \int_0^{x_0} \frac{1}{\sqrt{1 - \sin^2(x(t)/2)}}\, dx.$$

Use this result to show that

$$\lim_{x_0 \to \pi} \int_0^{x_0} \frac{1}{\sqrt{\sin^2(x_0/2) - \sin^2(x(t)/2)}}\, dx = \infty.$$

Explain how this relates to (6.45).

4. Expand the integrand in (6.46) as a power series in $\sin^2 u$, and use the fact that

$$\int_0^{\pi/2} \sin^{2n} u\, du = \frac{1 \cdot 3 \cdot 5 \cdots (2n-1)}{2 \cdot 4 \cdot 6 \cdots 2n} \frac{\pi}{2}$$

to find

$$T = 2\pi \sqrt{\frac{h}{g}} \left[1 + \left(\frac{1}{2}\right)^2 k^2 + \left(\frac{1 \cdot 3}{2 \cdot 4}\right)^2 k^4 + \left(\frac{1 \cdot 3 \cdot 5}{2 \cdot 4 \cdot 6}\right)^3 k^6 + \cdots \right],$$

where $k = \sin(x_0/2)$. Use this result with $x_0 = \pi/180$ [so that $k = \sin(\pi/360)$] to estimate T. Compare your result with the linearized period of $T = 2\pi\sqrt{h/g} \approx 6.2832$. As x_0 increases, do you expect T to get closer to or farther from the linearized period of $T = 2\pi\sqrt{h/g}$? As $x_0 \to \pi$, what do you expect to happen to T? Is it possible for the same pendulum to have every period from 0 to ∞, just by adjusting the initial angle?

5. A SECONDS PENDULUM CLOCK is one that ticks every second at the end of each swing of the pendulum, so it has a period of 2 seconds. We want to construct such a frictionless clock.

   (a) We decide to use the linearized model for the pendulum. How long should we make the pendulum, if $g = 32.2$ feet/sec^2? What has this to do with the size of grandfather clocks?

   (b) We build the clock according to the results from part (a) and start it from rest at an initial angle of $10°$. We know that the correct frictionless model is not the linearized model but is governed by $x'' + \lambda^2 \sin x = 0$. Will our clock run fast or slow? Over a 24-hour period, how inaccurate is our clock? How often should we restart the clock so it is never more than a minute wrong?

   (c) I have a pendulum clock that has a pendulum 9 inches long. At what angle should I start it from rest for it to tick every second in the nonlinear case?

6. The ELLIPTIC INTEGRAL OF THE FIRST KIND is defined by

$$F(z, k) = \int_0^z \frac{du}{\sqrt{1 - k^2 \sin^2 u}},$$

for $k^2 < 1$. Prove that

$$\int_0^{\pi/2} \frac{dx}{\sqrt{\sin x}} = \sqrt{2} F\left(\frac{\pi}{2}, \frac{1}{\sqrt{2}}\right).$$

(Hint: Let $x = \pi/2 - y$ and then let $\cos y = \cos^2 u$.)

7. The ELLIPTIC INTEGRAL OF THE SECOND KIND is defined by

$$E(z, k) = \int_0^z \sqrt{1 - k^2 \sin^2 u}\, du,$$

for $k^2 < 1$.

(a) Show that the perimeter of the ellipse $x^2/a^2 + y^2/b^2 = 1$, $(a > b)$ in the first quadrant, is

$$\int_0^a \sqrt{\frac{a^2 - e^2 x^2}{a^2 - x^2}}\, dx,$$

where $e = \sqrt{a^2 - b^2}/a$. By letting $x = a \sin u$, show that the perimeter of an ellipse is $4a E(\pi/2, e)$.

(b) What is the length of $\sin x$, from $x = 0$ to $x = 2\pi$?

8. Consider the differential equation

$$\left(\frac{dx}{dt}\right)^2 = (1 - x^2)(1 - k^2 x^2),$$

where $k^2 < 1$. Prove that this differential equation has periodic solutions with period

$$4 \int_0^1 \frac{1}{\sqrt{(1 - x^2)(1 - k^2 x^2)}}\, dx.$$

7. IDEAS FOR CHAPTER 7 — "SECOND ORDER LINEAR DIFFERENTIAL EQUATIONS WITH FORCING FUNCTIONS"

7.1 Page by Page Comments

Chapter 7 requires 3 to 4 class meetings of 50 minutes each.

Page 279. Equation (7.2), $x'' + x' = e^t$, could also be solved by the substitution $y = x'$, giving the first order linear equation $y' + y = e^t$.

Page 291. Students often miss the point regarding which part of a trial solution to multiply by t when the forcing function contains a sum, one term of which satisfies the associated homogeneous differential equation. It might be worth working an example in class to reinforce the proper technique. We usually give them a quiz asking students for the proper form for the trial solution of (for example) $x'' + x' - 6x = 2e^{2t} + 3e^{-2t} + \sin 2t + t^2$.

Page 293. Exercises 6(a) and (b) of Section 7.2 use results from Exercises 5(a) and (b). Exercises 6(c) and (d) of Section 7.2 use results from Exercises 5(g) and (i).

Page 294. Exercise 22 of Section 7.2 is worth assigning.

Page 301. Note the footnote.

Page 303. Example 7.17 could be replaced by BUMPY ROADS on page 229 of this manual.

Page 309. Exercise 11 of Section 7.3 is interesting, and the result surprises some students. It is solved in a different way in Exercise 12 of Section 6.5.

Page 310. Students enjoy Exercise 12 of Section 7.3.

Page 310. Table 7.2 is in TWIDDLE. It is called LIZARD.DTA and is in the subdirectory DD-07.

Page 311. Table 7.3 is in TWIDDLE. It is called LEWIS.DTA and is in the subdirectory DD-07.

Page 312. Table 7.4 is in TWIDDLE. It is called BEAM.DTA and is in the subdirectory DD-07.

Page 312. There are additional exercises for Section 7.3 on page 228 of this manual.

7.2 Additional Exercises

Section 7.3

Page 312. Add exercises after the current Exercise 14.

15. According to Robert Ehrlich[1] "You can find the resonant frequency of a hand-held spring from which a hanging weight is suspended, by gently shaking the top of the spring at different frequencies and seeing how the amplitude varies."

    (a) Explain how this experiment will find the resonant frequency.

    (b) Find the resonant frequency of a Slinky spring, by performing this experiment.

16. Consider the general damped spring-mass system with a sinusoidal forcing function $mx'' + bx' + kx = q\cos\omega t$.

    (a) Show that the general solution has the form $x(t) = C_1 e^{\alpha t}\sin\beta t + C_2 e^{\alpha t}\cos\beta t + A\sin\omega t + B\cos\omega t$ where C_1 and C_2 are arbitrary constants, $\alpha + i\beta = -b/(2m) + i\sqrt{4mk - b^2}/(2m)$, $A = b\omega q/[(\omega b)^2 + (k - \omega^2 m)^2]$, and $B = (k - \omega^2 m)q/[(\omega b)^2 + (k - \omega^2 m)^2]$.

    (b) Show that the maximum value of the amplitude of the steady state solution in part (a) is given when the frequency of the forcing function is given by $\omega = \sqrt{k/m - b^2/(2m^2)}$.

    (c) Show that the maximum value of the amplitude of the velocity of the steady state solution in part (a) is given by $\omega = \sqrt{k/m}$.

17. Consider a vertical slender metal rod, embedded in a solid base, with a mass at the free end. This rod will remain straight when the mass is very small. However, for larger masses, this equilibrium may be unstable, where a small displacement from equilibrium will cause the rod-mass combination to move to a new equilibrium point. If x denotes the displacement from equilibrium and t the distance along the rod, then the shape of the rod $x = x(t)$ is governed (for small values of x) by the differential equation $EIx'' = mg(x_0 - x) + (L - t)f$, where EI is the flexural rigidity, m is the mass at the end of the rod, g is the gravitational constant, x_0 is the displacement at the end of the rod, L is the length the rod, and f is the horizontal force on the mass. One end of the rod is like a built-in beam, so appropriate boundary conditions are $x(0) = x'(0) = 0$.

    (a) Find the solution of this initial value problem.

    (b) For a displacement of x_0, compute the resultant force f on the mass.

    (c) The critical mass for this rod is defined as the mass for which the force in part (b) is zero. Find the value of this critical mass.

    (d) Note that for this critical mass there are three equilibrium positions, two of them stable and one of them unstable. Explain why this is so.

18. **Projectiles.** See the section called PROJECTILES on page 231 of this manual.

---

[1] "Turning the World Inside Out" by R. Ehrlich, Princeton University Press, 1990, page 93.

7.3 Additional Materials

Bumpy Roads

Dirt roads frequently develop bumps or corrugations. A common driving practice when driving on such roads is to go as fast as possible. This reduces the up-and-down motion of a vehicle and is learned by driving over bumpy roads. We wish to examine this phenomenon.[2]

The vertical motion of a stationary vehicle can be modeled by the differential equation

$$mx'' + bx' + kx = 0,$$

where m, b, and k are positive constants, and $x(t)$ is the distance from the equilibrium position — the road. If we think of the vehicle moving with constant positive horizontal velocity ω from the viewpoint of someone inside the car, then this equation will also be valid if the road is level. However, if the road is not level then there will be a forcing function corresponding to the shape of the road. If we assume that the bumpy road is approximated by a *sine* function, with amplitude a and distance between the bumps c, then

$$f(t) = a\sin\left(\frac{2\pi\omega t}{c}\right),$$

and the differential equation becomes

$$mx'' + bx' + kx = a\sin\left(\frac{2\pi\omega t}{c}\right).$$

where m, b, k, a, c, and ω are positive constants. Notice that, if the road is level ($a = 0$), or the vehicle is not moving ($\omega = 0$), then $f(t) = 0$.

For illustrative purposes, we use the values of $b/m = 2$, $k/m = 17$, $a/m = 1$, and $2\pi/c = 1$ giving

$$x'' + 2x' + 17x = \sin\omega t. \tag{7.1}$$

We are interested in how the amplitude of the vertical distance, x, changes as we vary the horizontal velocity, ω.

We know that our solution will consist of two parts. The first part is the solution of the homogeneous differential equation associated with (7.1). This leads to the characteristic equation $r^2 + 2r + 17 = 0$, with solution $r = -1 \pm 4i$, so $x_h(t) = C_1 e^{-t}\sin 4t + C_2 e^{-t}\cos 4t$.

The particular solution will have the form $x_p(t) = A\sin\omega t + B\cos\omega t$. Substituting this form for x_p into (7.1) gives

$$-\omega^2(A\sin\omega t + B\cos\omega t) + 2\omega(A\cos\omega t - B\sin\omega t) + 17(A\sin\omega t + B\cos\omega t) = \sin\omega t.$$

We equate coefficients of like terms in this equation to obtain

$$\begin{aligned} -\omega^2 A - 2\omega B + 17A &= 1, \\ -\omega^2 B + 2\omega A + 17B &= 0. \end{aligned}$$

From the second equation we have that $A = (\omega^2 - 17)B/(2\omega)$. Using this fact in the first equation yields $[-(17-\omega^2)^2/(2\omega) - 2\omega]B = 1$, giving A and B as

$$\begin{aligned} A &= (17 - \omega^2)/[4\omega^2 + (17 - \omega^2)^2], \\ B &= -2\omega/[4\omega^2 + (17 - \omega^2)^2]. \end{aligned} \tag{7.2}$$

---

[2] For a more extensive treatment of this topic, see "Mathematics in Action : Modelling the Real World Using Mathematics" by R. Beare, Chartwell-Bratt, 1997, pages 334 through 358.

Thus, our particular solution is given by

$$x_p(t) = A\sin\omega t + B\cos\omega t,$$

and our general solution of (7.1) is

$$x(t) = C_1 e^{-t}\sin 4t + C_2 e^{-t}\cos 4t + A\sin\omega t + B\cos\omega t, \tag{7.3}$$

where C_1 and C_2 are arbitrary constants, and A and B are given in (7.2). The values of C_1 and C_2 are specified once initial conditions are given, but regardless of their values, the first two terms in (7.3) become very small as t increases. Because this is so, these terms constitute the transient solution, and the remaining terms that do not decay give the steady state solution.

We now change the form of this steady state solution — denoted by $x_{ss}(t)$ — by using the result that (see Exercise 6 on page 249)

$$x_{ss}(t) = A\sin\omega t + B\cos\omega t = \sqrt{A^2+B^2}\cos(\omega t + \phi), \tag{7.4}$$

where

$$\cos\phi = \frac{B}{\sqrt{A^2+B^2}}, \qquad \sin\phi = \frac{-A}{\sqrt{A^2+B^2}}.$$

Notice that the amplitude of our steady state solution, $\sqrt{A^2+B^2}$, depends on the value of the forcing frequency and, using (7.2), is given by

$$\sqrt{A^2+B^2} = \frac{1}{\sqrt{4\omega^2 + (17-\omega^2)^2}}. \tag{7.5}$$

The graph of this amplitude as a function of the forcing frequency, ω, is given in Figure 7.1, and shows that the amplitude has a maximum value. To find this maximum value we take the derivative of the amplitude with respect to ω and set the result to zero. Performing this calculation shows $\omega = \sqrt{15}$ is the forcing frequency that gives only the maximum value of the steady state amplitude (see Exercise 9 on page 309). Furthermore, as $\omega \to \infty$, the amplitude of the steady state solution $\sqrt{A^2+B^2} \to 0$. Thus, as the vehicle's speed, ω, approaches $\sqrt{15}$ the oscillations increase in magnitude, but after that they decrease to zero. Thus, if $\omega > \sqrt{15}$ the greater the velocity, ω, the smaller the amplitude, so driving faster is better than driving slower.

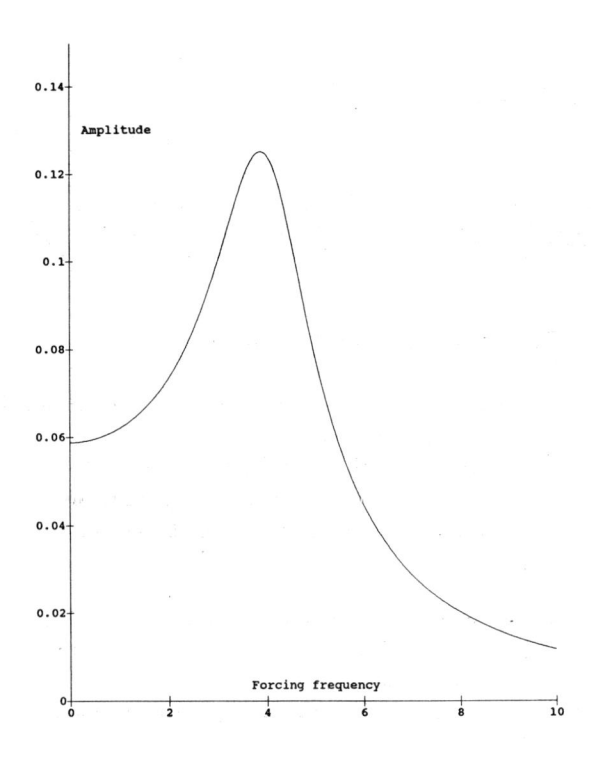

Figure 7.1 Amplitude of the steady state solution versus the forcing frequency

Projectiles

In earlier chapters we dealt with the trajectories of particles in one-dimension. Here we generalize this to two dimensions where a typical position of a particle of mass m has coordinates $(x(t), y(t))$ in the xy-plane. Two typical applications of this are to firing projectiles[3] and to skydiving.

No Drag (No Air Resistance)

If gravity is the only force acting on the particle — and it acts downward, parallel to the y-axis — then the particle's trajectory is governed by the two differential equations

$$mx'' = 0$$

and

$$my'' = -mg.$$

These equations are independent of each other and are solved separately giving

$$x'(t) = u_0,$$

$$x(t) = u_0 t + x_0,$$

---

[3] "The Mathematics of Projectiles in Sport" by N. de Mestre, Australian Mathematical Society Lecture Series 6, Cambridge University Press, 1990.

ADDITIONAL MATERIALS

$$y'(t) = -gt + v_0,$$

and

$$y(t) = -\frac{1}{2}gt^2 + v_0 t + y_0,$$

where u_0, x_0, v_0, and y_0 are constants. There is a simple interpretation for these constants. They correspond to the position (x_0, y_0) and velocity (u_0, v_0) of the particle at time $t = 0$.

Comments about this motion:

- Notice that neither $x(t)$ nor $y(t)$ depend on m. This means that if we take two particles with different masses at the same initial height y_0 and simultaneously release them from rest (so the initial velocities are $u_0 = v_0 = 0$) then they will fall the same distance y in the same time. A **simple experiment** to demonstrate this consists of taking two tennis balls, one loaded with lead shot, and dropping them from the same height. They hit the floor at the same time.[4]

- Notice that $y(t)$ does not depend on x_0 or u_0. This means that if we take two particles at the same initial height y_0 and simultaneously release them with different initial horizontal velocities u_0 then they will fall the same distance y in the same time. A **simple experiment** to demonstrate this consists of taking two large marbles and holding them side-by-side between thumb and first finger of one hand. Make sure that the balls are the same height above the ground. Now strike one of the balls horizontally with the other hand. If done correctly, the other marble drops vertically. They both have the same initial vertical velocities — namely, zero — but different initial horizontal velocities. They hit the floor at the same time.[5] A consequence of this is if a bullet is dropped at the same time that a rifle is fired horizontally then the two bullets will strike the earth at the same time.

The speed of the particle $w(t) = \sqrt{(x')^2 + (y')^2}$ at any time t is

$$w(t) = \sqrt{u_0^2 + (-gt + v_0)^2}.$$

The initial speed is $w_0 = w(0) = \sqrt{u_0^2 + v_0^2}$. There is no terminal speed because $\lim_{t \to \infty} w(t) = \infty$.

If we define the angle α by $u_0 = w_0 \cos \alpha$ and $v_0 = w_0 \sin \alpha$, where $-\pi/2 \leq \alpha \leq \pi/2$ — which means that the particle initially has a speed of w_0 in the direction α — then we can write

$$x(t) = (w_0 \cos \alpha) t + x_0,$$

$$y(t) = (w_0 \sin \alpha) t - \frac{1}{2} g t^2 + y_0,$$

and

$$w(t) = \sqrt{w_0^2 - 2gt w_0 \sin \alpha + g^2 t^2}.$$

---

[4] "Doing physics" by J. Bozovsky, The Physics Teacher, December 1983, pages 611 through 612.

[5] "A demonstration to show the independence of horizontal and vertical motion" by J. Hoskins and L. Lonney, The Physics Teacher, November 1983, page 525.

It is sometimes convenient to consider the situation where initially the particle is at the origin of the coordinate system. We can do this by using a translation from (x,y) to (X,Y) which places the particle's initial position at the origin of the XY-plane — namely,

$$X = x - x_0,$$

and

$$Y = y - y_0.$$

This gives the trajectory

$$X(t) = (w_0 \cos \alpha) t, \qquad (7.6)$$

$$Y(t) = (w_0 \sin \alpha) t - \frac{1}{2} g t^2. \qquad (7.7)$$

A natural question to ask is what is the shape of the trajectory followed by the particle in the XY-plane assuming that $u_0 \neq 0$. (What happens if $u_0 = 0$?) This we can obtain by eliminating t between the $X(t)$ and $Y(t)$ equations. Thus,

$$t = \frac{X}{w_0 \cos \alpha},$$

and so

$$Y = (\tan \alpha) X - \frac{1}{2} g \left(\frac{1}{w_0 \cos \alpha} \right)^2 X^2.$$

Thus, Y is a quadratic function of X and so the trajectory is a parabola opening downwards in the XY-plane. Typical trajectories are shown in Figure 7.2 for $\alpha > 0$, $\alpha = 0$, and $\alpha < 0$ — namely, $g = 32$, $w_0 = 200$, and $\alpha > 45^\circ$, $\alpha = 0^\circ$, and $\alpha < -45^\circ$.

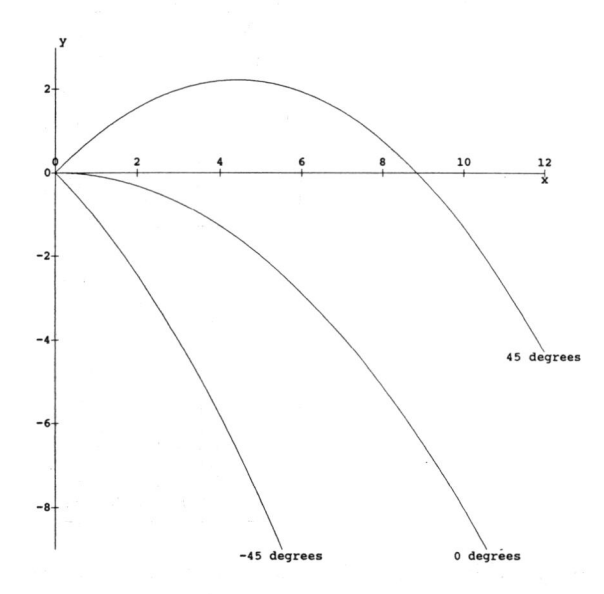

Figure 7.2 Trajectories with no drag for $\alpha > 0$, $\alpha = 0$, and $\alpha < 0$

The maximum height of the particle will occur when $dY/dX = 0$ and, because $dY/dX = Y'/X'$, this occurs when $Y' = 0$, that is

$$t = \frac{w_0}{g} \sin \alpha.$$

For this to occur for $t > 0$ we need $0 < \alpha \leq \pi/2$, which agrees with intuition, and at this time the particle is at
$$X = \frac{w_0^2}{2g} \sin 2\alpha,$$
when the maximum height H is
$$H = \frac{w_0^2}{2g} \sin^2 \alpha. \tag{7.8}$$

We notice that $Y(t) = 0$ when $t = 0$ and when $t = t_h$, where
$$t_h = \frac{2w_0}{g} \sin \alpha. \tag{7.9}$$

The time t_h is when the particle hits the X-axis — if the particle were a football we would call this the **hang time**. If we want $t_h > 0$, this requires that $0 < \alpha \leq \pi/2$ which also agrees with intuition. Notice it is twice the time taken to attain the maximum height, which is to be expected on the grounds of symmetry. In this case the particle hits the X-axis at a distance
$$R = \frac{w_0^2}{g} \sin 2\alpha \tag{7.10}$$
from where is was released, called the **range**.

Table 7.1 and Figure 7.3 show the results of an experiment recording the range R as a function of the initial angle of elevation α of a projectile.[6] Is this consistent with (7.10)? [Table 7.1 is already in TWIDDLE. It is called RANGE.DTA and is in the subdirectory DD-07.]

Table 7.1 The range R as a function of the elevation α

Angle (degrees)	Range (cms)
10.2	17.60
16.8	26.95
22.7	33.28
30.0	42.35
32.2	45.24
35.1	47.44
40.2	48.95
45.1	49.78
48.9	48.95
52.0	47.30
57.8	42.63
65.1	35.48
68.0	31.63
76.8	20.21

For a given positive range R and an initial velocity w_0 the initial angle α can be computed from (7.10). Apart for $\alpha = \pi/4$, this always has two solutions for $0 < \alpha \leq \pi/2$ — namely
$$\alpha_1 = \frac{1}{2} \arcsin\left(\frac{Rg}{w_0^2}\right)$$
and $\alpha_2 = \pi/2 - \alpha_1$. Thus, there are two different initial angles that have identical ranges for the same w_0 but different trajectories. The angle α_1 is less than $\pi/4$ while α_2 is greater. We

---

[6] "A study of the trajectories of projectiles" by A. Ruari Grant, Physics Education, 25, 1990, pages 288 through 292. Figure 4, page 289.

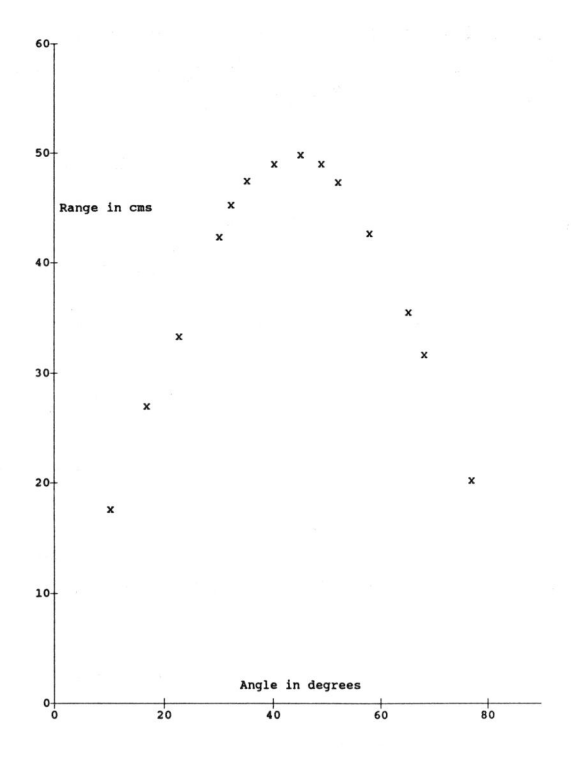

Figure 7.3 The range R as a function of the elevation α

can calculate the hang times t_1 and t_2 for each of these trajectories from (7.9), and we find

$$t_1 = \frac{2w_0}{g}\sin\alpha_1 \qquad t_2 = \frac{2w_0}{g}\sin\left(\frac{\pi}{2} - \alpha_1\right) = \frac{2w_0}{g}\cos\alpha_1.$$

The difference in flight time is

$$t_2 - t_1 = \frac{2w_0}{g}\left(\cos\alpha_1 - \sin\alpha_1\right).$$

This means that if we release two particles — say, snowballs — with the same w_0, one at an initial angle of α_2 and the other at an angle α_1 at time $t_2 - t_1$ later, they will both arrive at the range at the same time. Imagine you were throwing snowballs with an initial speed w_0 of 72 ft/sec at an person 100 feet away. In this case $\alpha_1 \approx 19°$ and $\alpha_2 \approx 71°$, and $t_2 - t_1 \approx 2.75$ sec. Thus throwing a high snowball followed by a low one 2.75 seconds later will cause confusion![7]

The particle which was launched with an initial angle of α will land at an angle of $\pi - \alpha$. One way to confirm this is to compute $dY/dX = Y'/X'$ at $t = (2w_0/g)\sin\alpha$, which will give the slope of the trajectory at the time of impact. In this way we find

$$\frac{dY}{dX} = -\tan\alpha.$$

If w_0 (the initial speed) is fixed then changing α will change the range. The maximum range occurs when $\sin 2\alpha = 1$, that is, when $\alpha = \pi/4$.

If w_0 (the initial speed) is fixed then changing α will change the height of the projectile. The maximum height occurs when $\sin\alpha = 1$, that is, when $\alpha = \pi/2$. This occurs when the particle is propelled straight up — not a recommended practice for firing projectiles.

---

[7] "Snowball Fighting: A Study in Projectile Motion" by P. N. Henriksen, The Physics Teacher, January 1975, page 43.

ADDITIONAL MATERIALS

If we observe the trajectory of a particle, there are two natural measurements we can make — the range R and the hang time t_h. If we measure these two parameters, can we identify the trajectory? From (7.9) we have

$$w_0 \sin \alpha = \frac{1}{2} g t_h, \tag{7.11}$$

which, when used in (7.10), gives

$$w_0 \cos \alpha = \frac{R}{t_h}. \tag{7.12}$$

If we substitute these into (7.6) and (7.7) we find

$$X(t) = \frac{R}{t_h} t, \tag{7.13}$$

$$Y(t) = \frac{1}{2} g (t_h - t) t, \tag{7.14}$$

so we can express the trajectory in terms of R and t_h.

It is sometimes useful to have the main features of a trajectory expressed in terms of R and t_h. From (7.11) and (7.12) we find that the initial speed is

$$w_0 = \sqrt{\left(\frac{1}{2} g t_h\right)^2 + \left(\frac{R}{t_h}\right)^2}$$

and the initial angle is

$$\alpha = \arctan\left(\frac{g t_h^2}{2R}\right).$$

From (7.9) and (7.8) we find that the maximum height H is

$$H = \frac{1}{8} g t_h^2. \tag{7.15}$$

Thus, a knowledge of the range and hang time will determine the trajectory completely.

Example 7.1 : Simple Experiment[8]

While watching a football game estimate and record the hang time and the range of passes and kicks.[9] This will usually require two people, one with a stop watch to record the hang time, and the other to estimate the range. For each of these trajectories determine the maximum height, the initial speed, and the initial angle of the ball. For example, a punt might have a hang time of 4.9 sec and a range of 25 yards. How high did the punt go? What was its initial speed and angle? □

Example 7.2 : Kicking a Football[10]

When kicking a football the kicker usually tries to maximize the range and to maximize the hang time. What initial angle α should the kicker use to accomplish this?

The range is maximized when $\sin 2\alpha$ is maximum — that is, when $\sin 2\alpha = 1$ — so $\alpha = \pi/4$. The hang time is maximized when $\sin \alpha$ is maximum — that is, when $\sin \alpha = 1$ — so $\alpha = \pi/2$. Thus, it is not possible to simultaneously maximize the range and hang time. This is sometimes called the "kicker's dilemma." □

---

[8] "Television, football, and physics: Experiments in kinematics" by A. A. Bartlett, The Physics Teacher, September 1984, pages 386 through 387.

[9] The range of a football is not necessarily the same as the distance quoted by announcers. Why?

[10] "The physics of kicking a football" by P. J. Brancazio, The Physics Teacher, **23**, 1985, pages 403 through 407.

Example 7.3 : Investigating Traffic Accidents[11]

When an investigating officer arrives at the scene of a traffic accident, one of the most important pieces of information is to estimate the speed of the vehicles involved. In a single car accident where a vehicle has left the road and crashed some distance below the road, the formula that investigators use to determine the speed w_0 at which the vehicle left the road at an angle α and fell a vertical distance d_v and a horizontal distance d_h is

$$w_0 \approx \frac{2.74 d_h}{\sqrt{d_h \tan \alpha - d_v}} \text{ mph.}$$

Here d_v and d_h are measurements (in feet) of the center of mass of the vehicle. (They are measured to where the vehicle landed, and not to where the vehicle eventually came to rest.) This formula is used when the vehicle is not traveling very fast and does not fall a great distance, so that air resistance can be neglected, and so there is no drag. Where does this formula come from?

From (7.6) and (7.7) we have these horizontal and vertical distances given by

$$d_h = (w_0 \cos \alpha) t,$$

and

$$d_v = (w_0 \sin \alpha) t - \frac{1}{2} g t^2,$$

where t is the time of impact. Eliminating t between these equations yields

$$d_v = d_h \tan \alpha - \frac{1}{2} g \left(\frac{d_h}{w_0 \cos \alpha} \right)^2.$$

Solving for w_0 gives

$$w_0 = \frac{d_h}{\cos \alpha} \sqrt{\frac{g}{2 (d_h \tan \alpha - d_v)}}.$$

This is the exact calculation for w_0. With $g = 32.2$ ft/sec^2 this is

$$w_0 = 4.01 \frac{d_h}{\cos \alpha \sqrt{d_h \tan \alpha - d_v}}$$

in ft/sec and

$$w_0 = \frac{2.74 d_h}{\cos \alpha \sqrt{d_h \tan \alpha - d_v}} \text{ mph.}$$

If α is small, so that $\cos \alpha \approx 1$, we have

$$w_0 \approx \frac{2.74 d_h}{\sqrt{d_h \tan \alpha - d_v}} \text{ mph.}$$

□

Example 7.4 : Simple Experiment[12]

Take a large coffee can and drill 5 holes of about 5 mm diameter equally spaced down one side of the can. Place the can near a sink and cover the holes. Fill the can with water. We are going to open the holes, keep the water level constant, and see which of the 5 streams travels the greatest horizontal distance. Before opening the holes, what does your intuition

---

[11] "Physics in accident investigations" by M. L. Brake, The Physics Teacher, January 1981, pages 26 through 29.

[12] "'Canned' Physics" by H. Kruglak, The Physics Teacher, **30**, 1992, pages 392 through 396, and "The Water Can Paradox" by L. G. Paldy, The Physics Teacher, **1**, 1963, page 126.

tell you? Now open the holes and keep the water level constant. Look at the 5 streams, and where they intersect the plane on which the can is standing. Which stream travels the greatest horizontal distance? Does the result of this experiment agree with your intuition? Surprised? We can explain this by thinking of the water stream as a succession of particles leaving a hole with a horizontal velocity due to the pressure of the water above it.

To do this, we let the can have height ℓ and consider one of the holes which is a distance h below the water level. We can find the water stream's trajectory by thinking of a particle leaving this hole with a horizontal velocity due to the head of water h. By Torricelli's law, this velocity is proportional to $\sqrt{h}$. Thus, we can model this situation in the following way.

A particle is fired horizontally from a height of $\ell - h$ above the ground with an initial horizontal velocity of $b\sqrt{h}$, where h, b, and ℓ are positive constants. Find the horizontal distance traveled when the particle hits the ground. What value of h makes this distance the greatest?

Here we have $u_0 = b\sqrt{h}$, $v_0 = 0$, $x_0 = 0$, $y_0 = \ell - h$, so

$$x(t) = b\sqrt{h}\, t$$

and

$$y(t) = -\frac{1}{2}gt^2 + \ell - h.$$

The time T when the particle hits the ground is determined from $y(T) = 0$ and is

$$T = \frac{1}{g}\sqrt{2(\ell - h)}.$$

The horizontal distance traveled in this time is

$$x(T) = \frac{b}{g}\sqrt{2(\ell - h)h}.$$

If we think of h as a parameter and plot $\sqrt{2(\ell - h)h}$ versus h, we see that this has a maximum when $h = \ell/2$. Thus, $x(T)$ is maximum when $h = \ell/2$. □

Now repeat the above experiment, but first place the coffee can on a raised platform. Does the result of this experiment agree with your intuition? Is it possible to adjust the height of the platform so that the water streams from middle and lowest holes travel the same horizontal distance? Use the above model to verify your conclusions. Based on this, write a paragraph explaining this phenomena — and in particular why it does not violate intuition — in language that a nonscientist could understand.

For applications to other areas, see

- Basketball. "Physics of basketball" by P. J. Brancazio, American Journal of Physics, **49**, 1981, pages 356 through 365. "Kinematics of the free throw in basketball" by A. Tan and G. Miller, American Journal of Physics, **49**, 1981, pages 542 through 544.

- Shot Put. "Maximizing the range of the shot put" by D. B. Lichtenberg and J. G. Wills, American Journal of Physics, **46**, 1978, pages 546 through 549.

- Shooting. "A puzzle in elementary ballistics" by O. A. Haugland, The Physics Teacher, April 1983, pages 246 through 248.

Linear Drag (Air Resistance Linear in Velocity)

Now let's include air resistance which is a force that opposes the motion. For slow velocities this force will be proportional to (x', y'), so our equations become

$$mx'' = -kx'$$

and

$$my'' = -mg - ky',$$

where k is a positive constant. Once more these equations are independent of each other and are solved separately giving

$$x(t) = x_0 + \frac{u_0}{\lambda}\left(1 - e^{-\lambda t}\right)$$

and

$$y(t) = y_0 - \frac{g}{\lambda}t + \left(\frac{v_0}{\lambda} + \frac{g}{\lambda^2}\right)\left(1 - e^{-\lambda t}\right),$$

where $\lambda = k/m$.

We can see whether these are reasonable answers by checking their behavior for small λ. Expanding $e^{-\lambda t}$ in powers of λt, namely,

$$e^{-\lambda t} = 1 - \lambda t + \frac{1}{2!}(\lambda t)^2 - \frac{1}{3!}(\lambda t)^3 + \cdots,$$

and collecting like terms, we find that x and y can be written

$$x(t) = x_0 + u_0 t - \lambda t^2 u_0 \left(\frac{1}{2!} - \frac{1}{3!}\lambda t + \cdots\right)$$

and

$$y(t) = y_0 + v_0 t - \lambda t^2 (v_0 + g\lambda) \left(\frac{1}{2!} - \frac{1}{3!}\lambda t + \cdots\right).$$

This reduces to the no drag case when $\lambda = 0$. Thus, these are reasonable equations.

The associated velocities are

$$x'(t) = u_0 e^{-\lambda t}$$

and

$$y'(t) = -\frac{g}{\lambda} + \left(v_0 + \frac{g}{\lambda}\right) e^{-\lambda t}.$$

As $t \to \infty$ we see that $x'(t) \to 0$ and $y'(t) \to -g/\lambda$. Also, because $x''(t) = -\lambda u_0 e^{-\lambda t} < 0$ and $y''(t) = -\lambda(v_0 + g/\lambda) e^{-\lambda t} < 0$ both x' and y' decrease with time. Thus as time increases the horizontal component of the velocity decreases to zero, while the vertical component decreases towards a terminal velocity of $-W$, where $W = g/\lambda = gm/k$. (Notice that we can obtain the same conclusion directly from the differential equations by looking for equilibrium solutions.) We can express $y(t)$ in terms of W as follows:

$$y(t) = y_0 - Wt + \frac{1}{\lambda}(v_0 + W)\left(1 - e^{-\lambda t}\right).$$

The speed of the particle $w(t) = \sqrt{(x')^2 + (y')^2}$ at any time t is

$$w(t) = \sqrt{u_0^2 e^{-2\lambda t} + [W - (v_0 + W)e^{-\lambda t}]^2}.$$

Notice that $\lim_{t \to \infty} w(t) = W$, so W is the terminal speed.

If we define the angle α by $u_0 = w_0 \cos\alpha$ and $v_0 = w_0 \sin\alpha$, where $-\pi/2 \leq \alpha \leq \pi/2$ and $w_0 = w(0)$ — which means that the particle initially has a speed of w_0 in the direction α — then we can write

$$x(t) = x_0 + \frac{w_0 \cos\alpha}{\lambda}\left(1 - e^{-\lambda t}\right),$$

and
$$y(t) = y_0 - Wt + \frac{1}{\lambda}(w_0 \sin\alpha + W)(1 - e^{-\lambda t}).$$

It is sometimes convenient to consider the situation where initially the particle is at the origin of the coordinate system. We can do this by using a translation from (x, y) to (X, Y) which places the particle's initial position at the origin of the XY-plane — namely,
$$X = x - x_0,$$
and
$$Y = y - y_0.$$
This gives the trajectory
$$X(t) = \frac{w_0 \cos\alpha}{\lambda}(1 - e^{-\lambda t}),$$
$$Y(t) = -Wt + \frac{1}{\lambda}(w_0 \sin\alpha + W)(1 - e^{-\lambda t}).$$

As $t \to \infty$ we see that $X(t) \to (w_0 \cos\alpha)/\lambda$ and $Y(t) \to -\infty$. Thus, this model predicts that there is a limit to the horizontal distance the particle can travel.

A natural question to ask is what is the shape of the trajectory followed by the particle in the XY-plane assuming that $u_0 \neq 0$. (What happens if $u_0 = 0$?) This we can obtain by eliminating t between the $X(t)$ and $Y(t)$ equations. Thus, if $u_0 \neq 0$ we have
$$(1 - e^{-\lambda t}) = \frac{\lambda X}{w_0 \cos\alpha},$$
from which we find
$$t = -\frac{1}{\lambda}\ln\left(1 - \frac{\lambda X}{w_0 \cos\alpha}\right),$$
and so
$$Y = \frac{W}{\lambda}\ln\left(1 - \frac{\lambda X}{w_0 \cos\alpha}\right) + \left(\tan\alpha + \frac{W}{w_0 \cos\alpha}\right)X.$$

Remembering that $W = g/\lambda$, and that the value of g is known, we see that the previous equation has three parameters in it, namely λ, w_0, and α.

To compare this to the case of no drag we rewrite it in the form
$$Y = \frac{W}{\lambda}\left[\ln\left(1 - \frac{\lambda X}{w_0 \cos\alpha}\right) + \frac{\lambda X}{w_0 \cos\alpha}\right] + (\tan\alpha)X,$$
and expand the quantity in brackets using the Taylor series expansion for $\ln(1 + z)$, namely,
$$\ln(1 + z) = z - \frac{1}{2}z^2 + \frac{1}{3}z^3 + \cdots,$$
with $z = -\lambda X/(w_0 \cos\alpha)$. We thus find
$$Y = \frac{W}{\lambda}\left[-\frac{1}{2}\left(\frac{\lambda X}{w_0 \cos\alpha}\right)^2 - \frac{1}{3}\left(\frac{\lambda X}{w_0 \cos\alpha}\right)^3 + \cdots\right] + (\tan\alpha)X,$$
or
$$Y = -\frac{1}{2}g\left(\frac{1}{w_0 \cos\alpha}\right)^2 X^2 - \frac{1}{3}g\lambda\left(\frac{1}{w_0 \cos\alpha}\right)^3 X^3 + \cdots + (\tan\alpha)X.$$

In this form we can see that the air resistance comes into play at the X^3 and higher terms.

Exercise: There is another way of obtaining this last equation directly from the differential equations $X'' = -\lambda X$, $Y'' = -g - \lambda Y'$, subject to the initial conditions $X(0) =$

$Y(0) = 0$, $X'(0) = w_0 \cos \alpha$, $Y'(0) = w_0 \tan \alpha$, by assuming that Y has a power series expansion in X, that is, $Y(t) = \sum_{n=0}^{\infty} a_n X^n(t)$. Show that $Y'(t) = \sum_{n=0}^{\infty} a_n n X^{n-1} X'$, $Y''(t) = \sum_{n=0}^{\infty} a_n n(n-1) X^{n-2} (X')^2 + \sum_{n=0}^{\infty} a_n n X^{n-1} X''$, and $Y'''(t) = \sum_{n=0}^{\infty} a_n n(n-1)(n-2) X^{n-3} (X')^3 + \sum_{n=0}^{\infty} 3 a_n n(n-1) X^{n-2} X' X'' + \sum_{n=0}^{\infty} a_n n X^{n-1} X'''$. From these, the differential equations, and the initial conditions, show that $a_0 = 0$, $a_1 = \tan \alpha$, $a_2 = -g/(2w_0^2 \cos^2 \alpha)$, and $a_3 = -g\lambda/(3w_0^3 \cos^3 \alpha)$.

The maximum height of the particle will occur when $dY/dX = 0$ and, because $dY/dX = Y'/X'$, this occurs when $Y' = 0$, that is

$$e^{-\lambda t} = \frac{W}{w_0 \sin \alpha + W},$$

or

$$t = \frac{1}{\lambda} \ln \left(1 + \frac{w_0}{W} \sin \alpha \right).$$

For this to occur for $t > 0$ we need $0 < \alpha \leq \pi/2$, which agrees with intuition, and at this time the particle is at

$$X = \frac{w_0^2 \sin \alpha \cos \alpha}{\lambda (w_0 \sin \alpha + W)},$$

when the maximum height H is

$$H = -W \frac{1}{\lambda} \ln \left(1 + \frac{w_0}{W} \sin \alpha \right) + \frac{1}{\lambda} w_0 \sin \alpha.$$

We notice that $Y(t) = 0$ when $t = 0$ and $t = t_h$ where

$$W t_h = \frac{1}{\lambda} (w_0 \sin \alpha + W) \left(1 - e^{-\lambda t_h}\right)$$

or

$$1 - e^{-\lambda t_h} = \frac{gW}{w_0 \sin \alpha + W} t_h.$$

Solving this for t_h by analytically techniques is beyond us. However, for given constants λ, g, w_0, α, and W, we can solve this either numerically or graphically by plotting the line $\{gW/(w_0 \sin \alpha + W)\} T$ and the curve $1 - e^{-T}$ as functions of T to find the point of intersection.[13] There are no values of T if $gW/(w_0 \sin \alpha + W) > 1$. (Why?)

Having obtained a numerical value for the time of flight t, the range R is computed from

$$R = \frac{gW w_0 \cos \alpha}{w_0 \sin \alpha + W} t_h,$$

while the angle of impact is determined from

$$\frac{dY}{dX} = \frac{-W + (w_0 \sin \alpha + W) e^{-\lambda t_h}}{w_0 \cos \alpha e^{-\lambda t_h}}.$$

Example 7.5 : The Flight of a Bullet with Drag

We illustrate these results by analyzing the flight of a bullet fired from a .17 Remington rifle. Table 7.2 shows the height of the bullet in inches above the muzzle of the rifle as a function of the horizontal distance in yards from the muzzle.[14]

---

[13] "The Mathematics of Projectiles in Sport" by N. de Mestre, Australian Mathematical Society Lecture Series 6, Cambridge University Press, 1990, page 29.

[14] Data from "Armed and Dangerous", by M. Newton, Writer's Digest Books, 1990, page 174. (This book also contains similar data for many other rifles.)

Table 7.2 Height of bullet as a function of horizontal distance

Distance (yards)	Height (inches)
0	0.0
100	2.1
150	2.5
200	1.9
250	0.0
300	-3.4
400	-17.0
500	-44.3

According to the previous model, X and Y should be related by the equation

$$Y = \frac{W}{\lambda}\left[\ln\left(1 - \frac{\lambda X}{w_0 \cos\alpha}\right) + \frac{\lambda X}{w_0 \cos\alpha}\right] + (\tan\alpha)X.$$

Using the program Kalkulator to perform a nonlinear regression, we find $\lambda = 1.8927$, $w_0 = 46478$, and $\alpha = 0.001057$, after converting the horizontal measurements to inches. This predicts a muzzle velocity of 3873 feet/sec and that the bullet will travel a maximum horizontal distance of 2046 feet. These are plausible predictions. Figure 7.4 shows the bullet's trajectory together with the theoretical curve.

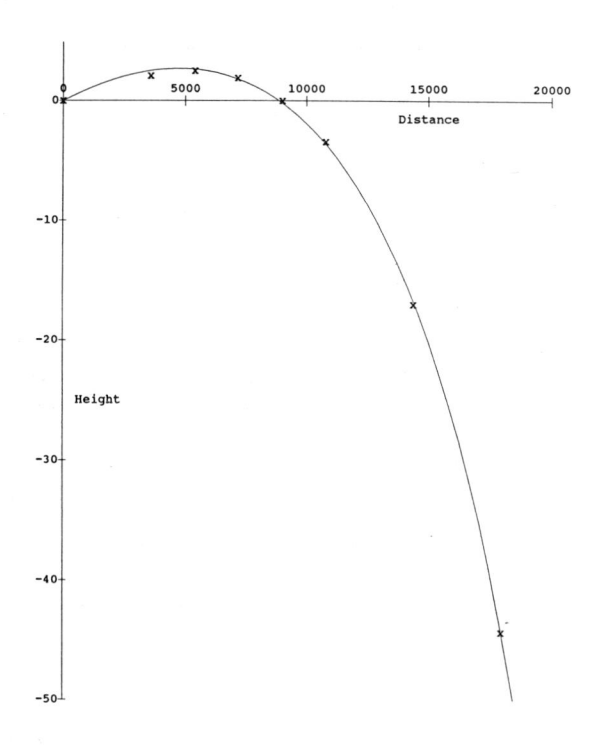

Figure 7.4 Bullet's trajectory together with the theoretical curve

Example 7.6 : The Flight of an Arrow with Drag

We also illustrate these results by analyzing the flight of an arrow fired from a Precision Shooting Equipment 30 inch Mach 7 bow with a 60 pound draw. Table 7.3 shows the height of the arrow in inches above the initial height of the arrow as a function of the horizontal distance in feet.[15]

Table 7.3 Height of arrow as a function of horizontal distance

Distance (feet)	Height (inches)
0	0.0
14	1.54
29	1.52
44	-1.35

According to the previous model, X and Y should be related by the equation

$$Y = \frac{W}{\lambda}\left[\ln\left(1 - \frac{\lambda X}{w_0 \cos\alpha}\right) + \frac{\lambda X}{w_0 \cos\alpha}\right] + (\tan\alpha)X.$$

Using the program Kalkulator to perform a nonlinear regression, we find $\lambda = 3.3212$, $w_0 = 3295$, and $\alpha = 0.01254$, after converting the horizontal measurements to inches. This predicts an initial velocity of 274.58 feet/sec and that the arrow will travel a maximum horizontal distance of 82.66 feet. These are plausible predictions. Figure 7.5 shows the arrow's trajectory together with the theoretical curve.

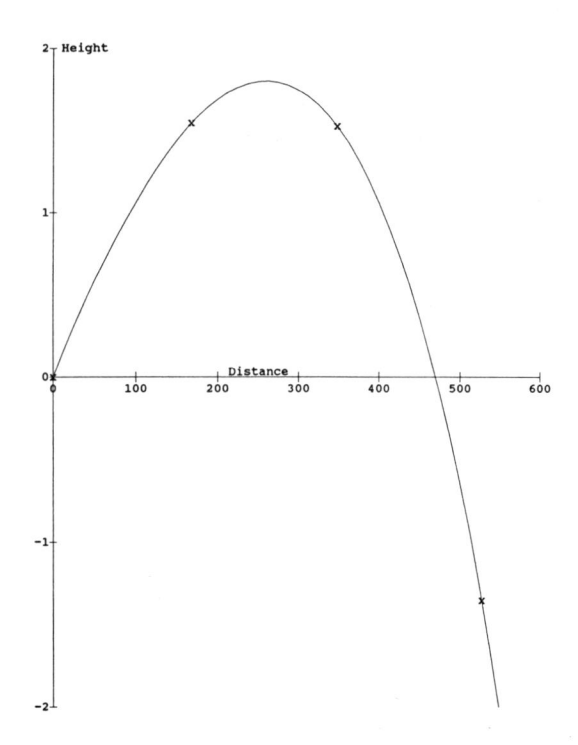

Figure 7.5 Arrow's trajectory together with the theoretical curve

Example 7.7 : The Flight of a Golf Ball with Drag [16]

---

[15] We would like to thank Doug Marcoux of Precision Shooting Equipment, Tucson, Arizona, for his help in conducting this experiment.

[16] "Maximum projectile range with drag and lift, with particular application to golf" by H. Erichson, American Journal of Physics, **51**, 1983, pages 357 through 362.

We also illustrate these results by analyzing the flight of a well-struck golf ball. Experiments have determined that for a golf ball the value of λ is given by $\lambda = k/m \approx 0.25$ sec^{-1}, and a well-struck golf ball has an initial velocity of $w_0 = 200$ ft/sec (≈ 136 mph). With $g = 32$ ft/sec^2 we find $W = 128$ ft/sec (≈ 87 mph.).

Under these circumstances, the trajectory of the golf ball with initial angle α is

$$X(t) = 800 \cos \alpha \left(1 - e^{-t/4}\right),$$

$$Y(t) = -128t + 4(200 \sin \alpha + 128)\left(1 - e^{-t/4}\right),$$

and

$$Y = 512 \ln\left(1 - \frac{X}{800 \cos \alpha}\right) + \left(\tan \alpha + \frac{128}{200 \cos \alpha}\right) X.$$

The speed at any time is

$$w(t) = \sqrt{(200 \cos \alpha)^2 e^{-t/2} + \left[128 - (200 \sin \alpha + 128) e^{-t/4}\right]^2}.$$

The hang time satisfies

$$1 - e^{-t_h/4} = \frac{32 \cdot 32}{200 \sin \alpha + 128} t_h,$$

and the range is

$$R = \frac{32 \cdot 128 \cdot 200 \cos \alpha}{200 \sin \alpha + 128} t_h.$$

In Figure 7.6 we show the (X,Y) trajectory for $\alpha = 15°$ to $\alpha = 65°$ in intervals of $10°$. Notice that for these values of α the greatest range occurs when $\alpha = 35°$. In Figure 7.7 we compare the no drag case to the linear drag case for $\alpha = 35°$. As any golfer will tell you, air resistance does play a role! In Table 7.4 we show the numerical computed values for t_h for $\alpha = 30°$ to $\alpha = 35°$ in intervals of $1°$ together with the range. For these values of α the greatest range occurs when $\alpha = 32$ and it is nearly 504 ft.

Table 7.4 Hang time and range for different initial angles

Angle (degrees)	Hang Time (seconds)	Range (feet)
30	5.167	502.43
31	5.300	503.45
32	5.431	503.91
33	5.560	503.84
34	5.686	503.16
35	5.810	501.98

Computer Experiment: Plot Y as a function of X for different values of α, and try to find the angle that gives the maximum range.

In Figure 7.8 we show the speed $w(t)$ as a function of time for $\alpha = 15°$ to $\alpha = 65°$ in intervals of $10°$. As t increases all speeds tend to the terminal speed of $W = 128$. But, wait a minute. In every case there are times when $w(t)$ attains a minimum value after which $w(t)$ increases to its terminal speed! Does this occur while the ball is in flight? In Figure 7.9 we show $w(t)$ and $y(t)$ as a function of time for the case $\alpha = 35°$. Notice that the time where $w(t)$ attains a maximum is just after the ball starts to fall back to earth. As the ball hits the earth its speed is increasing towards its terminal speed.

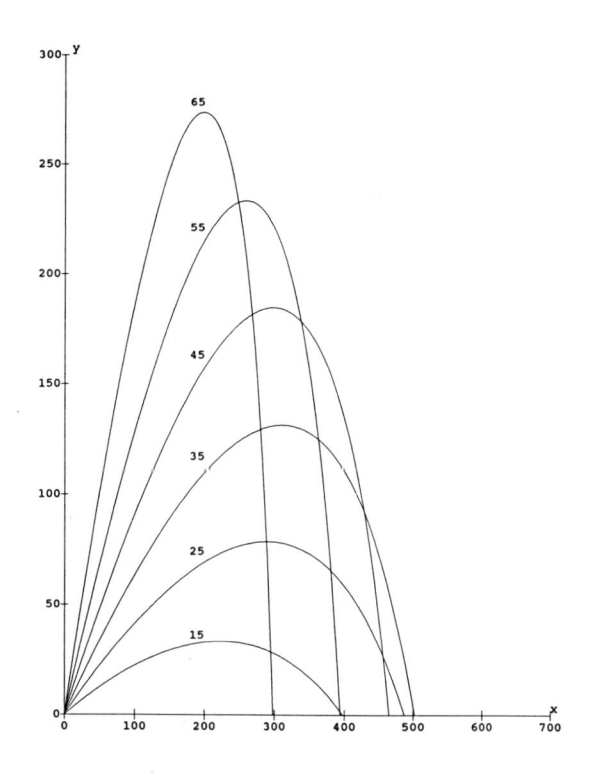

Figure 7.6 Golf ball trajectories with drag for $\alpha = 15°$ to $\alpha = 65°$

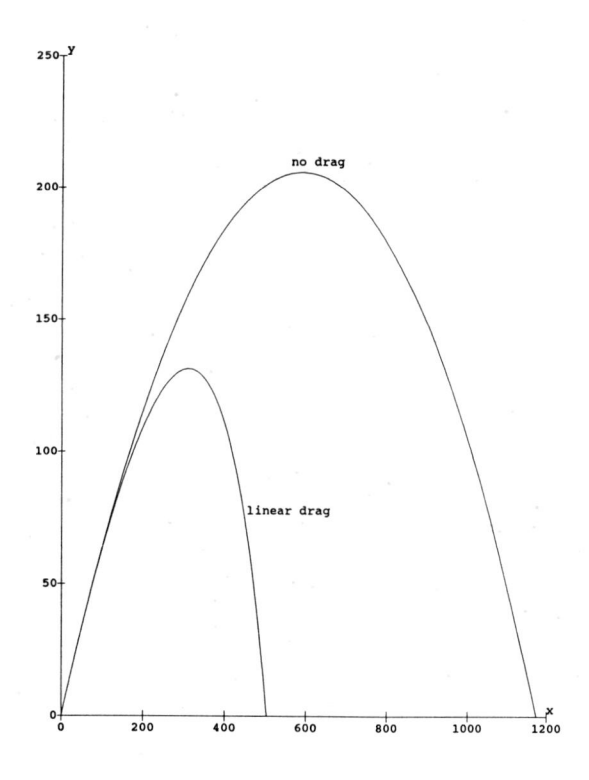

Figure 7.7 Golf ball trajectories with and without drag for $\alpha = 35°$

ADDITIONAL MATERIALS

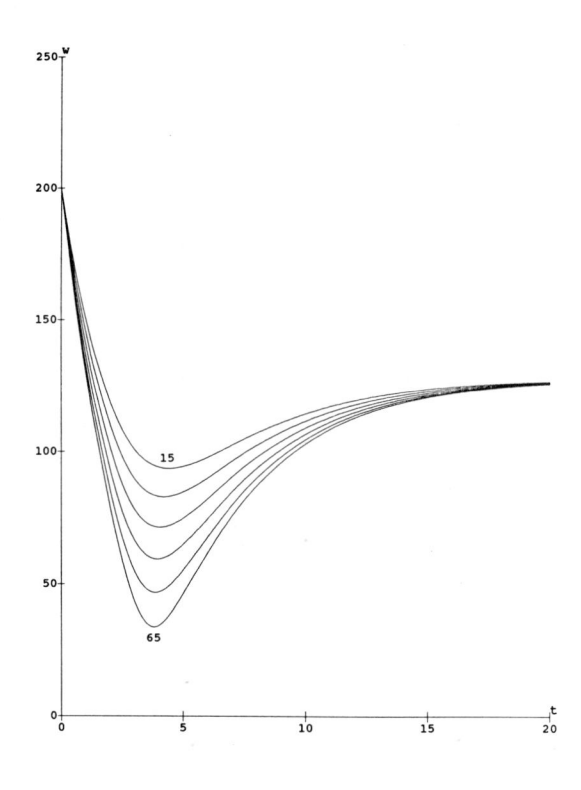

Figure 7.8 The speed $w(t)$ of a golf ball as a function of time for $\alpha = 15°$ to $\alpha = 65°$

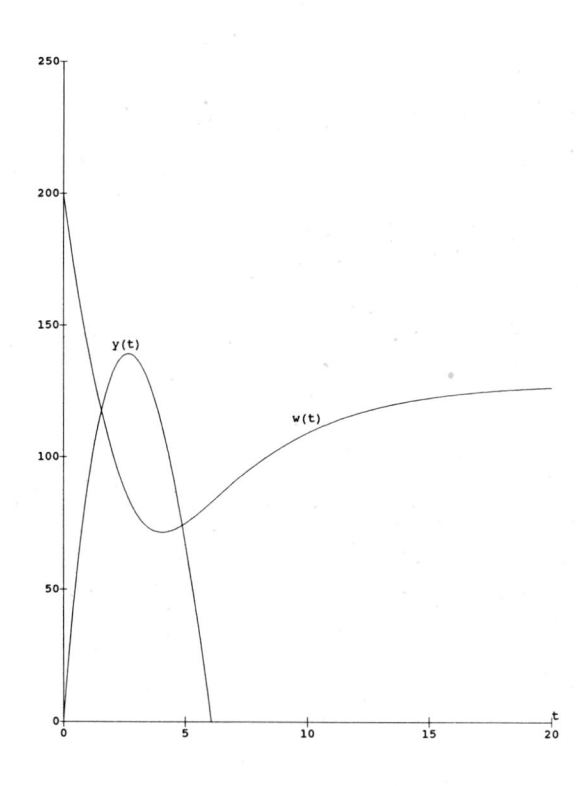

Figure 7.9 The speed $w(t)$ and height $y(t)$ of a golf ball as a function of time for $\alpha = 35°$

Computer Experiment: Find the largest angle α where $w(t)$ does not attain its minimum value while the ball is in the air. Find the angle α where $w(t)$ attains its minimum value at the time when the ball is at its maximum height in the air.

□

Based on the previous example it is natural to ask under what circumstances $w(t)$ attains a minimum. To answer this we differentiate

$$w^2(t) = u_0^2 e^{-2\lambda t} + \left[W - (v_0 + W)e^{-\lambda t}\right]^2$$

and find

$$2ww' = -2\lambda u_0^2 e^{-2\lambda t} + 2\left[W - (v_0 + W)e^{-\lambda t}\right]\lambda(v_0 + W)e^{-\lambda t}.$$

The right-hand side is zero when

$$-u_0^2 e^{-\lambda t} + \left[W - (v_0 + W)e^{-\lambda t}\right](v_0 + W) = 0,$$

which we can solve for $e^{-\lambda t}$ obtaining

$$e^{-\lambda t} = \frac{W(v_0 + W)}{u_0^2 + (v_0 + W)^2},$$

or

$$t = \frac{1}{\lambda} \ln \left[\frac{u_0^2 + (v_0 + W)^2}{W(v_0 + W)}\right].$$

There will be a time $t > 0$ where $w'(t) = 0$ if

$$\frac{u_0^2 + (v_0 + W)^2}{W(v_0 + W)} > 1,$$

that is, if

$$u_0^2 + v_0^2 + v_0 W > 0.$$

At this time the speed will be

$$w = \frac{u_0 W}{\sqrt{u_0^2 + (v_0 + W)^2}}.$$

We can show that

$$\frac{u_0 W}{\sqrt{u_0^2 + (v_0 + W)^2}} < W$$

and

$$\frac{u_0 W}{\sqrt{u_0^2 + (v_0 + W)^2}} < w(0),$$

which shows that there is a time when the particle attains a speed smaller than the terminal speed and smaller than the initial speed.[17] The values of X and Y at which this minimum speed is attained are

$$X(t) = \frac{u_0}{\lambda}\left[1 - \frac{W(v_0 + W)}{u_0^2 + (v_0 + W)^2}\right]$$

and

$$Y(t) = -W\frac{1}{\lambda}\ln\left[\frac{u_0^2 + (v_0 + W)^2}{W(v_0 + W)}\right] + \frac{1}{\lambda}(v_0 + W)\left[1 - \frac{W(v_0 + W)}{u_0^2 + (v_0 + W)^2}\right].$$

---

[17] Based on "Guide to Mathematical Modelling" by D. Edwards and M. Hamsom, Macmillan, 1989, pages 240 through 242.

There is another force that we have not yet considered, **lift**, due to backspin in the case of a golf ball and the position of the skis in the case of a ski jumper.[18] If this force is linear in the velocity then the equations of motion — including the linear drag terms — are

$$mx'' = -kx' - cy'$$

and

$$my'' = -mg - ky' + cx',$$

where k and c are positive constants, where as before k is associated with the drag, and now c is a measure of the lift. Unlike the previous cases, these equations are not independent. However, we can make some progress at this stage, because each of them can be integrated as they stand, yielding

$$x' = -\lambda x - \mu y + C_1$$

and

$$y' = -gt - \lambda y + \mu x + C_2,$$

where $\lambda = k/m$ and $\mu = c/g$. If we have the initial conditions $x(0) = y(0) = 0$ and $x'(0) = w_0 \cos\alpha$, $y'(0) = w_0 \sin\alpha$ then $C_1 = w_0 \cos\alpha$ and $C_2 = w_0 \sin\alpha$, and so

$$x' = -\lambda x - \mu y + w_0 \cos\alpha \tag{7.16}$$

and

$$y' = -gt - \lambda y + \mu x + w_0 \sin\alpha. \tag{7.17}$$

We can use the method of elimination discussed in Chapter 7 of the text to solve (7.16) and (7.17) by converting them to second order differential equations in one independent variable. If we differentiate (7.16) and then substitute for y' from (7.17) we find

$$x'' = -\lambda x' - \mu y' = -\lambda x' - \mu\left(-gt - \lambda y + \mu x + w_0 \sin\alpha\right).$$

We now eliminate y from this equation by using (7.16) in the form

$$\mu y = -x' - \lambda x + w_0 \cos\alpha \tag{7.18}$$

obtaining

$$x'' = -\lambda x' - \mu\left(-gt + \mu x + w_0 \sin\alpha\right) + \lambda\left(-x' - \lambda x + w_0 \cos\alpha\right),$$

or

$$x'' + 2\lambda x' + \left(\mu^2 + \lambda^2\right) x = \mu g t + w_0 \left(\lambda \cos\alpha - \mu \sin\alpha\right).$$

This has the solution

$$x(t) = e^{-\lambda t}\left(A \sin\mu t + B \cos\mu t\right) + Ct + D, \tag{7.19}$$

where

$$C = \frac{\mu g}{\mu^2 + \lambda^2},$$

$$D = \frac{1}{\left(\mu^2 + \lambda^2\right)^2}\left[w_0\left(\lambda\cos\alpha - \mu\sin\alpha\right)\left(\mu^2 + \lambda^2\right) - 2\lambda\mu g\right],$$

$$B = -D,$$

and

$$A = \frac{1}{\mu}\left(w_0 \cos\alpha - D\lambda - C\right).$$

---

[18] "Maximum projectile range with drag and lift, with particular application to golf" by H. Erichson, American Journal of Physics, 51, 1983, pages 357 through 362 and "The Flight of a Ski Jumper" by E. True, CoODEoE, Spring 1993, pages 5 through 8.

For the case $\mu \neq 0$, (7.18) and (7.19) yield

$$y(t) = e^{-\lambda t}(B\sin\mu t - A\cos\mu t) + \frac{1}{\mu}[-C - \lambda(Ct + D) + w_0\cos\alpha]. \qquad (7.20)$$

We leave it as an exercise to compute the velocity $w(t)$, and to find equations for the hang time t_h and range R for a particle experiencing both linear drag and lift.

Exercises

1. Does the particle experience a terminal velocity in the x- and y-directions? Does the particle experience a terminal velocity?

2. **Computer Experiment:** Does the particle experience a velocity lower than the terminal velocity?

Example 7.8 : The Flight of a Golf Ball with Drag and Lift[19]

For a golf ball the values of λ and μ are $\lambda = 0.25$ sec^{-1} and $\mu = 0.247$ sec^{-1}. To two decimal places these are the same, so we will consider the case where $\lambda = \mu = 0.25$, $w_0 = 200$ ft/sec, and $g = 32$ ft/sec^2. In this case (7.19) and (7.20) become

$$x(t) = e^{-t/4}(A\sin t/4 + B\cos t/4) + Ct + D, \qquad (7.21)$$

$$y(t) = e^{-t/4}(B\sin t/4 - A\cos t/4) - 4C - D - Ct + 800\cos\alpha, \qquad (7.22)$$

where

$$C = 64,$$
$$D = 400(\cos\alpha - \sin\alpha) - 256,$$
$$B = -D,$$

and

$$A = 400(\cos\alpha + \sin\alpha).$$

In Figure 7.10 we show the (X, Y) trajectory for $\alpha = 5°$ to $\alpha = 30°$ in intervals of $5°$. Notice that for these values of α the greatest range occurs when $\alpha = 10°$ and is a little over 600 ft. □

Exercises

1. **Computer Experiment:** Plot the trajectory of a golf ball for $\alpha = 5°$ to $\alpha = 15°$ in intervals of $1°$. Which value of α gives the greatest range?

2. We have looked at the flight of a golf ball with drag and no lift. Consider what happens if there is lift but no drag. Experiment with different values of μ to see whether it is possible to obtain unrealistic trajectories.

Quadratic Drag (Air Resistance Quadratic in Velocity)

[19] "Maximum projectile range with drag and lift, with particular application to golf" by H. Erichson, American Journal of Physics, **51**, 1983, pages 357 through 362.

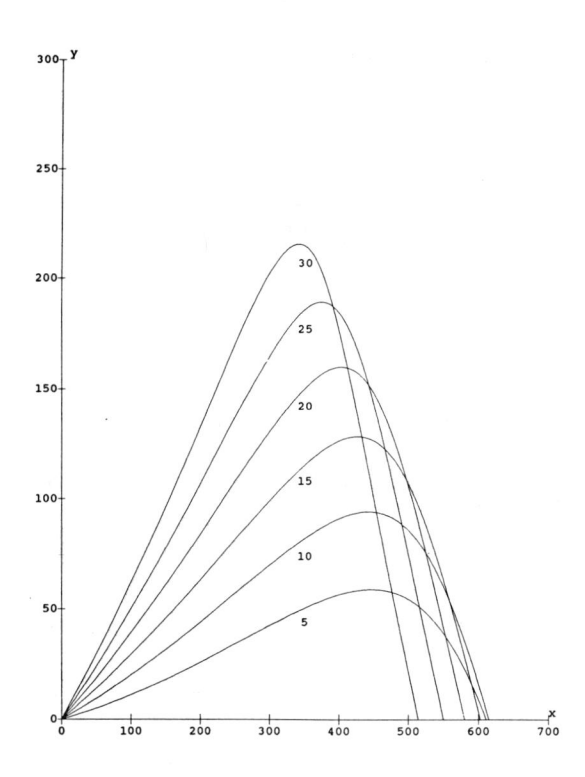

Figure 7.10 Golf ball trajectories with drag and lift for $\alpha = 5°$ to $\alpha = 30°$

Whereas linear drag and lift may be a good model for a golf ball, it is not a good model for the trajectory of many other objects which experiences quadratic drag, such as a baseball. The differential equations for quadratic drag are[20]

$$mx'' = -hwx'$$

and

$$my'' = -mg - hwy',$$

where h is a positive constant and

$$w = \sqrt{(x')^2 + (y')^2}.$$

These have not been solved analytically, so numerical solutions are the standard way to obtain information on the trajectories.

However, it is possible to rewrite these equations in a way that makes them amenable to standard techniques of numerical integration, and which give us additional insight.[21] We define the angle θ by

$$\sin\theta = \frac{y'}{w} \qquad \cos\theta = \frac{x'}{w},$$

so that

$$\tan\theta = \frac{y'}{x'}.$$

---

[20] "The Mathematics of Projectiles in Sport" by N. de Mestre, Australian Mathematical Society Lecture Series 6, Cambridge University Press, 1990, Chapter 3.

[21] "The Mathematics of Projectiles in Sport" by N. de Mestre, Australian Mathematical Society Lecture Series 6, Cambridge University Press, 1990, Chapter 3; "Principles of Mechanics" by J. L. Synge and B. A. Griffith, McGraw-Hill, 1949, Section 6.2; and "Shuttlecock Trajectories in Badminton" by A. Tan, Mathematical Spectrum, 19, 1987, pages 33 through 36.

In this case the differential equations can be expressed in the form

$$w' = -g\sin\theta - \mu w^2$$

and

$$\theta' = -g\frac{\cos\theta}{w},$$

where $\mu = h/m$. We can eliminate the t dependence between these two equations to obtain

$$\frac{dw}{d\theta} + w\tan\theta = -\frac{\mu}{g}w^3\sec\theta.$$

This is a Bernoulli differential equation with solution

$$\frac{\sec^2\theta}{w^2} = -\frac{\mu}{g}[\ln(\sec\theta + \tan\theta) + \sec\theta\tan\theta] + C.$$

Imposing the initial conditions $w(0) = w_0$ and $\theta(0) = \alpha$ at $t = 0$, we find

$$w(\theta) = \left\{\frac{1}{w_0^2}\frac{\cos^2\theta}{\cos^2\alpha} - \frac{\mu}{g}\cos^2\theta\left[\ln\left(\frac{\sec\theta + \tan\theta}{\sec\alpha + \tan\alpha}\right) + \frac{\sin\theta}{\cos^2\theta} - \frac{\sin\alpha}{\cos^2\alpha}\right]\right\}^{-1/2}$$

The quantity in brackets is negative if $-\pi/2 < \theta < \alpha$, so $0 < w(\theta) < \infty$, and $\lim_{\theta \to -\pi/2} = \sqrt{g/\mu}$, the terminal speed. This is one way in which μ could be measured. For example, the terminal speed of a baseball is about 140 feet/sec, giving $\mu \approx 0.0016$, in excellent agreement with published results obtained using different arguments.[22]

Because

$$\theta' = -g\frac{\cos\theta}{w},$$

we see, for $-\pi/2 < \theta < \pi/2$, that θ decreases from α as t increases from 0. Thus as $t \to \infty$, we have $\theta \to -\pi/2$. We also have

$$\frac{dt}{d\theta} = -\frac{1}{g}w\sec\theta,$$

from which we find

$$t(\theta) = -\frac{1}{g}\int_\alpha^\theta w(\phi)\sec\phi\,d\phi.$$

In the same way, from $dx/d\theta = x'dt/d\theta$ and $dy/d\theta = y'dt/d\theta$, we have

$$x(\theta) = -\frac{1}{g}\int_\alpha^\theta w^2(\phi)\,d\phi.$$

$$y(\theta) = -\frac{1}{g}\int_\alpha^\theta w^2(\phi)\tan\phi\,d\phi.$$

Thus, t, x, and y can be integrated numerically as functions of θ.

As $\theta \to -\pi/2$, we can show that $y \to -\infty$, while x is bounded. Thus, this model also predicts that there is a limit to the horizontal distance the particle can travel.

For an application of this to baseball see "Looking into Chapman's Homer: The physics of judging a fly ball" by P. J. Brancazio, American Journal of Physics, **53**, 1985, pages 849 through 855.

---

[22] "Looking into Chapman's Homer: The physics of judging a fly ball" by P. J. Brancazio, American Journal of Physics, **53**, 1985, pages 849 through 855.

There is a way of obtaining some information of how y varies with x directly from the differential equations $x'' = -\mu w x'$, $y'' = -g - vwy'$, subject to the initial conditions $x(0) = y(0) = 0$, $x'(0) = w_0 \cos\alpha$, $y'(0) = w_0 \tan\alpha$, by assuming that y has a power series expansion in x, that is,

$$y(t) = \sum_{n=0}^{\infty} a_n x^n(t).$$

From

$$y'(t) = \sum_{n=0}^{\infty} a_n n x^{n-1}(t) x',$$

$$y''(t) = \sum_{n=0}^{\infty} a_n n(n-1) x^{n-2}(t) (x')^2 + \sum_{n=0}^{\infty} a_n n x^{n-1}(t) x'',$$

and

$$y'''(t) = \sum_{n=0}^{\infty} a_n n(n-1)(n-2) x^{n-3}(t)(x')^3 + \sum_{n=0}^{\infty} 3a_n n(n-1) x^{n-2}(t) x' x'' + \sum_{n=0}^{\infty} a_n n x^{n-1}(t) x''',$$

the differential equations, and the initial conditions, we can show that $a_0 = 0$, $a_1 = \tan\alpha$, $a_2 = -g/(2w_0^2 \cos^2\alpha)$, and $a_3 = -g\mu w_0/(3w_0^3 \cos^3\alpha)$. Thus,

$$y = x\tan\alpha - \frac{1}{2}g\left(\frac{1}{w_0 \cos\alpha}\right)^2 x^2 - \frac{1}{3}g\mu w_0 \left(\frac{1}{w_0 \cos\alpha}\right)^3 x^3 + \cdots.$$

This are the same as we obtained for the case when air resistance is linear in the velocity, if we set $\lambda = \mu w_0$. However, in this case the terminal speed is $\sqrt{g/\mu}$, whereas in that it is $g/\lambda = g/(\mu w_0)$.

There is another way to analyze these equations, due to Siacci,[23] which was in common use for estimating the trajectory of a bullet before World War I. He introduced the "pseudovelocity" u as a new independent variable, where

$$w \cos\theta = u \cos\alpha.$$

In this case $x' = w\cos\theta = u\cos\alpha$, and $y' = w\sin\theta = w\cos\theta\tan\theta = u\cos\alpha\tan\theta$. We also have $x'' = u'\cos\alpha$, and $y'' = \frac{d}{du}(y')u'$, so that $x'' = -\mu w x'$ becomes

$$u' = -\mu u w,$$

or

$$\frac{dt}{du} = \frac{-1}{\mu u w}.$$

Thus as t increases, u decreases. Now $x' = u\cos\alpha$, so that $\frac{dx}{du} = x'\frac{dt}{du}$, or

$$\frac{dx}{du} = -\frac{\cos\alpha}{\mu w}.$$

Now $y' = u\cos\alpha\tan\theta$, so that $\frac{dy}{du} = y'\frac{dt}{du}$, or

$$\frac{dy}{du} = \frac{-\cos\alpha\tan\theta}{\mu w}.$$

---

[23] "Mathematics for Exterior Ballistics" by G.A. Bliss, Wiley, 1944, Chapter 3.

From $y'' = -g - \mu w y'$ we have $-\mu u w \frac{d}{du}(u \cos\alpha \tan\theta) = -g - \mu w u \cos\alpha \tan\theta$, or

$$\frac{d}{du}\tan\theta = \frac{g \sec\alpha}{\mu u^2 w}.$$

Because $w = u \cos\alpha \sec\theta$, these four first order equations for t, x, y, and θ as functions of u, are nonlinear and coupled. The initial conditions are $t = 0$, $x = 0$, $y = 0$, $\theta = \alpha$, and $u = w_0$. These equations cannot be solved explicitly.

However, Siacci made the approximation that

$$w \approx u.$$

It is believed that this approximation is reasonable in the case of small initial angles of elevation, for that part of the trajectory extending from the origin to the point where the bullet recrosses the horizontal line through the origin. It is therefore appropriate for low, flat trajectories, such as rifle fire, firing between aircraft, and anti-aircraft fire. It is inappropriate for trajectories such as that of a well-hit baseball.

Using this approximation the differential equations become

$$\frac{dt}{du} = \frac{-1}{\mu u^2},$$

$$\frac{dx}{du} = -\frac{\cos\alpha}{\mu u},$$

$$\frac{dy}{du} = \frac{-\cos\alpha \tan\theta}{\mu u},$$

and

$$\frac{d}{du}(\tan\theta) = \frac{g \sec\alpha}{\mu u^3}.$$

We can integrate the first two and the fourth, obtaining

$$t(u) = \frac{1}{\mu}\left(\frac{1}{u} - \frac{1}{w_0}\right),$$

$$x(u) = -\frac{\cos\alpha}{\mu}(\ln u - \ln w_0),$$

and

$$\tan\theta = \tan\alpha - \frac{g \sec\alpha}{2\mu}\left(\frac{1}{u^2} - \frac{1}{w_0^2}\right).$$

Substituting this into the third equation gives

$$\frac{dy}{du} = -\frac{\sin\alpha}{\mu u} + \frac{g}{2\mu^2}\left(\frac{1}{u^3} - \frac{1}{w_0^2 u}\right),$$

from which we find

$$y(u) = -\frac{\sin\alpha}{\mu}(\ln u - \ln w_0) + \frac{g}{2\mu^2}\left(-\frac{1}{2u^2} - \frac{1}{w_0^2}\ln u + \frac{1}{2w_0^2} + \frac{1}{w_0^2}\ln w_0\right).$$

We can find the trajectory by eliminating u from $x(u)$ and $y(u)$. From $x(u) = -\frac{\cos\alpha}{\mu}(\ln u - \ln w_0)$, we have $\ln u - \ln w_0 = -\mu x \sec\alpha$, so $u = w_0 \exp(-\mu x \sec\alpha)$, which when substituted in $y(u)$ gives

$$y = x \tan\alpha + \frac{g \sec\alpha}{2\mu w_0^2}x + \frac{g}{4\mu^2 w_0^2}[1 - \exp(2\mu x \sec\alpha)].$$

If we expand $\exp(2\mu x \sec\alpha)$ in a Taylor series we find

$$y = x \tan\alpha - \frac{g}{2(w_0 \cos\alpha)^2}x^2 - \frac{1}{3}\frac{g\mu}{w_0^2 \cos^3\alpha}x^3 + \ldots,$$

ADDITIONAL MATERIALS

in agreement with the first four terms of the exact model.

We can also find $x(t)$ and $y(t)$ by solving $t(u) = \frac{1}{\mu}\left(\frac{1}{u} - \frac{1}{w_0}\right)$ for

$$\frac{1}{u} = \mu t + \frac{1}{w_0} = \frac{1 + w_0 \mu t}{w_0},$$

giving $-\ln u + \ln w_0 = \ln(1 + w_0 \mu t)$, so that

$$x(t) = \frac{\cos \alpha}{\mu} \ln(1 + w_0 \mu t),$$

and

$$y(t) = \frac{\sin \alpha}{\mu} \ln(1 + w_0 \mu t) + \frac{g}{2\mu^2}\left(-\frac{1}{2}\mu^2 t^2 - \frac{\mu}{w_0} t + \frac{1}{w_0^2}\ln(1 + w_0 \mu t)\right).$$

Notice that $\lim_{t \to \infty} x(t) = \infty$, so the Siacci approximation to the model predicts that there is no limit to the horizontal distance the particle can travel, in stark contrast to the exact model.

Quadratic Drag with Atmospheric Density

If it is assumed that the atmospheric density changes with altitude in an exponential way, then the differential equation governing an object falling vertically from rest from an initial height h is[24]

$$y'' = -g + ce^{-ay}(y')^2, \qquad (7.23)$$

where a and c are positive constants, with typical values $a = 0.00014$ m^{-1} and $c = 0.001$ m^{-1}. In these units $g \approx 9.82$ m/sec^2. For what values of y is this differential equation valid?

Computer Experiment: Obtain plots of the vertical velocity y' versus t for various initial heights ranging from $h = 3,000$ m (≈ 1.86 miles) to $h = 30,000$ m (≈ 18.64 miles). A reasonable window is $0 \leq t \leq 150$ sec, $-400 \leq y' \leq 0$ m/sec. What do you notice when compared with the $a = 0$ situation? Now simultaneously plot y versus t. A reasonable window is $0 \leq t \leq 150$ sec, $0 \leq y \leq 30,000$ m. Does the phenomenon you noticed for the velocity occur before the object hits the ground ($y = 0$)? From the differential equation (7.23) show that if V is the minimum velocity that the object attains when dropped from a height h, then it will occur at the height H m, where

$$H = \frac{1}{a}\ln\left(\frac{cV^2}{g}\right). \qquad (7.24)$$

Confirm that the height predicted by this formula agrees with what you see on the computer screen.

We are now going to obtain more information from the differential equation (7.23), subject to the initial conditions $y(0) = h$, $y'(0) = 0$. We introduce the variable $v = y'$ and use the fact that

$$y'' = v\frac{dv}{dy}$$

to write the differential equation in the form

$$\frac{dv}{dy} = ce^{-ay}v - \frac{g}{v}.$$

This is a Bernoulli differential equation, so we change the variable from v to w where

$$w = v^2.$$

---

[24] "Terminal Speed and Atmospheric Density" by N. M. Shea, The Physics Teacher, March 1993, page 176.

We thus obtain
$$\frac{dw}{dy} - 2ce^{-ay}w = -2g.$$

This is a linear differential equation with integrating factor
$$\mu = \exp\left(\int -2ce^{-ay}\,dy\right) = e^{\alpha(y)},$$

where
$$\alpha(y) = 2\frac{c}{a}e^{-ay}. \tag{7.25}$$

The solution of this linear differential equation, using the fact that $w = v^2 = 0$ and $y = h$ when $t = 0$, can be written
$$w(y) = 2ge^{-\alpha(y)}\int_y^h e^{\alpha(u)}\,du,$$

so that
$$(y')^2 = 2ge^{-\alpha(y)}\int_y^h e^{\alpha(u)}\,du, \tag{7.26}$$

where α is given by (7.25). This is the general expression that relates the velocity of the object to its height and the initial height.

Equation (7.27) enables us to evaluate the velocity v_{term} of the object when it hits the ground ($y = 0$) as
$$v_{term} = -\sqrt{2ge^{-\alpha(y)}\int_0^h e^{\alpha(u)}\,du}, \tag{7.27}$$

which for $h = 30,000$ gives $v_{term} \approx 103$ m/sec (≈ 230 mph).

Equation (7.27) also enables us to answer the following question. If the object is to attain its minimum velocity at $H = 15,000$ m, from what height h should it be dropped? From (7.24) we know that the minimum velocity V will occur when
$$V^2 = \frac{g}{c}e^{-aH}.$$

If we substitute this in (7.27) we find that h must satisfy
$$\int_H^h e^{\alpha(u)}\,du = \frac{1}{a}\frac{e^\alpha}{\alpha}.$$

Using the previous values for the constants and the fact that $H = 15,000$ in this equation, we are faced with the problem of finding h for which
$$\int_{15000}^h e^{\alpha(u)}\,du = 23482,$$

where $\alpha(u) = 14.2857e^{-0.00014y}$. By using a numerical integration package, such as INTEGRAL, we try various values for h starting at say $h = 25,000$. It is not difficult to find that $h \approx 22,550$ m.

8. IDEAS FOR CHAPTER 8 — "SECOND ORDER LINEAR DIFFERENTIAL EQUATIONS — QUALITATIVE AND QUANTITATIVE ASPECTS"

8.1 Page by Page Comments

Chapter 8 requires about 8 class meetings of 50 minutes each.

Sections 8.5 through 8.8 are not critical for later sections.

You could go to Chapter 11 after Section 8.4. Chapter 11 does not use any material from Chapters 9 and 10.

You could go to Chapter 12 after Section 8.4.

Page 315. Section 8.1, *Qualitative Behavior of Solutions*, may be new to you. It shows how we can sometimes obtain the qualitative behavior of solutions even if we cannot find an explicit solution. Theorem 8.2 deals with $x'' + Q(t)x = 0$, and Theorem 8.3 extends these results to $x'' + p(t)x' + q(t)x = 0$.

Page 317. Theorem 8.2 of the text may be motivated by rewriting the differential equation $x'' + Q(t)x = 0$ in the form

$$x'' = -Q(t)x$$

and thinking about concavity.

Let's first consider a solution $x = x_1(t)$ of this differential equation for the case where $Q(t) \leq 0$ for $t \geq t_0$. Thus, $x'' \geq 0$ when $x_1 > 0$ and $x'' \leq 0$ when $x_1 < 0$. This means that generally $x_1(t)$ is concave up when x_1 is above the t-axis, and concave down when x_1 is below the t-axis. One possibility is that $x_1(t)$ is not zero to the right of t_0, in which case the maximum number of times $x_1(t)$ can be zero to the right of t_0 is zero. Another possibility is that there is a point $t_1 > t_0$ at which $x_1(t_1) = 0$. Then generally the function x_1 will have been either decreasing and concave up or increasing and concave down just to the left of t_1. If we concentrate on the case when the curve is decreasing and concave up then the curve could just touch the t-axis at $t = t_1$ but not cross into the lower half-plane — in which case, because it is concave up in the upper half-plane, the curve must continue to increase and so will never again be zero — or the curve could cross into the lower half-plane — in which case, because it is concave down in the lower half-plane, the curve must continue to decrease and so will never again be zero. A similar argument applies to the case when the curve is increasing and concave down. Thus, the maximum number of times $x_1(t)$ can be zero to the right of t_0 is one.

Now we consider a solution $x = x_1(t)$ of the differential equation for the case where $Q(t) \geq 0$ for $t \geq t_0$. Thus, $x'' \leq 0$ when $x_1 > 0$ and $x'' \geq 0$ when $x_1 < 0$. This means that generally $x_1(t)$ is concave down when x_1 is above the t-axis, and concave up when x_1 is below the t-axis. One possibility is that $x_1(t)$ oscillates about the t-axis and so is zero at an infinite number of points. However, another possibility is a curve which is decreasing and concave down in the upper half-plane, that crosses the t-axis, and then continues to decrease, even though it is concave up. This is an example of why $Q(t) \geq 0$ is insufficient to guarantee that x_1 has an infinite number of zeros.

Page 318. The differential equation in Example 8.1 is solved in Exercise 1(a) of Section 8.5.

Page 319. The differential equation in Example 8.3, Airy's equation, is solved by the power series method as Example 12.6.

Page 321. The differential equation in Example 8.4 is solved in Example 8.13 of Section 8.5. This same example is used to motivate the reduction of order technique in the next section, Section 8.2.

Page 321. The differential equation in Example 8.5 is solved by the power series method as Example 12.3.

Page 322. Exercise 7 of Section 8.1 asks the student to confirm the solution is the one given. It relates to a comment on page 318. Exercise 5 of Section 8.5 asks the student to solve the same differential equation.

Page 323. There are additional exercises for Section 8.1 on page 260 of this manual.

Page 323. This treatment of *Reduction of Order* may be new to you. We use reduction of order not only to solve homogeneous equations, but also nonhomogeneous equations.

Page 324. The differential equation in Example 8.6 is the same as the one used in Example 8.4.

Page 324. In Example 8.6 the change of variable is $x(t) = tz(t)$. Some students mistakenly think that $x(t) = tz(t)$ is the substitution to make in all reduction of order problems.

Page 328. Example 8.8 could also be done in Section 8.5 — *Solving Cauchy-Euler Equations*.

Page 329. Note in Figures 8.9 and 8.10, that the vertical axis passes through $t = 1$, not $t = 0$.

Page 329. Note in Figure 8.9, $C_2 = 2$ because $x(1) = 0$.

Page 329. Note in Figure 8.10, $C_2 = 3 - C_1$ because $x'(1) = 0$.

Page 329. In Figures 8.10, it appears that all solutions for which $x'(1) = 0$ pass through a common point. This is true. They pass through the point $(e, 3e - e^2)$.

Page 329. You may want to demonstrate, or have your students demonstrate, that it is usually better to use the method of undetermined coefficients than to use reduction of order, if you have a choice. You could use the same example — namely, $x'' + 16x = 16t^2$, see Example 8.10 on page 335 — that we will use to demonstrate that it is usually better to use the method of undetermined coefficients than to use variation of parameters. (We have already solved this by the method of undetermined coefficients, see Example 7.6 on page 287.) Here is an outline using reduction of order.

The solution of the associated homogeneous equation, $x'' + 16x = 0$, is $x_h(t) = C_1 \sin 4t + C_2 \cos 4t$. If we choose $x_1(t) = \sin 4t$, then the substitution $x(t) = x_1(t) z(t) = z \sin 4t$, makes $x'' + 16x = 16t^2$ become $z'' \sin 4t + 8z' \cos 4t = 16t^2$. The integrating factor is $\sin^2 4t$ so $(z' \sin^2 4t)' = 16t^2 \sin 4t$. Because $\int 16t^2 \sin 4t \, dt = -4t^2 \cos 4t + 2t \sin 4t + \frac{1}{2} \cos 4t + C_1$, we have $z' \sin^2 4t = -4t^2 \cos 4t + 2t \sin 4t + \frac{1}{2} \cos 4t + C_1$, so

$$z' = \left(-4t^2 \cos 4t + 2t \sin 4t + \frac{1}{2} \cos 4t + C_1\right) / \sin^2 4t.$$

By noting that $\left(-4t^2 \cos 4t + 2t \sin 4t\right) / \sin^2 4t = \left(t^2 / \sin 4t\right)'$, we find $z = \frac{t^2}{\sin 4t} - \frac{1}{8 \sin 4t} - \frac{\cos 4t}{4 \sin 4t} C_1 + C_2$ so $x = z \sin 4t = t^2 - \frac{1}{8} - \frac{1}{4} C_1 \cos 4t + C_2 \sin 4t$.

Page 330. There are additional exercises for Section 8.2 on page 261 of this manual.

Page 332. Example 8.9 could also be done by reduction of order.

Page 335. Example 8.10 demonstrates that it is usually better to use the method of undetermined coefficients than to use variation of parameters, if you have a choice.

Page 336. Exercise 6 of Section 8.3 is an excellent example to test whether students understand the three different methods for finding particular solutions.

Page 336. There are additional exercises for Section 8.3 on page 261 of this manual.

Page 336. Table 8.1 compares the three different methods for finding particular solutions.

Table 8.1 A comparison of the three methods

	UC	RO	VP
Restrictions on a_2, a_1, a_0?	Constants	No	No
Requires $x_h(t)$?	Yes	No, any nontrivial solution x_1	Yes
Restrictions on $f(t)$?	Yes	No	No
Technique always works?	Yes	Up to integration	Up to integration
Technique	Educated guess	$x_p = x_1 z$	$x_p = x_1 z_1 + x_2 z_2$

Page 340. The fact that if x_1 and x_2 are linearly dependent then they will have the same zeros, is used in some of the exercises.

Page 342. Exercise 8 shows that we need only solve two initial value problems to find the general solution.

Page 342. Exercise 10 of Section 8.4 is very useful if you want to construct second order linear differential equations with specific solutions. This technique is demonstrated in Exercise 11. (We used this to construct a number of the differential equations in this book.)

Page 343. Exercise 12 of Section 8.4 is used to motivate the definition of a regular singular points in Chapter 12, page 602.

Page 343. There are additional exercises for Section 8.4 on page 261 of this manual.

Page 348. A flowchart showing how to solve second order equations can be found on page 265 in this manual.

Page 349. Note that Exercise 1(a) of Section 8.5 is the one used in Example 8.1 on page 318.

Page 349. Exercise 5 of Section 8.5 asks the student to solve the differential equation that was mentioned in Section 8.1 (page 318, and Exercise 7 of Section 8.1).

Page 349. The differential equation in Exercise 8 of Section 8.5 is the same as the one in Exercise 18 of Section 6.2.

Page 349. There are additional exercises for Section 8.5 on page 262 of this manual.

Page 350. Section 8.6, *Boundary Value Problems and the Shooting Method*, is optional. It is a nontraditional introduction to Boundary Value Problems. Such problems are not discussed in the sequel.

Page 355. Section 8.7, *Solving Higher Order Homogeneous Differential Equations*, is optional.

Page 367. Table 8.1 is in TWIDDLE. It is called PENDULUM.DTA and is in the subdirectory DD-08.

Page 367. There are additional exercises for Section 8.7 on page 262 of this manual.

Page 368. Section 8.8, *Solving Higher Order Nonhomogeneous Differential Equations*, is optional.

Page 376. There are additional exercises for Section 8.8 on page 263 of this manual.

8.2 Additional Exercises

Section 8.1

Page 323. Add exercises following the current Exercise 16.

17. **The Riccati Equation I.** Show that any solution of the Riccati equation $x' = a_0(t) + a_1(t)x + a_2(t)x^2$ ($a_2(t) \neq 0$) is the solution of a second order linear differential equation by the following steps.

    (a) Define a new variable $u = \exp\left(-\int a_2(t)x(t)\,dt\right)$ and show that
    $$x = -\frac{u'}{a_2 u}$$
    and
    $$x' = -\frac{u''}{a_2 u} + a_2\left(\frac{u'}{a_2 u}\right)^2 + \frac{a_2'}{a_2^2}\left(\frac{u'}{u}\right).$$

    (b) Substitute these into the equation $x' = a_0(t) + a_1(t)x + a_2(t)x^2$ to find
    $$-\frac{u''}{a_2 u} + \frac{a_2'}{a_2^2}\left(\frac{u'}{u}\right) = a_0 + a_1\left(-\frac{u'}{a_2 u}\right),$$
    which is the second order linear differential equation
    $$a_2 u'' - (a_2 a_1 + a_2')\,u' + a_2^2 a_0 u = 0.$$

(c) Convert the Riccati equation $x' = t^2 + x^2$ to the second order linear differential equation.
$$u'' + t^2 u = 0.$$

18. **The Riccati Equation II.** Show that any nontrivial solution of the second order linear differential equation $a_2(t)x'' + a_1(t)x' + a_0(t)x = 0$ $(a_2(t) \neq 0)$ is the solution of a Riccati equation by the following steps.

    (a) Define a new variable $u = x'/x$ and show that
    $$u' = \frac{x''}{x} - \left(\frac{x'}{x}\right)^2.$$

    (b) Solve the preceding equation for x'' obtaining
    $$x'' = x\left(u' + u^2\right).$$

    (c) Substitute this and $x' = ux$ into the equation $a_2(t)x'' + a_1(t)x' + a_0 x = 0$ to find
    $$a_2\left(u' + u^2\right) + a_1 u + a_0 = 0$$
    which is the Riccati equation
    $$u' = -\frac{a_0}{a_2} - \frac{a_1}{a_2} u - u^2.$$

    (d) Convert the differential equation $x'' + x = 0$ to a Riccati equation.

Section 8.2

Page 330. Add exercise following the current Exercise 7.

8. Solve $x'' + 16x = 16t^2$ by reduction of order using the fact that $x_1(t) = \sin 4t$ is a solution of $x'' + 16x = 0$. Compare the work involved in this exercise with that required to solve it by the method of undetermined coefficients (Example 7.6 on page 287.) Is it better to use the method of undetermined coefficients than to use reduction of order, if you have a choice?

Section 8.3

Page 336. Add exercise following the current Exercise 6.

7. Solve $x'' - 3x' + 2x = -e^{2t}/(e^t + 1)$ by reduction of order by noting that $x = e^t$ is a solution of the corresponding homogeneous equation. Compare your answer to the one obtained in Example 8.9.

Section 8.4

Page 343. Add exercise following the current Exercise 15.

16. Consider a spring-mass system with friction, and we conduct the following two experiments.

    (a) First, the mass is lowered one unit below its equilibrium position and then released from rest. Assume that the solution of this initial value problem is $x_1(t)$. What other initial value problems for the same spring-mass system can you immediately solve? What does this correspond to physically. [Hint: See Exercise 8.]

(b) Second, the mass is at its equilibrium position and is given an initial velocity of one unit downward. Assume that the solution of this initial value problem is $x_2(t)$. What other initial value problems for the same spring-mass system can you immediately solve? What does this correspond to physically. [Hint: See Exercise 8.]

Section 8.5

Page 349. Add exercise following the current Exercise 8.

9. Solve $x^2 y'' - x y' + y = x^2 \ln x$. Compare your answer to the one obtained in Example 8.8.

10. Solve $x^2 y'' + x y' - y = 2/(x+1)$ by reduction of order by noting that $y = x$ is a solution of the corresponding homogeneous equation. Compare your answer to the one obtained in Example 8.15.

11. Solve $x^2 y'' + x y' - y = 2/(x+1)$ by first making the substitution $x = e^t$ to find $d^2 y/dt^2 - y = 2/(e^t + 1)$ and then solving the later by reduction of order or by variation of parameters. (Why can't you use the method of undetermined coefficients?) Compare your answer to the one obtained in Example 8.15.

Section 8.7

Page 367. Add exercise following the current Exercise 12.

13. **Reduction of Order.** Show that if $x(t) = x_1(t)$ is a solution of the homogeneous differential equation

$$a_n(t) \frac{d^n x}{dt^n} + a_{n-1}(t) \frac{d^{n-1} x}{dt^{n-1}} + \cdots + a_2(t) \frac{d^2 x}{dt^2} + a_1(t) \frac{dx}{dt} + a_0(t) x = 0,$$

then the substitution $x(t) = x_1(t) z(t)$ reduces the differential equation to one which is of order $n - 1$ in $v = z'$.

14. **Coupled Pendulums II — A Simple Experiment.** Construct the following piece of apparatus, which is a coupled pendulum. Suspend two identical pendulums (same length wire, same mass) and join the masses with a light spring. Make sure that the points of suspension of the pendulums are the same distance apart as the unstretched spring. We are first going to perform two experiments, and then we are going to explain what we have found.

   (a) Perform the following experiment. Displace the two pendulums through the same small initial angle and simultaneously release both from rest. Describe what happens.

   (b) Perform the following experiment. Displace the two pendulums through the same small initial angle, but in opposite directions, and simultaneously release both from rest. Describe what happens.

   (c) Perform the following experiment. Displace one of the pendulums through a small initial angle holding the other pendulum vertical. Now simultaneously release both from rest. Describe what happens.

   (d) Show that if $x(t)$ and $y(t)$ are the angles displaced by the two pendulums from vertical, and $k > 0$ is the spring constant, then the linearized differential equations governing the subsequent motion, neglecting air resistance, are $x'' + \lambda^2 x = k(y-x)$, $y'' + \lambda^2 y = k(x - y)$. Solve these differential equations subject to $x(0) = x_0$, $x'(0) = u_0$, $y(0) = y_0$, $y'(0) = v_0$.

(e) What is the solution of the differential equations in part (d) subject to the initial conditions $x(0) = y(0) = x_0$ and $x'(0) = y'(0) = 0$? How does this agree with part (a)?

(f) What is the solution of the differential equations in part (d) subject to the initial conditions $x(0) = x_0$, $y(0) = -x_0$ and $x'(0) = y'(0) = 0$? How does this agree with part (b)?

(g) What is the solution of the differential equations in part (d) subject to the initial conditions $x(0) = x_0$, $y(0) = 0$ and $x'(0) = y'(0) = 0$? How does this agree with part (c)?

15. Write down a homogenous third order linear differential equation with constant coefficients that has 1, e^t, and e^{2t} as solutions.

16. Write down a homogenous third order linear differential equation with constant coefficients that has e^t, te^t, and $t^2 e^t$ as solutions.

17. Write down a homogenous fourth order linear differential equation with constant coefficients that has $\sin t$ and $\cos 2t$ as solutions. Is it possible to write down a homogenous second order linear differential equation with constant coefficients, that has $\sin t$ and $\cos 2t$ as solutions?

Section 8.8

Page 393. Add exercises following the current Exercise 3.

4. **Sagging Bookshelf.** The shape of a sagging bookshelf of length L can be determined by solving
$$\frac{d^4 y}{dx^4} = \frac{a}{EI},$$
where a, E, and I are constants, subject to $y(0) = 0$, $y'(0) = 0$, $y(L) = 0$, and $y'(L) = 0$. What is the shape?

5. Two cars are stationary behind each other at a stoplight at time $t = 0$. The first car's position is characterized by $x(t)$ and the second by $y(t)$. A simple model for the motion of these cars[1] (called Pipes' Model) is given by the system of equations
$$\begin{aligned} x'' &= a - x' \\ y'' &= x' - y', \end{aligned}$$
where a is a positive constant. It is said that the second driver adjusts his acceleration to be exactly equal in magnitude to the relative difference in velocity between his vehicle and the vehicle ahead. Explain why this is plausible. Write down a similar statement about the first driver's acceleration. What does a represent? Explain why the initial conditions $x(0) = x'(0) = 0$, $y(0) = y_0$, $y'(0) = 0$, are reasonable. What does y_0 represent? Solve this system of equations for $x(t)$ and $y(t)$. Is there any time at which the cars collide?

6. **Glucose**[2] A simple model used in studies of an oral glucose tolerance test involves the concentration of glucose in the blood, G, and the net effective hormonal concentration, H. If G_0 and H_0 are the nominal values of glucose and hormone in a body, and we

---

[1] "Car-Following Models" by R.L. Baker, Jr., in "Modules in Applied Mathematics, Volume 1, Differential Equations — Models" edited by M. Braun, C.S. Courteny, and D.A. Drew, Springer-Verlag 1983, Chapter 12

[2] This exercise is based on "Blood Glucose Regulation and Diabetes" by E. Ackerman et al., as reported in "Concepts and Models of Biomathematics" ed. F. Heinmets, Marcel Dekker, Inc. 1969, pages 131 - 156.

introduce new dependent variables, $g(t)$ and $h(t)$, by $g = G - G_0$, and $h = H - H_0$, the appropriate system of differential equations is

$$\begin{aligned} g' &= -m_1 g - m_2 h + J(t), \\ h' &= m_3 g - m_4 h, \end{aligned}$$

where m_1, m_2, m_3, and m_4 are positive constants. $J(t)$ is the rate of glucose into the system.

(a) Find an appropriate second order differential equation for $h(t)$

(b) Solve the equation from part (a) if $J(t) = J_0 e^{-t}$, where J_0 is a positive constant. Use the bottom equation to determine $g(t)$.

To find an analytical solution of a
$$a_2(t)y'' + a_1(t)y' + a_0(t)y = f(t)$$

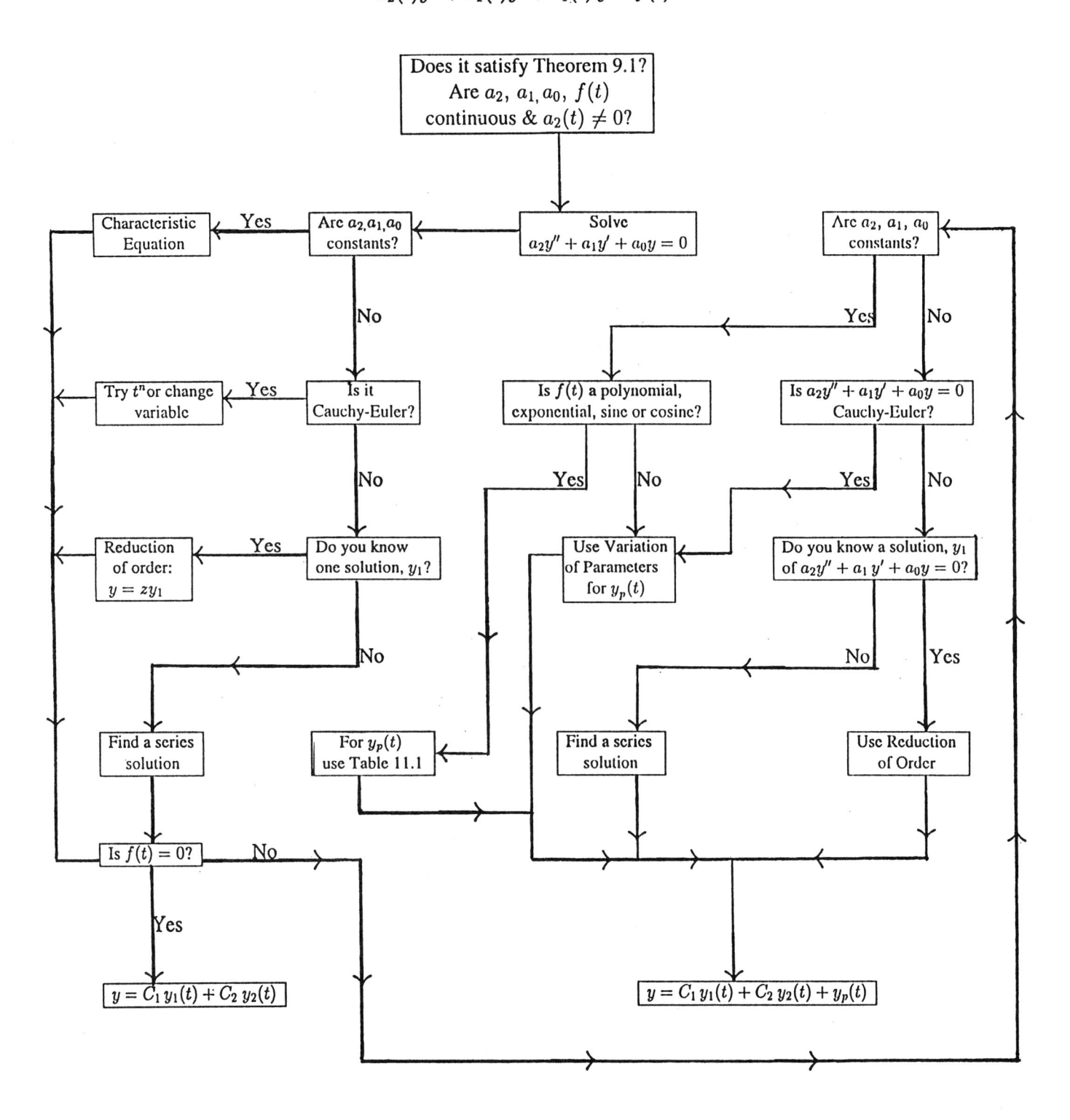

```
                    Does it satisfy Theorem 9.1?
                    Are a₂, a₁, a₀, f(t)
                    continuous & a₂(t) ≠ 0?
                              │
                              ▼
  Characteristic  ◄──Yes── Are a₂,a₁,a₀ ◄──── Solve ────► Are a₂, a₁, a₀ ──No──►
  Equation                 constants?       a₂y'' + a₁y' + a₀y = 0   constants?
                              │ No                  │                    │ Yes
                              ▼                     ▼                    ▼
  Try tⁿ or change  ◄──Yes── Is it          Is f(t) a polynomial,   Is a₂y'' + a₁y' + a₀y = 0
  variable                 Cauchy-Euler?    exponential, sine or cosine?   Cauchy-Euler?
                              │ No             Yes │ No                Yes │ No
                              ▼                    │                       ▼
  Reduction         ◄──Yes── Do you know           │              Do you know a solution, y₁
  of order:                  one solution, y₁?     │              of a₂y'' + a₁y' + a₀y = 0?
  y = zy₁                       │ No               │                    No │ Yes
                                ▼                  ▼                       ▼
  Find a series             For yₚ(t)       Use Variation          Use Reduction
  solution                  use Table 11.1  of Parameters          of Order
                                             for yₚ(t)
       │                                          │                       │
       ▼                                          ▼                       ▼
  Is f(t) = 0? ──No──►                     Find a series
       │ Yes                               solution
       ▼                                          │
  ┌──────────────────────┐                        ▼
  │ y = C₁y₁(t) + C₂y₂(t) │        ┌──────────────────────────────┐
  └──────────────────────┘         │ y = C₁y₁(t) + C₂y₂(t) + yₚ(t) │
                                    └──────────────────────────────┘
```

Figure 8.1 Flowchart

ADDITIONAL EXERCISES

9. IDEAS FOR CHAPTER 9 — "LINEAR AUTONOMOUS SYSTEMS"

9.1 Page by Page Comments

Chapter 9 requires 5 class meetings of 50 minutes each.
Chapters 9 follows from Chapter 7. It does not use any material from Chapter 8.

Page 388. There are additional exercises for Section 9.1 on page 268 of this manual.

Page 401. The Parabolic Classification Scheme is used in Chapter 10. A sophisticated student may ask why only two constants, namely, $D = ad - bc$ and $T = a + d$, are required in order to classify the equilibrium point of a system that has four arbitrary constants, a, b, c, and d. The answer is contained in the additional Exercise 5 for Section 9.2 on page 268 of this manual.

Page 403. Exercise 2 of Section 9.2 is time consuming to work through all possible cases for this exercise, but very illuminating.

Page 403. There are additional exercises for Section 9.2 on page 268 of this manual.

Page 403. In Section 9.5, we show that straight-line orbits are eigenvectors.

Page 403. It is worth pointing out that a straight-line orbit is really three orbits. See Footnote 7.

Page 403. Because the orbits satisfy a first order differential equation with homogeneous coefficients, zooming in or out on the phase plane will not change reveal anything new — the picture remains the same.

Page 411. It is worth stressing — as is pointed out in the third comment — that a nullcline analysis cannot by itself distinguish between a node, a center, and a focus.

Page 431. Inulin is the correct word.

Page 433. Table 9.1 is in TWIDDLE. It is called CALVES.DTA and is in the subdirectory DD-09.

9.2 Additional Exercises

Section 9.1

Page 388. Add exercises following the current Exercise 5.

6. Two cars are stationary behind each other at a stoplight at time $t = 0$. The first car's position is characterized by $x(t)$ and the second by $y(t)$, with velocities $u(t) = x'$ and $v(t) = y'$, respectively. A simple model for the motion of these cars[1] (called Pipes' Model) is given by the two equations

$$\begin{aligned} u' &= a - u \\ v' &= u - v, \end{aligned}$$

where a is a positive constant. It is said that the second driver adjusts her acceleration to be exactly equal in magnitude to the relative difference in velocity between her vehicle and the vehicle ahead. Explain why this is plausible. Write down a similar statement about the first driver's acceleration. What does a represent? Explain why the initial conditions $x(0) = u(0) = 0$, $y(0) = y_0$, $v(0) = 0$, are reasonable. What does y_0 represent? Solve this system for $u(t)$ and $v(t)$. Now solve $x' = u$, $y' = v$ for $x(t)$ and $y(t)$. Is there any time at which the cars collide?

Section 9.2

Page 403. Add exercise following the current Exercise 4.

5. **(a)** Show that the equilibrium points of

$$\begin{aligned} x' &= ax + by \\ y' &= cx + dy \end{aligned} \quad \text{and} \quad \begin{aligned} x' &= a\lambda x + b\lambda y \\ y' &= c\lambda x + d\lambda y \end{aligned},$$

where λ is any positive constant, have the same stability — that is, are classified the same under the parabolic classification scheme. By considering the transformation $t \to \lambda t$, explain how the first system can be converted into the second but, in general, they will have different explicit solutions.

(b) Show that the equilibrium points of

$$\begin{aligned} x' &= ax + by \\ y' &= cx + dy \end{aligned} \quad \text{and} \quad \begin{aligned} x' &= dx + cy \\ y' &= bx + ay \end{aligned}$$

have the same stability — that is, are classified the same under the parabolic classification scheme. By considering the transformation $x \longleftrightarrow y$, explain how the first system can be converted into the second but, in general, they will have different explicit solutions.

(c) Using parts (a) and (b), explain why only **two** constants, namely, $D = ad - bc$ and $T = a + d$, are required in order to classify the equilibrium point of the system

$$\begin{aligned} x' &= ax + by \\ y' &= cx + dy \end{aligned}$$

which has **four** arbitrary constants, a, b, c, and d.

---

[1] "Car-Following Models" by R.L. Baker, Jr., in "Modules in Applied Mathematics, Volume 1, Differential Equations — Models" edited by M. Braun, C.S. Courteny, and D.A. Drew, Springer-Verlag 1983, Chapter 12

10. IDEAS FOR CHAPTER 10 — "NONLINEAR AUTONOMOUS SYSTEMS"

10.1 Page by Page Comments

Chapter 10 requires 4 to 6 class meetings of 50 minutes each.
   Sections 10.5 through 10.7 are not critical for later sections.
   Chapter 10 follows from Section 9.4. It is not used in later chapters.

Page 439. We have the students draw the equilibrium points on Figure 10.1.

Page 441. When we talk about closed curves, we mean simple closed curves — curves that do not intersect, i.e. they look like circles as opposed to eights.

Page 442. We have the students draw the equilibrium points on Figure 10.2.

Page 444. We have the students draw the equilibrium points on Figure 10.4.

Page 445. We have the students draw the equilibrium points on Figure 10.6.

Page 448. There are additional exercises for Section 10.1 on page 270 of this manual.

Page 448. It is worthwhile having the students think about constructing a phase portrait consistent with the information at the bottom of the page.

Page 451. We have the students draw the equilibrium points on Figure 10.9.

Page 452. It is worth stressing — as is pointed out in the Important point — that a nullcline analysis cannot distinguish between a node, center, or focus.

Page 453. We have the students draw the equilibrium points on Figure 10.11.
Page 455. There are additional exercises for Section 10.2 on page 270 of this manual.

Page 456. It is worthwhile having the students think about constructing a phase portrait consistent with the information at the top of the page.

Page 460. We have the students draw the equilibrium points on Figure 10.16.

Page 460. It is worth stressing — as is pointed out in the Important point — that a linearization analysis cannot distinguish between a center and a spiral.

Page 462. We have the students draw the equilibrium points on Figure 10.18.

Page 464. We have the students draw the equilibrium points on Figures 10.19 and 10.20.

Page 465. It is also worth pointing out that the phase plane does not tell the whole story. See BE CAREFUL INTERPRETING TRAJECTORIES IN THE PHASE PLANE on page 274 of this manual.

Page 469. We have the students draw the equilibrium points on Figure 10.22.

Page 469. We have the students draw the equilibrium points on Figure 10.24.

Page 474. We have the students draw the equilibrium points on Figure 10.26.

Page 476. Footnote 4 is interesting.

Page 484. There are additional exercises for Section 10.4 on page 271 of this manual.

Page 485. Section 10.5, *Bungee Jumping*, is optional.

Page 491. Table 10.4 is in TWIDDLE. It is called RUBBER.DTA and is in the subdirectory DD-10.

Page 492. Section 10.6, *Linear Versus Nonlinear Differential Equations*, is optional.

Page 498. There are additional exercises for Section 10.6 on page 273 of this manual.

Page 498. Section 10.7, *Autonomous Versus Nonautonomous Differential Equations*, is optional.

10.2 Additional Exercises

Section 10.1

Page 448. Add exercises following the current Exercise 9.

10. Assuming that orbits have the properties described in the summary on page 446 of the text, explain why the curves in Figure 10.1 are not orbits.

11. Is it possible for an orbit of $x' = P(x, y)$, $y' = Q(x, y)$ to oscillate backward and forward on the x-axis between $(-1, 0)$ and $(1, 0)$?

Section 10.2

Page 455. Add exercises following the current Exercise 4.

5. Consider the system of equations $x' = y - x + y^2 - xy$, $y' = -x + y^2$.

    (a) Determine all the equilibrium points.

    (b) Using a nullcline analysis, determine the stability (if possible) of these equilibrium points.

Figure 10.1 Curves, but not orbits

Section 10.4

Page 484. Add exercises following the current Exercise 6.

7. **S-I-R Model**[1] A simple S-I-R model that includes the possibility of reinfection (as with a sore throat) is $S' = -aSI + bI$, $I' = aSI - bI$, where a and b are positive constants.

    (a) Find the appropriate differential equation for the orbits in the phase plane. Contrast the equilibrium points and nullclines for this phase plane with those from Example 10.14.

    (b) Find an explicit solution using initial conditions $I(0) = n$, $S(0) = N - n$. What is the limit of your solution for large times?

    (c) Using the values $N = 261$, $a = 2.78/159$, and $b = 2.78$, draw some orbits for your phase plane and compare the results with those in Figure 10.32.

8. **Bacteria**[2] A model of bacteria growth that accounts for changes in a limiting nutrient as well as bacteria size can be written as $dP/dt = vNP/(k + N)$, $dN/dt = -vNP/[c(k + N)]$, where $P(t)$ and $N(t)$ are the concentration of cells and nutrients (respectively) as a function of time. The bacterial yield, c, saturation constant, k, and uptake velocity, v, are all positive constants. Typical initial conditions would be $P(0) = P_0$, $N(0) = N_0$.

    (a) Determine all the equilibrium points and do a nullcline analysis.

---

[1] This exercise is based on "Qualitative Analyses of Communicable Disease Models" by H. W. Hethcote, Mathematical Biosciences 28, 1976, pages 335 - 356.

[2] The exercise is based on "Mathematics in Medicine and the Life Sciences" by F.C. Hoppensteadt and C.S. Peskin, Springer-Verlag, 1992, page 27.

(b) By linearizing these equations about the equilibrium points, determine the stability of the equilibrium points as a function of the parameters.

(c) Find the differential equation for the orbits in the phase plane and find its solution.

(d) Use the result from part (b) to find $P(t)$ and $N(t)$ explicitly.

(e) What is the long term behavior of your solution from part (c)?

9. **Chemostat**[3] A chemostat is a continuous culture device used for growing and studying bacteria. Nutrient is added at a constant rate, say a N_0, to the growth chamber where living cells are stirred in the enriched media. The growth chamber is continually adjusted to keep a constant volume by removing liquid at the rate of a. Let $N(t)$ be the concentration of nutrient and $B(t)$ the concentration of bacteria as a function of time. A differential system that is used to model this chemostat is $dN/dt = a(N_0 - N) - vNB/[c(k+N)]$, $dB/dt = vNB/(k+N) - aB$, where a is the flow rate, v the maximum uptake rate, k the saturation constant of nutrient uptake and c is the yield of cells per unit nutrient extracted.

   (a) Find all equilibrium points for this system.

   (b) By linearizing these equations about the equilibrium points, determine the stability of the equilibrium points that occur on the vertical axis. Give relations between the parameters where the stability changes.

   (c) Find relations between the parameters so one equilibrium point is in the first quadrant.

   (d) For the hypothetical case $c = k = 1$, $v = 2$, $N_0 = 10$, use linearization to determine the stability of the equilibrium points as a function of the flow rate a. Does the behavior for large values of flow rate with your intuition? Why or why not?

10. Consider the system of equations $x' = y - x + y^2 - xy$, $y' = -x + y^2$.

    (a) Determine all the equilibrium points.

    (b) Using a nullcline analysis, determine the stability (if possible) of these equilibrium points.

    (c) By linearizing these equations about the equilibrium points, determine the stability of the equilibrium points

11. **Nonlinear Springs**. On page 194 of this manual you looked at an example of a nonlinear spring where
$$F(x) = \frac{x}{1-x^2}.$$

    [Here x is measured from the natural length of the spring. In that example x — which we shall now call X to avoid confusion — was measured from the top of the spring of length 1, so $X = x + 1$. Thus,
$$\frac{X-1}{X(2-X)} = \frac{x}{1-x^2}$$

    gives $F(x)$.]

    (a) Sketch $F(x)$ and compare it to Figure 10.36 of the text.

    (b) Expand $F(x)$ in a Taylor series about $x = 0$ and compare it to (10.46) of the text — $F(x) = kx + ax^3$ — for small x.

---

[3] The exercise is based on "Mathematics in Medicine and the Life Sciences" by F.C. Hoppensteadt and C.S. Peskin, Springer-Verlag, 1992, page 41.

(c) Based on your answers to parts (a) and (b) does this spring behave like a hard or a soft spring? Is this consistent with the conclusions you drew for the exercise on page 194 of this manual?

Section 10.6

Page 498. Add exercises following the current Exercise 12.

13. The Linear and Nonlinear Pendulum

(a) Solve the linearized pendulum equation $x'' + 9x = 0$, subject to (i) $x(0) = 1$, $x'(0) = 0$; (ii) $x(0) = 0$, $x'(0) = 1$; (iii) $x(0) = 1$, $x'(0) = 1$, to find $x(1)$ in each case. How is the solution you found in part (iii) related to the solutions in parts (i) and (ii)? Is this always the case? Why?

(b) **Computer Experiment.** Solve the nonlinear pendulum equation $x'' + 9\sin x = 0$, subject to (i) $x(0) = 1$, $x'(0) = 0$; (ii) $x(0) = 0$, $x'(0) = 1$; (iii) $x(0) = 1$, $x'(0) = 1$, to find $x(1)$ in each case. How is the solution you found in part (iii) related to the solutions in parts (i) and (ii)? Is this always the case? Why?

10.3 Additional Materials

Be Careful Interpreting Trajectories in the Phase Plane

If we consider the nonlinear system

$$x' = x(x+y)$$
$$y' = y(x+y)$$

for $x \geq 0$, $y \geq 0$, the only equilibrium point is $(0,0)$ and the phase plane equation is

$$\frac{dy}{dx} = \frac{y}{x},$$

with trajectories $y = cx$. These trajectories move away from the equilibrium point $(0,0)$ as t increases and are shown in Figure 10.2.

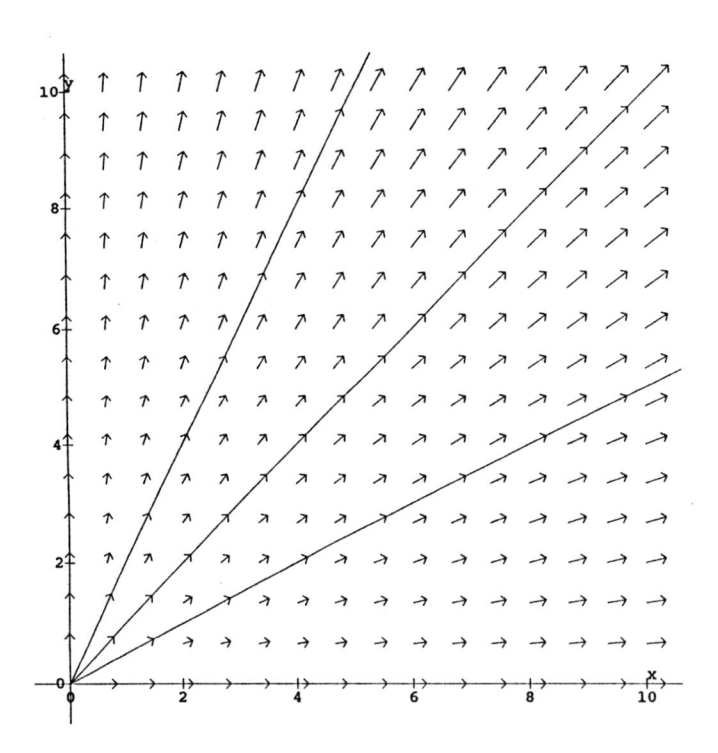

Figure 10.2 Phase plane for $x' = x(x+y)$, $y' = y(x+y)$

If we substitute $y = cx$ into $x' = x(x+y)$ we find $x' = (1+c)x^2$ with solution

$$x(t) = \frac{-1}{a + (1+c)t},$$

so

$$y(t) = \frac{-c}{a + (1+c)t}.$$

IDEAS FOR CHAPTER 10 — "NONLINEAR AUTONOMOUS SYSTEMS"

If we use the initial conditions $x(0) = x_0$, $y(0) = y_0$, we find

$$x(t) = \frac{x_0}{1 - (x_0 + y_0)t},$$

$$y(t) = \frac{y_0}{1 - (x_0 + y_0)t}.$$

Notice that for $x_0 > 0$, $y_0 > 0$ as $t \to 1/(x_0 + y_0)$ then $x \to \infty$ and $y \to \infty$. Thus, the trajectories that emanate from the origin reach infinity in a finite time $1/(x_0 + y_0)$. Thus, although the phase plane indicates that the origin is an unstable node (a source) in this case the trajectories go to infinity in finite time.

ADDITIONAL MATERIALS

11. IDEAS FOR CHAPTER 11 — "USING LAPLACE TRANSFORMS"

11.1   Page by Page Comments

Chapter 11 requires 5 to 7 class meetings of 50 minutes each.

This chapter follows from Section 8.4. It does not use any material from Chapters 9 and 10. It is not used in later chapters.

This is a very traditional treatment of the Laplace Transform, except the discussion of its existence is delayed until the final section, and the initial motivation is more in keeping with the philosophy of the book.

Pages 505 — 511. This is where we motivate the Laplace Transform in a nontraditional way.

Page 517. We could solve (11.29) by writing it as

$$\frac{s^{n+1}}{n!}\mathcal{L}\{t^n\} = \frac{s^n}{(n-1)!}\mathcal{L}\{t^{n-1}\}.$$

By repeated use of this identity we find

$$\frac{s^{n+1}}{n!}\mathcal{L}\{t^n\} = \frac{s^n}{(n-1)!}\mathcal{L}\{t^{n-1}\} = \frac{s^{n-1}}{(n-2)!}\mathcal{L}\{t^{n-2}\} = \cdots = \frac{s^2}{1!}\mathcal{L}\{t\}.$$

But $\mathcal{L}\{t\} = 1/s^2$, so $\mathcal{L}\{t^n\} = n!/s^{n+1}$, which is (11.30).

Page 529. Not all the entries of Table 11.6 are derived in the text. Those that are not make excellent exercises.

Page 546. The inverse Laplace transform of $1/\left[(s-k)s^2\right]$ can also be obtained from the final entry in Table 11.6 and the Shifting Theorem.

Page 546. The inverse Laplace transform of $1/\left[s(s+1)^2\right]$ can also be obtained from the final entry in Table 11.6.

Page 549. Students find Footnote 7 interesting.

Page 560. There are additional exercises for Section 11.6 on page 278 of this manual.

Page 565. The inverse Laplace transform of $2/\left[s^2(s+4)\right]$ can also be obtained from the final entry in Table 11.6 and the Shifting Theorem.

Page 566. There are additional exercises for Section 11.7 on page 278 of this manual.

11.2 Additional Exercises

Section 11.6

Page 560. Add exercise after the current Exercise 7.

8. **Glucose**[1] A simple model used in studies of an oral glucose tolerance test involves the concentration of glucose in the blood, G, and the net effective hormonal concentration, H. If G_0 and H_0 are the nominal values of glucose and hormone in a body, and we introduce new dependent variables, $g(t)$ and $h(t)$, by $g = G - G_0$, and $h = H - H_0$, the appropriate system of differential equations is

$$\begin{aligned} g' &= -m_1 g - m_2 h + J(t), \\ h' &= m_3 g - m_4 h, \end{aligned}$$

where m_1, m_2, m_3, and m_4 are positive constants. $J(t)$ is the rate of glucose into the system.

(a) Find an appropriate second order differential equation for $h(t)$.

(b) Solve the equation from part (a) if $J(t) = J_0 \delta(t)$, where J_0 is a positive constant. (The case for an injection.) Use the bottom equation to determine $g(t)$.

(c) Repeat (b) if $J(t) = J_0[u(t) - u(t-1)]$. (The case for an intravenous dose.)

Section 11.7

Page 566. Add exercises following the current Exercise 7.

8. **Glucose**[2] A simple model used in studies of an oral glucose tolerance test involves the concentration of glucose in the blood, G, and the net effective hormonal concentration, H. If G_0 and H_0 are the nominal values of glucose and hormone in a body, and we introduce new dependent variables, $g(t)$ and $h(t)$, by $g = G - G_0$, and $h = H - H_0$, the appropriate system of differential equations is

$$\begin{aligned} g' &= -m_1 g - m_2 h + J(t), \\ h' &= m_3 g - m_4 h, \end{aligned}$$

where m_1, m_2, m_3, and m_4 are positive constants. $J(t)$ is the rate of glucose into the system.

(a) Solve this system if $J(t) = J_0 \delta(t)$, where J_0 is a positive constant. (The case for an injection.)

(b) Repeat (a) if $J(t) = J_0[u(t) - u(t-1)]$. (The case for an intravenous dose.)

---

[1] This exercise is based on "Blood Glucose Regulation and Diabetes" by E. Ackerman et al., as reported in "Concepts and Models of Biomathematics" ed. F. Heinmets, Marcel Dekker, Inc. 1969, pages 131 - 156.

[2] This exercise is based on "Blood Glucose Regulation and Diabetes" by E. Ackerman et al., as reported in "Concepts and Models of Biomathematics" ed. F. Heinmets, Marcel Dekker, Inc. 1969, pages 131 - 156.

12. IDEAS FOR CHAPTER 12 — "USING POWER SERIES"

12.1 Page by Page Comments

Chapter 12 requires 4 or 5 class meetings of 50 minutes each.
Chapter 12 follows from Section 8.4.

Page 580. Example 12.3 is the differentiated form of $y' = 1 - 2xy$, which occurs in Exercise 10, page 92; Exercise 2, page 117; Example 5.8, page 205; Exercise 27, page 211. It is also the same example as Example 12.5 on page 589.

Page 584. Notice that the case $m = 0$ is included in (12.20).

Page 585. Notice that the case $m = 0$ is included in (12.21).

Page 589. Example 12.3 is the differentiated form of $y' = 1 - 2xy$, which occurs in Exercise 10, page 92; Exercise 2, page 117; Example 5.8, page 205; Exercise 27, page 211. It is also the same example as Example 12.3 on page 580.

Page 590. Notice that the case $m = 0$ is included in the equations for c_{2m} and c_{2m+1}.

Page 585. Notice that the case $m = 0$ is included in (12.21).

Page 586. The technique described for solving recurrence relations under the third bullet is used repeatedly in the sequel. Here are some exercises.

(a) Solve $(2m+3)(2m+2)c_{m+1} + c_m = 0$.
(b) Solve $(2m+2)(2m+1)c_{m+1} + c_m = 0$.
(c) Solve $(m+2)(m+1)c_{m+2} + c_m = 0$.
(d) Solve $m^2 c_{2m} + c_{2m-2} = 0$.
(e) Solve $m(m-2)c_m + (m-3)c_{m-1} = 0$.

Page 606. We frequently use Theorem 12.4 to find the indicial equation.

Page 608. Expressing x in terms of $x - 3$ by $x = (x-3) + 3$ and similar expansions — a technique needed for Example 12.17 and subsequent exercises — is foreign to many of our students.

Page 609. Example 12.18 could be replaced by other examples. For example see the section called A REPLACEMENT FOR EXAMPLE 12.18 on page 281 of this manual.

Page 615. Comment for Example 12.20.
The differential equation (12.65), namely,

$$3x^2 y'' + xy' - \frac{2}{2-x} y = 0,$$

can also be solved by writing it in the form

$$3(2-x)x^2 y'' + x(2-x) y' - 2y = 0.$$

In this case

$$3(2-x)x^2 y'' + x(2-x) y' - 2y$$
$$= 2(3s+1)(s-1) c_0 x^s + \sum_{n=0}^{\infty} [2c_{n+1}(3n+3s+4)(n+s) - c_n(3n+3s-2)(n+s)] x^{n+s+1},$$

giving the indicial equation

$$(3s+1)(s-1) = 0,$$

and recurrence relation

$$2c_{n+1}(3n+3s+4)(n+s) = c_n(3n+3s-2)(n+s), \quad n = 0, 1, 2, \cdots.$$

With $s = 1$, we find

$$c_{n+1} = \frac{3n+1}{2(3n+7)} c_n,$$

giving

$$c_1 = \frac{1}{14} c_0, \qquad c_2 = \frac{1}{70} c_0, \qquad c_3 = \frac{1}{260} c_0.$$

With $s = -1/3$, we find

$$c_{n+1} = \frac{n-1}{2(n+1)} c_n,$$

giving

$$c_1 = -\frac{1}{2} c_0, \qquad c_2 = 0, \qquad c_3 = 0.$$

Page 630. A flowchart showing how to solve second order equations can be found on page 284 in this manual.

12.2 Additional Materials

A Replacement for Example 12.18

Example 12.1 :

We now find the general solution of

$$x^2 y'' - 2xy' + (2 + x^2) y = 0 \tag{12.1}$$

as a series about the origin using the method of Frobenius. We see that $x = 0$ is a regular singular point, and using Theorem 10.1 we see that the series solution we obtain will converge for all values of x. If we substitute the series in (10.20), with $x_0 = 0$, into this differential equation, we obtain

$$\sum_{n=0}^{\infty}(n+s)(n+s-1)c_n x^{n+s} - \sum_{n=0}^{\infty} 2(n+s)c_n x^{n+s} + \sum_{n=0}^{\infty} 2c_n x^{n+s} + \sum_{n=0}^{\infty} c_n x^{n+s+2} = 0,$$

which, by combining the first three terms, we can rewrite as

$$\sum_{n=0}^{\infty} [(n+s)(n+s-1) - 2(n+s) + 2] c_n x^{n+s} + \sum_{n=0}^{\infty} c_n x^{n+s+2} = 0.$$

By noting that $(n+s)(n+s-1) - 2(n+s) + 2 = (n+s)(n+s) - (n+s) - 2(n+s) + 2 = (n+s)^2 - 3(n+s) + 2 = (n+s-1)(n+s-2)$ we finally have

$$\sum_{n=0}^{\infty} (n+s-1)(n+s-2) c_n x^{n+s} + \sum_{n=0}^{\infty} c_n x^{n+s+2} = 0.$$

We now raise the index of summation in the first term by 2 (that is, we let $n - 2 = m$) and get

$$(s-1)(s-2) c_0 x^s + s(s-1) c_1 x^{1+s} + \sum_{m=0}^{\infty} (m+s+1)(m+s) c_{m+2} x^{m+s+2} + \sum_{n=0}^{\infty} c_n x^{n+s+2} = 0. \tag{12.2}$$

The indicial equation is obtained by setting the coefficient of the lowest power of x to 0, giving

$$(s-1)(s-2) c_0 = 0.$$

Because $c_0 \neq 0$, we have 2 and 1 as our two values of s.

The recurrence relation is obtained by setting the coefficients of the remaining powers of x in (12.2) to 0. This gives

$$s(s-1) c_1 = 0$$

and

$$(m+s+1)(m+s) c_{m+2} + c_m = 0, \qquad m = 0, 1, 2, \cdots. \tag{12.3}$$

If we use the larger value of s, $s = 2$, in this recurrence relation, we obtain

$$c_1 = 0$$

and
$$(m+3)(m+2)c_{m+2} + c_m = 0, \quad m = 0, 1, 2, \cdots.$$

By putting $m = 1, 3, 5, \cdots$ into the recurrence relation, and using the fact that $c_1 = 0$ we find all the terms with odd subscripts are zero, that is,
$$c_{2m+1} = 0, \quad m = 0, 1, 2, \cdots.$$

We have no condition on c_0, so it may be chosen arbitrarily, and the rest of the even coefficients are determined in terms of c_0 by
$$(2m+3)(2m+2)c_{2m+2} + c_{2m} = 0, \quad m = 0, 1, 2, \cdots.$$

so that
$$\begin{aligned}
c_{2m+2} &= -1/[(2m+3)(2m+2)]\, c_{2m} \\
&= -1/[(2m+3)(2m+2)]\{-1/[(2m+1)(2m)]c_{2m-2}\} \\
&= (-1)^2/[(2m+3)(2m+2)(2m+1)(2m)]\, c_{2m-2} \\
&= (-1)^2/[(2m+3)(2m+2)(2m+1)(2m)]\{-1/[(2m-1)(2m-2)]c_{2m-4}\} \\
&= (-1)^3/[(2m+3)(2m+2)(2m+1)(2m)(2m-1)(2m-2)]\, c_{2m-4} \\
&= (-1)^{m+1}/[(2m+3)(2m+2)(2m+1)(2m)\cdots(3)(2)]\, c_0 \\
&= (-1)^{m+1}/[(2m+3)!]\, c_0, \quad m = 0, 1, 2, \cdots.
\end{aligned}$$

This gives the series corresponding to $s = 2$ as
$$c_0 + \sum_{m=0}^{\infty} \frac{(-1)^{m+1}}{(2m+3)!} x^{2m+2} c_0 = c_0 \sum_{m=0}^{\infty} \frac{(-1)^m}{(2m+1)!} x^{2m},$$

and so a solution of (12.1) may be written as
$$y_1(x) = x^2 \sum_{m=0}^{\infty} \frac{(-1)^m}{(2m+1)!} x^{2m},$$

where we have restricted ourselves to the case $x > 0$. This series can be written in the form
$$y_1(x) = x \sum_{m=0}^{\infty} \frac{(-1)^m}{(2m+1)!} x^{2m+1},$$

which we recognize converges to
$$y_1(x) = x \sin x.$$

Because we have found one solution, we can now use reduction of order techniques to obtain a second solution. Recall from Section 9.2 that when we know one solution, say $y_1(x)$, of
$$a_2(x)\frac{d^2y}{dx^2} + a_1(x)\frac{dy}{dx} + a_0(x)y = 0,$$

then the second one is
$$y(x) = y_1 \int \left[\frac{1}{y_1^2} \exp\left(-\int \frac{a_1}{a_2} dx\right)\right] dx.$$

In our example, $a_1(x)/a_2(x) = -2/x$, so $\int [a_1(x)/a_2(x)]\, dx = 2\ln x$. This means that
$$\begin{aligned}
y_2(x) &= y_1 \int \tfrac{1}{y_1^2} \exp(2\ln x)\, dx \\
&= y_1(x) \int \tfrac{x^2}{y_1^2}\, dx,
\end{aligned}$$

or
$$y_2(x) = x\sin x \int \frac{x^2}{x^2 \sin^2 x}\, dx = x\sin x \int \csc^2 x\, dx.$$

This is a standard integral, $\int \csc^2 x\, dx = -\cot x$, giving rise to

$$y_2(x) = -x \cos x.$$

Thus, the general solution may be written

$$y(x) = C_1 x \sin x + C_2 x \cos x.$$

Although this is the solution we seek, let's see what happens if we use the other solution of the indicial equation ($s = 1$) in the recurrence relation of (12.3), namely,

$$s(s-1)c_1 = 0$$

and

$$(m+s+1)(m+s)c_{m+2} + c_m = 0, \qquad m = 0, 1, 2, \cdots.$$

Doing so shows that c_1 is undetermined and

$$c_{m+2} = -\frac{1}{(m+2)(m+1)} c_m, \qquad m = 0, 1, 2, \cdots.$$

After some calculation we find

$$c_{2m+2} = \frac{(-1)^m}{(2m+2)!} c_0$$

and

$$c_{2m+1} = \frac{(-1)^m}{(2m+1)!} c_1.$$

Thus, we have the series solution corresponding to $s = 1$ as

$$y(x) = c_0 x \sum_{m=0}^{\infty} \frac{(-1)^m}{(2m)!} x^{2m} + c_1 x \sum_{m=0}^{\infty} \frac{(-1)^m}{(2m+1)!} x^{2m+1},$$

which we immediately recognize as the series expansion for the function $c_0 x \sin x + c_1 x \cos x$.

Figure 12.1 shows the graphs of $x \sin x$ and $x \cos x$. Notice how the roots are interlaced for $x > 0$. □

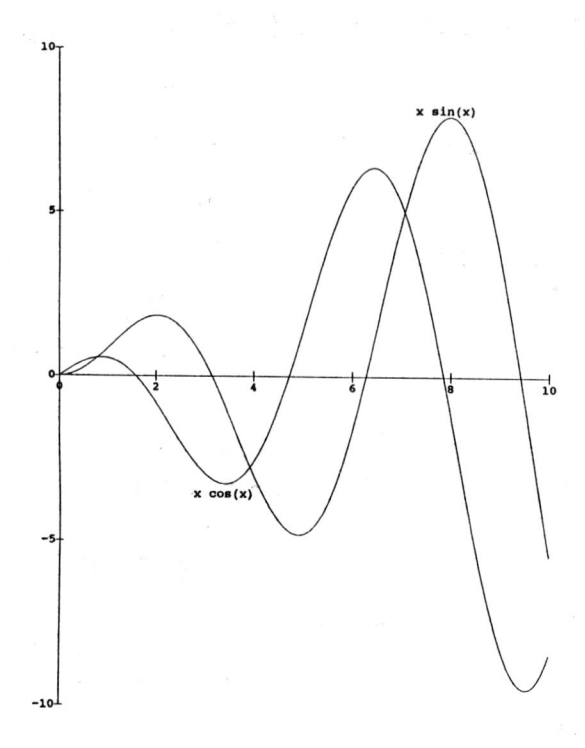

Figure 12.1 The functions $x \sin x$ and $x \cos x$

To find an analytical solution of a
$$a_2(t)y'' + a_1(t)y' + a_0(t)y = f(t)$$

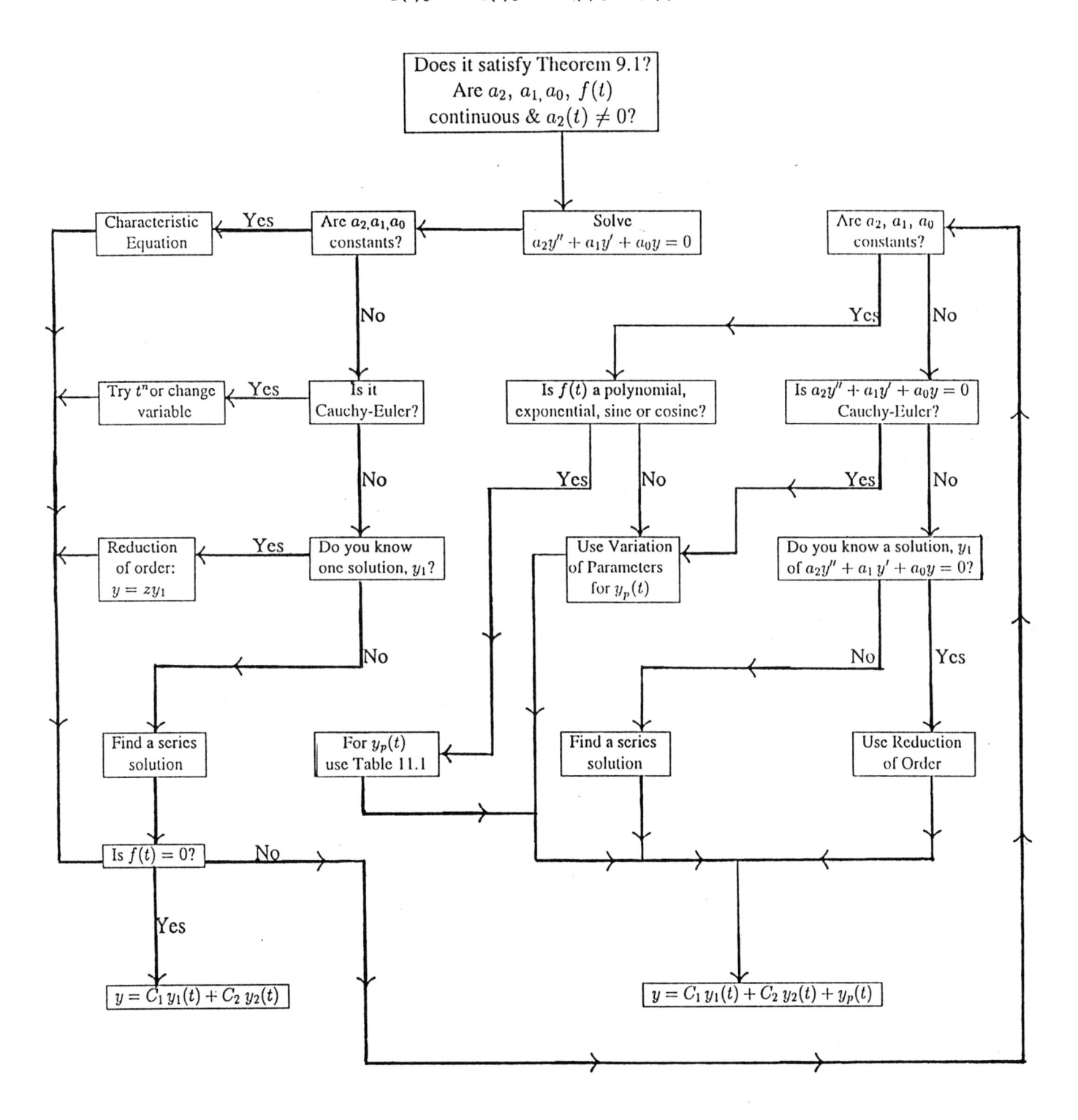

- Does it satisfy Theorem 9.1? Are $a_2, a_1, a_0, f(t)$ continuous & $a_2(t) \neq 0$?
- Solve $a_2 y'' + a_1 y' + a_0 y = 0$
- Are a_2, a_1, a_0 constants?
  - Yes: Characteristic Equation
  - No: Is it Cauchy-Euler?
    - Yes: Try t^n or change variable
    - No: Do you know one solution, y_1?
      - Yes: Reduction of order: $y = zy_1$
      - No: Find a series solution
- Is $f(t) = 0$?
  - Yes: $y = C_1 y_1(t) + C_2 y_2(t)$
  - No: Is $f(t)$ a polynomial, exponential, sine or cosine?
    - Yes: For $y_p(t)$ use Table 11.1
    - No: Are a_2, a_1, a_0 constants?
      - Yes: Use Variation of Parameters for $y_p(t)$
      - No: Is $a_2 y'' + a_1 y' + a_0 y = 0$ Cauchy-Euler?
        - Yes: Do you know a solution, y_1 of $a_2 y'' + a_1 y' + a_0 y = 0$?
          - Yes: Use Reduction of Order
          - No: Find a series solution
        - No: Do you know a solution, y_1 of $a_2 y'' + a_1 y' + a_0 y = 0$?
- $y = C_1 y_1(t) + C_2 y_2(t) + y_p(t)$

Figure 12.2 Flowchart

ADDITIONAL MATERIALS

13. USING MATRICES

Where We Are Going — and Why

In Section 9.5 we used eigenvalues, eigenvectors, and the notion of a fundamental matrix to solve systems of homogeneous linear differential equations with constant coefficients. In this chapter we extend these methods to the nonhomogeneous case by following the ideas developed in Chapter 8. We do this in two ways — first, by generalizing the method of undetermined coefficients; and second, by generalizing the method of variation of parameters. These methods for solving systems of two linear differential equations with constant coefficients are then extended to systems of higher order equations.

The methods of solution developed in this chapter are those of choice when analyzing complicated systems of differential equations with many dependent variables.

13.1 Method of Undetermined Coefficients Using Matrices

We now develop methods for solving nonhomogeneous systems of linear differential equations with constant coefficients, namely,

$$\begin{cases} x' = ax + by + f(t), \\ y' = cx + dy + g(t), \end{cases}$$

where a, b, c, and d are given constants and $f(t)$, $g(t)$, are given functions of t, which, using matrix notation, can be written as

$$\mathbf{X}' = \mathbf{M}\mathbf{X} + \mathbf{V},$$

where

$$\mathbf{X} = \begin{bmatrix} x \\ y \end{bmatrix}, \qquad \mathbf{M} = \begin{bmatrix} a & b \\ c & d \end{bmatrix}, \qquad \mathbf{V} = \begin{bmatrix} f(t) \\ g(t) \end{bmatrix}.$$

We will combine the matrix formulation from Section 9.5 with the method for solving nonhomogeneous second order linear differential equations in Chapter 7 — the Method of Undetermined Coefficients. This section will parallel Section 7.2, so we will consider forcing functions that are exponentials, polynomials, and sine and cosine functions.

We start with a slight modification of Example 9.1 — the Two-Container Mixture Problem — in which, instead of adding pure water to one of the containers, we add a salt solution.

Example 13.1 : Two-Container Mixture Problem Revisited.

A solution with concentration of 4 pounds of salt per gallon is entering container A at a rate of 3 gallons per minute, and the well-stirred mixture is leaving container B at the same rate. (See Figure 13.1.) There are 100 gallons in each container, and the well-stirred mixture flows from container A to container B at a rate of 4 gallons per minute and leaks back from container B to container A at a rate of 1 gallon per minute. Predict the amounts of salt in each container at any time.

We let $x(t)$ and $y(t)$ be the amounts of salt in containers A and B respectively at time t. Both x and y have units of pounds, and t has units of minutes. The appropriate differential

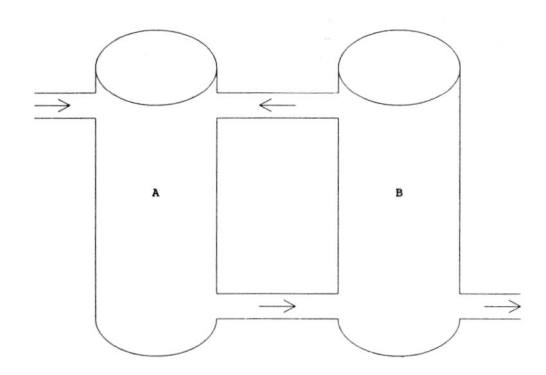

Figure 13.1 Two connected containers

equations governing this system are obtained from conservation considerations. For container A we have
$$x' = \text{input} - \text{output}$$
$$= (4)(3) + (y/100)(1) - (x/100)(4),$$
or
$$x' = 12 + \frac{1}{100}y - \frac{1}{25}x.$$

For container B we have
$$y' = \frac{x}{100}4 - \frac{y}{100}1 - \frac{y}{100}3 = \frac{1}{25}x - \frac{1}{25}y.$$

To find an explicit solution of this system of equations, we recast them in matrix form as
$$\mathbf{X}' = \mathbf{M}\mathbf{X} + \mathbf{V}, \tag{13.1}$$

where
$$\mathbf{X} = \begin{bmatrix} x \\ y \end{bmatrix}, \quad \mathbf{M} = \begin{bmatrix} -1/25 & 1/100 \\ 1/25 & -1/25 \end{bmatrix}, \quad \mathbf{V} = \begin{bmatrix} 12 \\ 0 \end{bmatrix}.$$

In Section 7.1 we found general solutions of nonhomogeneous linear differential equations as the sum of the general solution of the associated homogeneous differential equation and a particular solution. This is the situation here as well (see Exercise 2, page 298), so we first need to solve the associated homogeneous system,
$$\mathbf{X}'_\mathbf{h} = \mathbf{M}\mathbf{X}_\mathbf{h}. \tag{13.2}$$

In Section 9.5 we developed the general solution of (13.2) by considering the matrix equation
$$\begin{bmatrix} -1/25 - r & 1/100 \\ 1/25 & -1/25 - r \end{bmatrix} \begin{bmatrix} A \\ B \end{bmatrix} = \begin{bmatrix} 0 \\ 0 \end{bmatrix},$$

with a characteristic equation of $(-1/25 - r)(-1/25 - r) - 1/2500 = 0$, and eigenvalues $r = -1/25 \pm 1/50$, or $-1/50$ and $-3/50$. To find the eigenvector associated with $r = -1/50$, we solve
$$\begin{bmatrix} -1/50 & 1/100 \\ 1/25 & -1/50 \end{bmatrix} \begin{bmatrix} A \\ B \end{bmatrix} = \begin{bmatrix} 0 \\ 0 \end{bmatrix},$$

giving $B = 2A$, and eigenvector
$$\begin{bmatrix} 1 \\ 2 \end{bmatrix} A.$$

Associated with $r = -3/50$ we have

$$\begin{bmatrix} 1/50 & 1/100 \\ 1/25 & 1/50 \end{bmatrix} \begin{bmatrix} A \\ B \end{bmatrix} = \begin{bmatrix} 0 \\ 0 \end{bmatrix},$$

giving $B = -2A$, and eigenvector

$$\begin{bmatrix} 1 \\ -2 \end{bmatrix} A.$$

Letting $A = 1$ in the last two expressions gives $\mathbf{X_h}(t) = \mathbf{UC}$, where a fundamental matrix $\mathbf{U}$ is given by

$$\mathbf{U} = \begin{bmatrix} e^{-t/50} & e^{-3t/50} \\ 2e^{-t/50} & -2e^{-3t/50} \end{bmatrix},$$

and $\mathbf{C}$ is a vector of arbitrary constants.

Now we need to find a particular solution, $\mathbf{X_p}(t)$, of the original equation (13.1). Because $\mathbf{V}$ is a constant vector, we try a particular solution in the form of a constant vector. Thus, we try

$$\mathbf{X_p}(t) = \begin{bmatrix} K_1 \\ K_2 \end{bmatrix}.$$

Substituting this trial solution into (13.1) gives the set of algebraic equations

$$\begin{bmatrix} -1/25 & 1/100 \\ 1/25 & -1/25 \end{bmatrix} \begin{bmatrix} K_1 \\ K_2 \end{bmatrix} = \begin{bmatrix} -12 \\ 0 \end{bmatrix},$$

with solution $K_1 = K_2 = 400$, so

$$\mathbf{X_p}(t) = \begin{bmatrix} 400 \\ 400 \end{bmatrix}.$$

This gives the general solution of our original system of differential equations as

$$\mathbf{X}(t) = \begin{bmatrix} x(t) \\ y(t) \end{bmatrix} = \begin{bmatrix} e^{-t/50} & e^{-3t/50} \\ 2e^{-t/50} & -2e^{-3t/50} \end{bmatrix} \begin{bmatrix} C_1 \\ C_2 \end{bmatrix} + \begin{bmatrix} 400 \\ 400 \end{bmatrix}, \quad (13.3)$$

or in matrix form as

$$\mathbf{X}(t) = \mathbf{UC} + \mathbf{X_p}(t),$$

where

$$\mathbf{C} = \begin{bmatrix} C_1 \\ C_2 \end{bmatrix}.$$

Had this process started with no salt in either container, the initial conditions at $t = 0$ would be $x(0) = y(0) = 0$. If we substitute these values into (13.3), we obtain

$$\begin{bmatrix} 0 \\ 0 \end{bmatrix} = \begin{bmatrix} 1 & 1 \\ 2 & -2 \end{bmatrix} \begin{bmatrix} C_1 \\ C_2 \end{bmatrix} + \begin{bmatrix} 400 \\ 400 \end{bmatrix},$$

with solution $C_1 = -300$ and $C_2 = -100$. In this case the amount of salt in containers A and B at time t would be

$$\mathbf{X}(t) = \begin{bmatrix} x(t) \\ y(t) \end{bmatrix} = \begin{bmatrix} e^{-t/50} & e^{-3t/50} \\ 2e^{-t/50} & -2e^{-3t/50} \end{bmatrix} \begin{bmatrix} -300 \\ -100 \end{bmatrix} + \begin{bmatrix} 400 \\ 400 \end{bmatrix},$$

or

$$\mathbf{X}(t) = \begin{bmatrix} x(t) \\ y(t) \end{bmatrix} = \begin{bmatrix} 400 - 300e^{-t/50} - 100e^{-3t/50} \\ 400 - 600e^{-t/50} + 200e^{-3t/50} \end{bmatrix}.$$

Figure 13.2 shows the graphs of $x(t) = 400 - 300e^{-t/50} - 100e^{-3t/50}$, and $y(t) = 400 - 600e^{-t/50} + 200e^{-3t/50}$.

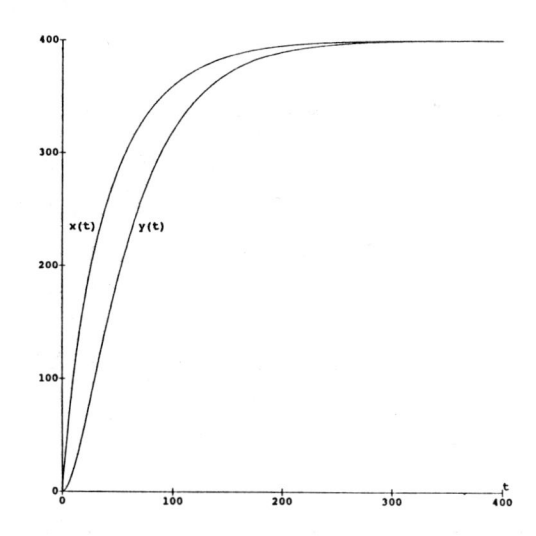

Figure 13.2 The functions $x(t) = 400 - 300e^{-t/50} - 100e^{-3t/50}$, and $y(t) = 400 - 600e^{-t/50} + 200e^{-3t/50}$

On the other hand, had this process started with no salt in container A, but 300 pounds in container B, the initial conditions would be $x(0) = 0$ and $y(0) = 300$. If we substitute these values into (13.3), we obtain

$$\begin{bmatrix} 0 \\ 300 \end{bmatrix} = \begin{bmatrix} 1 & 1 \\ 2 & -2 \end{bmatrix} \begin{bmatrix} C_1 \\ C_2 \end{bmatrix} + \begin{bmatrix} 400 \\ 400 \end{bmatrix},$$

with solution $C_1 = -225$ and $C_2 = -175$. In this case the amount of salt in containers A and B at time t would be

$$\mathbf{X}(t) = \begin{bmatrix} x(t) \\ y(t) \end{bmatrix} = \begin{bmatrix} e^{-t/50} & e^{-3t/50} \\ 2e^{-t/50} & -2e^{-3t/50} \end{bmatrix} \begin{bmatrix} -225 \\ -175 \end{bmatrix} + \begin{bmatrix} 400 \\ 400 \end{bmatrix},$$

or

$$\mathbf{X}(t) = \begin{bmatrix} x(t) \\ y(t) \end{bmatrix} = \begin{bmatrix} 400 - 225e^{-t/50} - 175e^{-3t/50} \\ 400 - 450e^{-t/50} + 350e^{-3t/50} \end{bmatrix}.$$

Figure 13.3 shows the graphs of $x(t) = 400 - 225e^{-t/50} - 175e^{-3t/50}$, and $y(t) = 400 - 450e^{-t/50} + 350e^{-3t/50}$.

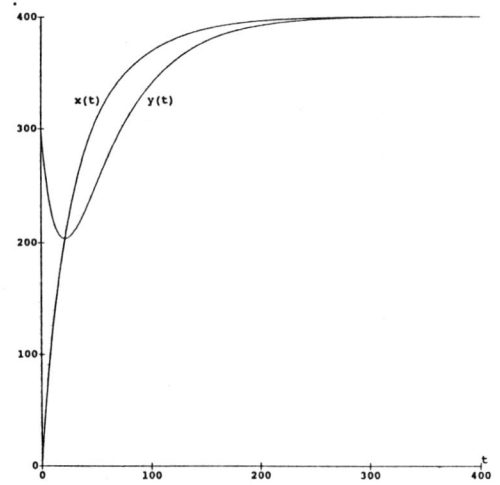

Figure 13.3 The functions $x(t) = 400 - 225e^{-t/50} - 175e^{-3t/50}$, and $y(t) = 400 - 450e^{-t/50} + 350e^{-3t/50}$

How much of the information contained in these explicit solutions could we have obtained from considerations in the phase plane? Looking at our original differential equations,

$$\begin{cases} x' = 12 + \frac{1}{100}y - \frac{1}{25}x, \\ y' = \frac{1}{25}x - \frac{1}{25}y, \end{cases}$$

we see that equilibrium points exist at values of x and y that satisfy

$$\begin{array}{rcl} 0 &=& 12 + \frac{1}{100}y - \frac{1}{25}x, \\ 0 &=& \frac{1}{25}x - \frac{1}{25}y. \end{array}$$

This gives $x = 400$, $y = 400$, as the only equilibrium point. The differential equation in the phase plane is

$$\frac{dy}{dx} = \frac{x/25 - y/25}{12 + y/100 - x/25} = \frac{4(x - y)}{1200 + y - 4x}, \tag{13.4}$$

so the y-nullclines are given by $y = x$ (horizontal tangents) and the x-nullclines by $y = 4x - 1200$ (vertical tangents). These nullclines are plotted in Figure 13.4 along with the direction field for the phase plane from (13.4). The direction of the arrows on the direction field suggests that the point $x = 400$, $y = 400$, is a stable equilibrium. The behavior of orbits in the phase plane for various initial conditions can be determined in a manner we now outline for the preceding two sets of initial conditions.

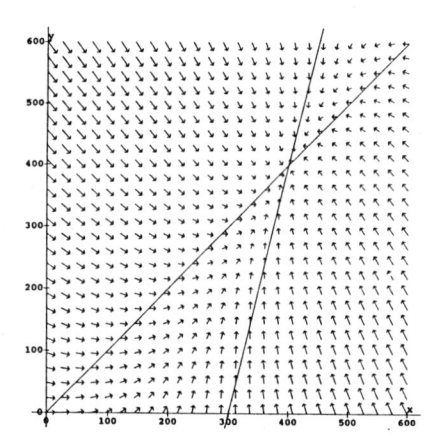

Figure 13.4 Direction field and nullclines for $dy/dx = 4(x - y)/(1200 + y - 4x)$

If we start with both containers free of the substance, we start at the origin ($x = y = 0$). This point is on the isocline for horizontal tangents, and because in the region $x - y > 0$ and $1200 + y - 4x > 0$, both x and y must be increasing, we must move to the right and up. However, because we cannot cross the isocline for vertical tangents (both x and y must increase in this region), we are forced to proceed to the equilibrium point at the intersection of these two isoclines.

On the other hand, if we had initial conditions $x = 0$, $y = 300$, we would start in a region where y must decrease and x must increase. This behavior would continue until we cross the isocline for horizontal tangents, where our previous reasoning takes over and we would end up at the equilibrium solution.

In fact, if we look at the form of the general solution in (13.3), we see that the values of C_1 and C_2 are determined by the initial conditions. However, the presence of $e^{-t/50}$ and $e^{-3t/50}$ means that in the limit as $t \to \infty$, x and y approach their equilibrium value regardless of the initial condition. Thus, the equilibrium point is stable. Figure 13.5 shows numerical solutions for the orbits in the phase plane for many initial conditions. Note how they all approach the equilibrium point. The orbit corresponding to the initial conditions of our solutions curves in Figures 13.2 and 13.3 are among those shown. □

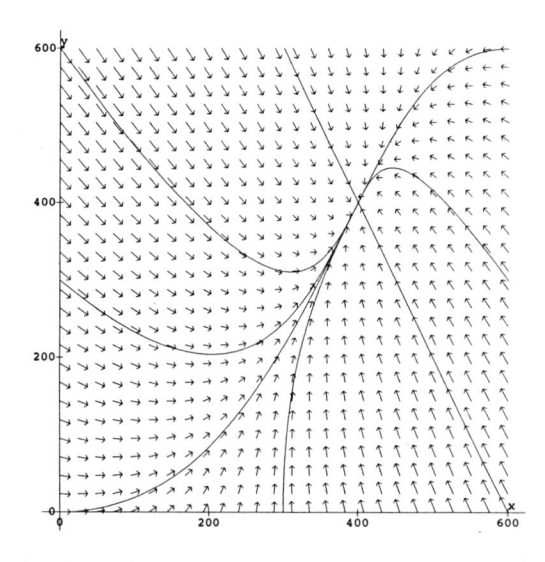

Figure 13.5 Orbits for $dy/dx = 4(x - y)/(1200 + y - 4x)$ for several initial conditions

Notice that had we made a change of variables from x and y to X and Y so that the equilibrium point was transformed from $(400, 400)$ to $(0, 0)$, we could have used the technique from Section 9.5. (See Exercise 7, page 299.)

Our next example has an exponential forcing function.

Example 13.2 :

Solve the system of nonhomogeneous first order differential equations

$$\begin{bmatrix} x \\ y \end{bmatrix}' = \begin{bmatrix} 1 & -2 \\ 1 & 4 \end{bmatrix} \begin{bmatrix} x \\ y \end{bmatrix} + \begin{bmatrix} 4e^{-t} \\ -4e^{-t} \end{bmatrix}. \tag{13.5}$$

Written in matrix form, this system of equations becomes $\mathbf{X}' = \mathbf{MX} + \mathbf{V}$, where

$$\mathbf{M} = \begin{bmatrix} 1 & -2 \\ 1 & 4 \end{bmatrix}$$

and $\mathbf{V}$ is the vector given by

$$\mathbf{V} = \begin{bmatrix} 4e^{-t} \\ -4e^{-t} \end{bmatrix} = \begin{bmatrix} 4 \\ -4 \end{bmatrix} e^{-t}.$$

Now the general solution of (13.5) is the sum of the general solution of the associated homogeneous differential equation and a particular solution of (13.5). Thus, we first consider the homogeneous system of differential equations, $\mathbf{X}'_\mathbf{h} = \mathbf{MX}_\mathbf{h}$, which leads to

$$\begin{bmatrix} 1-r & -2 \\ 1 & 4-r \end{bmatrix} \begin{bmatrix} A \\ B \end{bmatrix} = \begin{bmatrix} 0 \\ 0 \end{bmatrix}. \tag{13.6}$$

This gives the characteristic equation as $(1-r)(4-r) + 2 = r^2 - 5r + 6 = 0$, with solutions (eigenvalues) of 2 and 3.

Using $r = 2$ in (13.6) gives

$$\begin{bmatrix} -1 & -2 \\ 1 & 2 \end{bmatrix} \begin{bmatrix} A \\ B \end{bmatrix} = \begin{bmatrix} 0 \\ 0 \end{bmatrix},$$

so $A = -2B$, and the associated eigenvector is

$$\begin{bmatrix} -2 \\ 1 \end{bmatrix} B.$$

Using $r = 3$ in (13.6) gives

$$\begin{bmatrix} -2 & -2 \\ 1 & 1 \end{bmatrix} \begin{bmatrix} A \\ B \end{bmatrix} = \begin{bmatrix} 0 \\ 0 \end{bmatrix},$$

so $A = -B$, and the associated eigenvector is

$$\begin{bmatrix} -1 \\ 1 \end{bmatrix} B.$$

Thus, the solution of the associated homogeneous differential equation is $\mathbf{X_h}(t) = \mathbf{UC}$, where $\mathbf{C}$ is a vector of arbitrary constants and $\mathbf{U}$ is a fundamental matrix given by

$$\mathbf{U} = \begin{bmatrix} -2e^{2t} & -e^{3t} \\ e^{2t} & e^{3t} \end{bmatrix}. \tag{13.7}$$

Because the right-hand side of (13.5) contains a constant vector times e^{-t}, and the derivative of a constant times e^{-t} is a constant times e^{-t}, we try a particular solution of the form

$$\mathbf{X_p}(t) = \begin{bmatrix} K_1 \\ K_2 \end{bmatrix} e^{-t}.$$

(This is what we did for second order nonhomogeneous equations in Chapter 7.)

Substituting this expression into (13.5) gives

$$\begin{bmatrix} -K_1 \\ -K_2 \end{bmatrix} e^{-t} = \begin{bmatrix} 1 & -2 \\ 1 & 4 \end{bmatrix} \begin{bmatrix} K_1 \\ K_2 \end{bmatrix} e^{-t} + \begin{bmatrix} 4e^{-t} \\ -4e^{-t} \end{bmatrix}.$$

Because the expression e^{-t} is common to all terms and is never zero, we may divide through by it and rearrange the resulting system of algebraic equations as

$$\begin{bmatrix} -2 & 2 \\ -1 & -5 \end{bmatrix} \begin{bmatrix} K_1 \\ K_2 \end{bmatrix} = \begin{bmatrix} 4 \\ -4 \end{bmatrix}. \tag{13.8}$$

The solution of (13.8) is $K_1 = -1$, $K_2 = 1$, so we may write the general solution of (13.5) as $\mathbf{X}(t) = \mathbf{UC} + \mathbf{X_p}(t)$, where $\mathbf{U}$ is given by (13.7) and

$$\mathbf{X_p}(t) = \begin{bmatrix} -e^{-t} \\ e^{-t} \end{bmatrix}. \tag{13.9}$$

□

The method of finding a particular solution we used in these two examples works for many other situations. In fact, the particular solution for a system of nonhomogeneous differential equations $\mathbf{X}' = \mathbf{MX} + \mathbf{V}$, with $\mathbf{V}$ being a constant vector times $e^{\omega t}$, has the form

$$\mathbf{X_p}(t) = \begin{bmatrix} K_1 \\ K_2 \end{bmatrix} e^{\omega t}.$$

This is the proper form provided that ω is not an eigenvalue of the matrix $\mathbf{M}$. (Explain what would happen if we used such an $\mathbf{X_p}(t)$ with ω an eigenvalue of $\mathbf{M}$.) This result concerning the proper form of a particular solution is also of use for cases where the vector $\mathbf{V}$ in $\mathbf{X}' = \mathbf{MX} + \mathbf{V}$ contains a sine function or a cosine function. This use is not obvious at first, because a single derivative of a sine function gives a cosine function, and vice versa. However, the sine and cosine functions are related to the exponential function through Euler's identity, $e^{i\omega t} = \cos \omega t + i \sin \omega t$, and the derivative of $e^{i\omega t}$ is $i\omega e^{i\omega t}$.

The key result may be stated in the following manner:

- If **X** is a solution of
$$\mathbf{X}' = \mathbf{MX} + \mathbf{B}e^{i\omega t}, \tag{13.10}$$
where **M** and **B** contain only real numbers, then the real and imaginary parts of **X** — namely, $Re[\mathbf{X}]$ and $Im[\mathbf{X}]$ — satisfy the differential equations
$$\frac{d}{dt}Re[\mathbf{X}] = \mathbf{M}Re[\mathbf{X}] + \mathbf{B}\cos\omega t, \tag{13.11}$$
and
$$\frac{d}{dt}Im[\mathbf{X}] = \mathbf{M}Im[\mathbf{X}] + \mathbf{B}\sin\omega t. \tag{13.12}$$

In other words, if **X** satisfies (13.10), then the real part of **X** satisfies (13.11), and the imaginary part of **X** satisfies (13.12).

An example should make clear how to use this idea.

Example 13.3 :

Find the general solution of a nonhomogeneous system of differential equations
$$\mathbf{X}' = \mathbf{MX} + \begin{bmatrix} 2\cos t \\ 0 \end{bmatrix}, \text{ where } \mathbf{M} = \begin{bmatrix} -2 & 4 \\ 1 & 1 \end{bmatrix}. \tag{13.13}$$

We can write a three-step procedure as follows.

1. Express a companion differential equation as
$$\mathbf{X}' = \mathbf{MX} + \begin{bmatrix} 2 \\ 0 \end{bmatrix} e^{it}. \tag{13.14}$$

2. Solve the companion differential equation. The eigenvalues of **M** are determined by solving the quadratic equation obtained from the determinant of the matrix
$$\begin{bmatrix} -2-r & 4 \\ 1 & 1-r \end{bmatrix}.$$

This gives $(-2-r)(1-r) - 4 = r^2 + r - 6 = (r+3)(r-2) = 0$, so the eigenvalues are 2 and -3. Because the coefficient of t in the exponential forcing function is not an eigenvalue, we may try a particular solution of (13.14) of the form
$$\begin{bmatrix} K_1 \\ K_2 \end{bmatrix} e^{it}.$$

Substituting this expression into (13.14) yields
$$\begin{bmatrix} iK_1 \\ iK_2 \end{bmatrix} e^{it} = \begin{bmatrix} -2 & 4 \\ 1 & 1 \end{bmatrix} \begin{bmatrix} K_1 \\ K_2 \end{bmatrix} e^{it} + \begin{bmatrix} 2 \\ 0 \end{bmatrix} e^{it}.$$

Because the common factor e^{it} is not zero, we may divide by it and rearrange the result to obtain the system of algebraic equations
$$\begin{bmatrix} i+2 & -4 \\ -1 & i-1 \end{bmatrix} \begin{bmatrix} K_1 \\ K_2 \end{bmatrix} = \begin{bmatrix} 2 \\ 0 \end{bmatrix}. \tag{13.15}$$

The solution of (13.15) is
$$\begin{bmatrix} K_1 \\ K_2 \end{bmatrix} = \begin{bmatrix} (8-6i)/25 \\ (-7-i)/25 \end{bmatrix},$$

giving
$$\begin{bmatrix} K_1 \\ K_2 \end{bmatrix} e^{it} = \begin{bmatrix} (8-6i)/25 \\ (-7-i)/25 \end{bmatrix} (\cos t + i\sin t). \tag{13.16}$$

3. Take the real and imaginary parts of this particular solution as needed. In this case, we need the real part of (13.16), which gives

$$\mathbf{X_p}(t) = \begin{bmatrix} 8/25\cos t + 6/25\sin t \\ -7/25\cos t + 1/25\sin t \end{bmatrix}$$

as the appropriate particular solution of (13.13). □

Our final example in this section considers a nonhomogeneous system of differential equations describing an electrical circuit.

Example 13.4 : A Parallel RLC Circuit

In Chapter 6 we examined the discharge of a capacitor in a parallel RLC circuit. Here we consider the RLC circuit as shown in Figure 13.6, which contains a voltage source, $E(t)$.

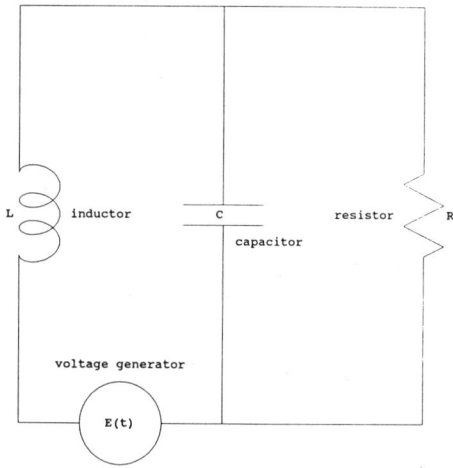

Figure 13.6 Parallel RLC circuit with a voltage source

If we consider the voltage drops around the outer loop of the circuit, we obtain

$$V_L + V_R = E(t). \tag{13.17}$$

(Recall that the algebraic sum of the voltages is zero for elements in series.) In the loop containing the resistor and capacitor, we have that the voltage drop across each element must be the same, so

$$V_R = V_C. \tag{13.18}$$

We now have two equations for our three unknowns. If we consider the voltage drops around the loop consisting of the inductor and the capacitor, we obtain $V_L + V_C = E(t)$, which is also the result of combining the first two equations.

To find a third, independent equation, we use Kirchhoff's current law for the junction in which wires from the three circuit elements meet — at the top of Figure 13.6. Kirchhoff's current law states that the current entering a junction must equal the current leaving the junction. (Thus, we see that this is a type of conservation equation.) If we consider the current to be positive for flow in a clockwise direction, we have that

$$I_L = I_R + I_C. \tag{13.19}$$

These three equations may be combined to obtain a system of differential equations in two unknowns as follows. Recalling that for inductors and resistors, voltage drops are given by LI' and RI, we rewrite (13.17) as

$$LI'_L + RI_R = E(t). \tag{13.20}$$

METHOD OF UNDETERMINED COEFFICIENTS USING MATRICES

Because the voltage drop across a capacitor is $(1/C)\int I(t)\,dt$, we differentiate (13.18) to obtain $RI'_R = (1/C)I_C$. Then we substitute the value of I_C from (13.19) into this result to obtain

$$RI'_R = \frac{1}{C}(I_L - I_R). \tag{13.21}$$

To use our usual notation, we let $x = I_L$, $y = I_R$, which gives the system of differential equations from (13.20) and (13.21) as

$$\begin{cases} x' = -(R/L)y + (1/L)E(t), \\ y' = 1/(RC)(x-y). \end{cases}$$

If we consider the situation in which $L = 1$, $R = 5$, $C = 1/20$, and $E(t) = 20\cos 2t$, this gives the system of differential equations

$$\begin{cases} x' = -5y + 20\cos 2t, \\ y' = 4x - 4y. \end{cases}$$

In matrix form this becomes $\mathbf{X'} = \mathbf{MX} + \mathbf{V}$, where

$$\mathbf{M} = \begin{bmatrix} 0 & -5 \\ 4 & -4 \end{bmatrix} \text{ and } \mathbf{V} = \begin{bmatrix} 20\cos 2t \\ 0 \end{bmatrix}.$$

We know that the general solution of such nonhomogeneous differential equations is given by $\mathbf{X}(t) = \mathbf{UC} + \mathbf{X_p}(t)$, where $\mathbf{U}$ is a fundamental matrix for the associated homogeneous system $\mathbf{X'_h} = \mathbf{MX_h}$, $\mathbf{C}$ is a vector of arbitrary constants, and $\mathbf{X_p}(t)$ is a particular solution.

Thus, our first step is to find a solution of this associated homogeneous system of differential equations, $\mathbf{X'_h} = \mathbf{MX_h}$. We recall that when $\mathbf{M}$ has constant coefficients, we always have solutions of such a system in the form

$$\mathbf{X_h}(t) = \begin{bmatrix} A \\ B \end{bmatrix} e^{rt},$$

where r, A, and B are determined from

$$\begin{bmatrix} -r & -5 \\ 4 & -4-r \end{bmatrix} \begin{bmatrix} A \\ B \end{bmatrix} = \begin{bmatrix} 0 \\ 0 \end{bmatrix}. \tag{13.22}$$

This gives rise to the characteristic equation $r^2 + 4r + 20 = 0$, and eigenvalues $r = -2 \pm 4i$. Substituting $r = -2 + 4i$ into the top equation in (13.22), $-rA - 5B = 0$, allows us to choose the components of our eigenvector as $A = 5$, $B = 2 - 4i$. Because our solution had the form of this eigenvector times e^{rt}, it may also be written as

$$\begin{bmatrix} 5 \\ 2-4i \end{bmatrix} e^{-2t+4it} = \begin{bmatrix} 5e^{-2t}\cos 4t \\ 2e^{-2t}\cos 4t + 4e^{-2t}\sin 4t \end{bmatrix} + i \begin{bmatrix} 5e^{-2t}\sin 4t \\ 2e^{-2t}\sin 4t - 4e^{-2t}\cos 2t \end{bmatrix}.$$

Our original system of differential equations had real coefficients, so both the real and imaginary parts of this vector must be solutions. Thus, we may form a fundamental matrix as

$$\mathbf{U} = \begin{bmatrix} 5e^{-2t}\cos 4t & 5e^{-2t}\sin 4t \\ 2e^{-2t}\cos 4t + 4e^{-2t}\sin 4t & 2e^{-2t}\sin 4t - 4e^{-2t}\cos 2t \end{bmatrix}. \tag{13.23}$$

Earlier we discovered that in finding particular solutions for trigonometric forcing functions, it was convenient to replace such functions with related complex valued exponential functions. (See also Example 13.3 on page 294.) Using that idea in this example means we replace $20\cos 2t$ with $20e^{2it}$ and seek a particular solution of

$$\mathbf{X'} = \mathbf{MX} + \begin{bmatrix} 20e^{2it} \\ 0 \end{bmatrix}. \tag{13.24}$$

The particular solution of our original differential equation will be the real part of the particular solution of (13.24).

Because $2i$ is not an eigenvalue of $\mathbf{M}$, we try a particular solution of (13.24) in the form

$$\begin{bmatrix} K_1 \\ K_2 \end{bmatrix} e^{2it}.$$

Substituting this expression into (13.24) gives

$$\begin{bmatrix} 2iK_1 \\ 2iK_2 \end{bmatrix} e^{2it} = \begin{bmatrix} 0 & -5 \\ 4 & -4 \end{bmatrix} \begin{bmatrix} K_1 \\ K_2 \end{bmatrix} e^{2it} + \begin{bmatrix} 20 \\ 0 \end{bmatrix} e^{2it},$$

which leads to the algebraic system of equations

$$2iK_1 = -5K_2 + 20, \qquad (2i+4)K_2 = 4K_1.$$

The solution of these two equations is $K_1 = 5$, $K_2 = 4 - 2i$, so our particular solution for the forcing function $20e^{2it}$ is

$$\begin{bmatrix} 5 \\ 4-2i \end{bmatrix} e^{2it} = \begin{bmatrix} 5\cos 2t \\ 4\cos 2t + 2\sin 2t \end{bmatrix} + i \begin{bmatrix} 5\sin 2t \\ 4\sin 2t - 2\cos 2t \end{bmatrix}.$$

We now have the particular solution of our original system as the real part of this vector, namely,

$$\mathbf{X_p}(t) = \begin{bmatrix} 5\cos 2t \\ 4\cos 2t + 2\sin 2t \end{bmatrix}. \tag{13.25}$$

This gives our general solution as $\mathbf{X}(t) = \mathbf{UC} + \mathbf{X_p}(t)$, where $\mathbf{U}$ is the fundamental matrix from (13.23), $\mathbf{C}$ is a vector of arbitrary constants, and $\mathbf{X_p}(t)$ is the particular solution from (13.25).

Note that at this stage, we have also found a particular solution for the differential equation

$$\mathbf{X'} = \begin{bmatrix} 0 & -5 \\ 4 & -4 \end{bmatrix} \mathbf{X} + \begin{bmatrix} 20\sin 2t \\ 0 \end{bmatrix}$$

as

$$\mathbf{X_p}(t) = \begin{bmatrix} 5\sin 2t \\ 4\sin 2t - 2\cos 2t \end{bmatrix}.$$

If we examine the behavior of this circuit subject to the initial conditions $x(0) = y(0) = 0$ — that is, there is no current in any part of the circuit at time $t = 0$ — we have $\mathbf{0} = \mathbf{U}(0)\mathbf{C} + \mathbf{X_p}(0)$, or

$$\begin{bmatrix} 0 \\ 0 \end{bmatrix} = \begin{bmatrix} 5 & 0 \\ 2 & -4 \end{bmatrix} \begin{bmatrix} C_1 \\ C_2 \end{bmatrix} + \begin{bmatrix} 5 \\ 4 \end{bmatrix}.$$

This gives $C_1 = -1$, $C_2 = 1/2$, and the solution of this initial value problem is

$$\mathbf{X}(t) = \begin{bmatrix} 5e^{-2t}\cos 4t & 5e^{-2t}\sin 4t \\ 2e^{-2t}\cos 4t + 4e^{-2t}\sin 4t & 2e^{-2t}\sin 4t - 4e^{-2t}\cos 4t \end{bmatrix} \begin{bmatrix} -1 \\ 1/2 \end{bmatrix} + \begin{bmatrix} 5\cos 2t \\ 4\cos 2t + 2\sin 2t \end{bmatrix},$$

or

$$\mathbf{X}(t) = \begin{bmatrix} 5e^{-2t}(-\cos 4t + 1/2\sin 4t) + 5\cos 2t \\ e^{-2t}(-4\cos 4t - 3\sin 4t) + 4\cos 2t + 2\sin 2t \end{bmatrix}. \tag{13.26}$$

We see that the terms involving e^{-2t} are the transient part of the solution, and that the terms containing $\cos 2t$ and $\sin 2t$ are the steady state part. Figure 13.7 shows the solution given in (13.26). The upper right-hand box shows $y(t)$, the lower right-hand box $x(t)$, and the upper left-hand box the phase plane. Notice that the amplitude of the steady state oscillations for $x(t)$ is 5, whereas that for $y(t)$ is $\sqrt{4^2 + 2^2} = \sqrt{20}$. Notice also that periodic steady state oscillations give rise to closed orbits in the phase plane. $\square$

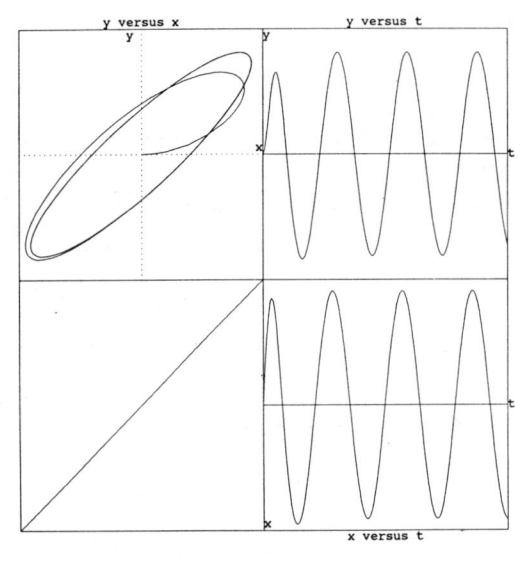

Figure 13.7 Currents in a parallel RLC circuit and the associated phase plane

Exercises

1. Find a particular solution of $\mathbf{X}' = \mathbf{MX} + \mathbf{V}$ for the following $\mathbf{M}$ and $\mathbf{V}$.

   (a) $\mathbf{M} = \begin{bmatrix} -2 & 4 \\ 1 & 1 \end{bmatrix}$ $\mathbf{V} = \begin{bmatrix} 7e^{3t} \\ -2e^{3t} \end{bmatrix}$

   (b) $\mathbf{M} = \begin{bmatrix} -2 & 4 \\ 1 & 1 \end{bmatrix}$ $\mathbf{V} = \begin{bmatrix} 7 \\ 1 \end{bmatrix}$

   (c) $\mathbf{M} = \begin{bmatrix} -2 & 4 \\ 1 & 1 \end{bmatrix}$ $\mathbf{V} = \begin{bmatrix} 0 \\ -\sin 2t \end{bmatrix}$

   (d) $\mathbf{M} = \begin{bmatrix} -2 & 4 \\ 1 & 1 \end{bmatrix}$ $\mathbf{V} = \begin{bmatrix} e^{-t} \\ -2e^{-2t} \end{bmatrix}$

   (e) $\mathbf{M} = \begin{bmatrix} -2 & 4 \\ 1 & 1 \end{bmatrix}$ $\mathbf{V} = 2\begin{bmatrix} 7e^{3t} \\ -2e^{3t} \end{bmatrix} + \pi \begin{bmatrix} 0 \\ -\sin 2t \end{bmatrix}$

   (f) $\mathbf{M} = \begin{bmatrix} 2 & 1 \\ -1 & 2 \end{bmatrix}$ $\mathbf{V} = \begin{bmatrix} e^{\pi t} \\ 3e^{\pi t} \end{bmatrix}$

   (g) $\mathbf{M} = \begin{bmatrix} 2 & 1 \\ -1 & 2 \end{bmatrix}$ $\mathbf{V} = \begin{bmatrix} 3e^{t} \\ -e^{t} \end{bmatrix}$

   (h) $\mathbf{M} = \begin{bmatrix} 2 & 1 \\ -1 & 2 \end{bmatrix}$ $\mathbf{V} = \begin{bmatrix} \cos t \\ 0 \end{bmatrix}$

   (i) $\mathbf{M} = \begin{bmatrix} 2 & 1 \\ -1 & 2 \end{bmatrix}$ $\mathbf{V} = \begin{bmatrix} 3e^{t} \\ -e^{t} \end{bmatrix} + 17 \begin{bmatrix} e^{\pi t} \\ 3e^{\pi t} \end{bmatrix}$

2. Show that if $\mathbf{X_p}(t)$ is a particular solution of

$$\mathbf{X}' = \mathbf{MX} + \mathbf{V}, \qquad (13.27)$$

   and $\mathbf{X}_1(t)$ and $\mathbf{X}_2(t)$ are solutions of the associated homogeneous differential equation, then $C_1\mathbf{X}_1(t) + C_2\mathbf{X}_2(t) + \mathbf{X_p}(t)$, where C_1 and C_2 are arbitrary constants, is a solution of (13.27). [Note that a fundamental matrix has $\mathbf{X}_1(t)$ and $\mathbf{X}_2(t)$ as its column vectors, and the solution is $\mathbf{UC} + \mathbf{X_p}(t)$.]

3. Consider $\mathbf{X}' = \mathbf{MX} + \mathbf{V}$, where $\mathbf{V} = \mathbf{Z}e^{rt}$, with $\mathbf{Z}$ a constant vector. If r is an eigenvector of $\mathbf{M}$, show that there are no particular solutions of the form

$$\mathbf{X_p}(t) = \begin{bmatrix} K_1 \\ K_2 \end{bmatrix} e^{rt}.$$

4. Show that if $\mathbf{X_{p_1}}$ and $\mathbf{X_{p_2}}$ are particular solutions of $\mathbf{X'} = \mathbf{MX} + \mathbf{V}_1$ and $\mathbf{X'} = \mathbf{MX} + \mathbf{V}_2$, respectively, and $\mathbf{X}_1$ and $\mathbf{X}_2$ are solutions of the associated homogeneous differential equation, then $C_1\mathbf{X}_1 + C_2\mathbf{X}_2 + \mathbf{X_{p_1}} + \mathbf{X_{p_2}}$, where C_1 and C_2 are arbitrary constants, is a solution of $\mathbf{X'} = \mathbf{MX} + \mathbf{V}_1 + \mathbf{V}_2$. This contains the Principle of Linear Superposition for systems of linear differential equations.

5. Show that if $\mathbf{X}_1$ and $\mathbf{X}_2$ are solutions of $\mathbf{X'} = \mathbf{MX}$, then $C_1\mathbf{X}_1 + C_2\mathbf{X}_2 + A\mathbf{X}_1$ is not a solution of $\mathbf{X'} = \mathbf{MX} + \mathbf{X}_1$ for any choice of the constant A.

6. Create a *How to Find Particular Solutions Using Complex Exponential Functions* by adding statements under Purpose, Process, and Comments that generalize the ideas given in Example 13.3.

7. Consider the initial value problem $x' = 12 - \frac{1}{25}x + \frac{1}{100}y$, $y' = \frac{1}{25}x - \frac{1}{25}y$, where $x(0) = y(0) = 0$.

    (a) Show that there is an equilibrium point at $(400, 400)$.

    (b) Make the change of variables from x and y to X and Y given by $x = X + a$, $y = Y + b$, and choose the constants a and b so the equilibrium point is transformed from $x = y = 400$ to $X = Y = 0$.

    (c) Find the general solution of this system of equations in X and Y using the results of Section 9.1. Choose the arbitrary constants in your solution so the initial conditions are satisfied. Transform back to the original variables and compare your answer with that found in Example 13.1.

8. Show that the methods of this section also apply to polynomial forcing functions by considering the differential equation $\mathbf{X'} = \mathbf{MX} + \mathbf{V}$, where

$$\mathbf{M} = \begin{bmatrix} 1 & -2 \\ 1 & 4 \end{bmatrix}, \quad \mathbf{V} = \begin{bmatrix} 6t \\ 7 \end{bmatrix},$$

by choosing a particular solution of the form

$$\mathbf{X_p}(t) = \begin{bmatrix} At + B \\ Ct + D \end{bmatrix}.$$

9. Consider the situation in which the concentration of salt in the solution entering container A in Example 13.1 is given by $4 - 2e^{-t}$. If all the other specifications of that example remain the same, solve the following initial value problem, $\mathbf{X'} = \mathbf{MX} + \mathbf{V}$, where

$$\mathbf{M} = \begin{bmatrix} -1/25 & 1/100 \\ 1/25 & -1/25 \end{bmatrix}, \quad \mathbf{V} = \begin{bmatrix} 12 \\ 0 \end{bmatrix} - \begin{bmatrix} 6e^{-t} \\ 0 \end{bmatrix},$$

subject to the initial condition

$$\mathbf{X}(0) = \begin{bmatrix} 0 \\ 0 \end{bmatrix}.$$

10. If the forcing function in Exercise 9 is changed to

$$\mathbf{V} = \begin{bmatrix} 12 - \sin t \\ 0 \end{bmatrix},$$

find the solution of this initial value problem.

11. Find the solution of the initial value problem for the circuit of Figure 13.6 for $L = 1/5$, $C = 1/4$, $R = 1/3$, and $E(t) = 4\sin 2t$. The system of differential equations is therefore $\mathbf{X'} = \mathbf{MX} + \mathbf{V}$, where

$$\mathbf{M} = \begin{bmatrix} 0 & -5/3 \\ 12 & -12 \end{bmatrix}, \quad \mathbf{V} = \begin{bmatrix} 20\sin 2t \\ 0 \end{bmatrix},$$

and the initial condition is $\mathbf{X}(0) = \mathbf{0}$.

12. Find the solution of the initial value problem for the circuit of Figure 13.6 for $L = 1$, $C = 1/20$, $R = 5$, and $E(t) = 4(1 - e^{-t})$. The system of differential equations is therefore $\mathbf{X}' = \mathbf{MX} + \mathbf{V}$, where

$$\mathbf{M} = \begin{bmatrix} 0 & -5 \\ 4 & -4 \end{bmatrix}, \quad \mathbf{V} = \begin{bmatrix} 4(1 - e^{-t}) \\ 0 \end{bmatrix},$$

and the initial condition is $\mathbf{X}(0) = \mathbf{0}$.

13. Let x denote the number of fish in a lake and y the number of people fishing at the lake. If the rate of stocking the lake is denoted by $S(t)$, the differential equations describing x and y are

$$\begin{cases} x' = ax - by + S(t), \\ y' = cx - dy, \end{cases} \quad (13.28)$$

where a, b, c, and d are positive constants. Here a is the difference between the natural birth and death rates of the fish, and b is the catch rate. The constants c and d are both positive because the rate of change of people fishing should increase with the number of fish present and decrease with the number of people fishing.

(a) If $a = c = 2$, $b = 8$, $d = 6$, and the stocking rate is approximated by $S(t) = S_0 [1 - \cos(t/10)]$, where S_0 is a constant, solve (13.28) with initial conditions $x(0) = A$, $y(0) = 0$.

(b) Note that the solution for part (a) has terms that decrease rapidly with time (the transient solutions) as well as terms that do not (the steady state solution). Using the steady state solution of this model, determine what values of S_0 will guarantee that the number of people fishing will always be less than 300.

(c) Solve (13.28) if $a = 4$, $c = 5$, $b = d = 2$, $S(t) = S_0 (1 - \cos t)$, $x(0) = A$, $y(0) = 0$.

13.2 Solutions by Variation of Parameters

The method we used in Section 13.1 for finding particular solutions of

$$\mathbf{X}' = \mathbf{MX} + \mathbf{V} \quad (13.29)$$

will not work if the constant r in $\mathbf{V} = \mathbf{Z}e^{rt}$, with $\mathbf{Z}$ a constant vector, is an eigenvalue of the matrix $\mathbf{M}$; or with some other forms for $\mathbf{V}$. (See Exercise 5 on page 299.) However, as with the nonhomogeneous second order differential equations in Section 8.3, we have the method of variation of parameters that we can use.

In this case we start with the general solution of the associated homogeneous differential system, which we write as $\mathbf{U}(t)\mathbf{C}$, where $\mathbf{U}$ is a fundamental matrix. Instead of having $\mathbf{C}$ as a constant vector, we allow it to be a function of time. Thus, we have $\mathbf{X_p}(t) = \mathbf{U}(t)\mathbf{C}(t)$, which we substitute into (13.29) to obtain $[\mathbf{U}(t)\mathbf{C}(t)]' = \mathbf{MU}(t)\mathbf{C}(t) + \mathbf{V}(t)$, or

$$\mathbf{U}(t)\mathbf{C}'(t) + \mathbf{U}'(t)\mathbf{C}(t) = \mathbf{MU}(t)\mathbf{C}(t) + \mathbf{V}(t). \quad (13.30)$$

(Exercise 2 on page 302 gives the product rule for differentiating a matrix times a vector.)

Also, because a fundamental matrix $\mathbf{U}(t)$ of the matrix $\mathbf{M}$ satisfies the differential equation $\mathbf{U}'(t) = \mathbf{MU}(t)$ (see Exercise 4, page 302), we may multiply both sides of this matrix equation on the right by $\mathbf{C}(t)$ to obtain $\mathbf{U}'(t)\mathbf{C}(t) = \mathbf{MU}(t)\mathbf{C}(t)$. Using this fact, (13.30) reduces to

$$\mathbf{U}(t)\mathbf{C}'(t) = \mathbf{V}(t). \quad (13.31)$$

Because a fundamental matrix always has an inverse (see Exercise 6 on page 302), we may multiply both sides of the expression in (13.31) by $\mathbf{U}^{-1}(t)$ to obtain $\mathbf{U}^{-1}(t)\mathbf{U}(t)\mathbf{C}'(t) = \mathbf{U}^{-1}(t)\mathbf{V}(t)$. Because $\mathbf{U}^{-1}(t)\mathbf{U}(t) = \mathbf{I}$, the identity matrix, this equation simplifies to $\mathbf{C}'(t) = \mathbf{U}^{-1}(t)\mathbf{V}(t)$.

Thus, if we integrate the vector $\mathbf{C}'(t)$, we obtain

$$\mathbf{C}(t) = \int \mathbf{U}^{-1}(t)\mathbf{V}(t)\,dt$$

and find

$$\mathbf{X_p}(t) = \mathbf{U}(t)\mathbf{C}(t) = \mathbf{U}(t)\int \mathbf{U}^{-1}(t)\mathbf{V}(t)\,dt.$$

We include no constant of integration, because we need only one particular solution. Sometimes it is convenient to write limits on the integration in the preceding equation from 0 to t, so the value of this integral is zero at $t = 0$. If we do this, our particular solution is

$$\mathbf{X_p}(t) = \mathbf{U}(t)\mathbf{C}(t) = \mathbf{U}(t)\int_0^t \mathbf{U}^{-1}(s)\mathbf{V}(s)\,ds.$$

Example 13.5 :

Let us use this technique to find a particular solution of the nonhomogeneous problem in Example 13.2, page 292. The pertinent system of differential equations is

$$\begin{bmatrix} x \\ y \end{bmatrix}' = \begin{bmatrix} 1 & -2 \\ 1 & 4 \end{bmatrix}\begin{bmatrix} x \\ y \end{bmatrix} + \begin{bmatrix} 4e^{-t} \\ -4e^{-t} \end{bmatrix},$$

or $\mathbf{X}' = \mathbf{MX} + \mathbf{V}$, where

$$\mathbf{V} = \begin{bmatrix} 4e^{-t} \\ -4e^{-t} \end{bmatrix}.$$

In the previous section we discovered that a fundamental matrix for the associated homogeneous equation was

$$\mathbf{U}(t) = \begin{bmatrix} -2e^{2t} & -e^{3t} \\ e^{2t} & e^{3t} \end{bmatrix}, \qquad (13.32)$$

which has an inverse of

$$\mathbf{U}^{-1}(t) = -e^{-5t}\begin{bmatrix} e^{3t} & e^{3t} \\ -e^{2t} & -2e^{2t} \end{bmatrix}.$$

Thus, a particular solution will be $\mathbf{U}(t)\mathbf{C}(t)$, where $\mathbf{U}(t)$ is given by (13.32), and where

$$\mathbf{C}(t) = \int \mathbf{U}^{-1}(t)\mathbf{V}(t)\,dt = \int -e^{-5t}\begin{bmatrix} e^{3t} & e^{3t} \\ -e^{2t} & -2e^{2t} \end{bmatrix}\begin{bmatrix} 4e^{-t} \\ -4e^{-t} \end{bmatrix}\,dt,$$

which we integrate to obtain

$$\mathbf{C}(t) = \int -e^{-5t}\begin{bmatrix} 0 \\ 4e^{t} \end{bmatrix}\,dt = \int \begin{bmatrix} 0 \\ -4e^{-4t} \end{bmatrix}\,dt = \begin{bmatrix} 0 \\ e^{-4t} \end{bmatrix}.$$

Performing the indicated multiplication gives our particular solution as

$$\mathbf{X_p}(t) = \mathbf{U}(t)\mathbf{C}(t) = \begin{bmatrix} -2e^{2t} & -e^{3t} \\ e^{2t} & e^{3t} \end{bmatrix}\begin{bmatrix} 0 \\ e^{-4t} \end{bmatrix} = \begin{bmatrix} -e^{-t} \\ e^{-t} \end{bmatrix},$$

in agreement with (13.9). $\square$

Exercises

1. Show that if $ad - bc \neq 0$, the inverse of the matrix
$$\begin{bmatrix} a & b \\ c & d \end{bmatrix}$$
is
$$\frac{1}{ad-bc}\begin{bmatrix} d & -b \\ -c & a \end{bmatrix}.$$

2. The derivative of a matrix
$$\mathbf{N}(t) = \begin{bmatrix} f(t) & g(t) \\ h(t) & k(t) \end{bmatrix}$$
is defined as
$$\mathbf{N}'(t) = \begin{bmatrix} f'(t) & g'(t) \\ h'(t) & k'(t) \end{bmatrix}.$$

Show that

(a) $[C\mathbf{N}(t)]' = C\mathbf{N}'(t)$.
(b) $[\mathbf{N}(t) + \mathbf{M}(t)]' = \mathbf{N}'(t) + \mathbf{M}'(t)$.
(c) $[z(t)\mathbf{N}(t)]' = z'(t)\mathbf{N}(t) + z(t)\mathbf{N}'(t)$.
(d) $[\mathbf{N}(t)\mathbf{X}(t)]' = \mathbf{N}'(t)\mathbf{X}(t) + \mathbf{N}(t)\mathbf{X}'(t)$.

3. Find a particular solution of $\mathbf{X}' = \mathbf{M}\mathbf{X} + \mathbf{V}$ for the following $\mathbf{M}$ and $\mathbf{V}$.

(a) $\mathbf{M} = \begin{bmatrix} -2 & 4 \\ 1 & 1 \end{bmatrix}$ $\quad \mathbf{V} = \begin{bmatrix} 7e^{-3t} \\ -2e^{-3t} \end{bmatrix}$

(b) $\mathbf{M} = \begin{bmatrix} -2 & 4 \\ 1 & 1 \end{bmatrix}$ $\quad \mathbf{V} = \begin{bmatrix} e^{2t} \\ -2e^{2t} \end{bmatrix}$

(c) $\mathbf{M} = \begin{bmatrix} 2 & 1 \\ -3 & -2 \end{bmatrix}$ $\quad \mathbf{V} = \begin{bmatrix} 3e^t \\ -e^t \end{bmatrix}$

(d) $\mathbf{M} = \begin{bmatrix} 2 & 1 \\ -3 & -2 \end{bmatrix}$ $\quad \mathbf{V} = \begin{bmatrix} e^{-t} \\ 2e^{-t} \end{bmatrix}$

(e) $\mathbf{M} = \begin{bmatrix} 2 & 1 \\ -3 & -2 \end{bmatrix}$ $\quad \mathbf{V} = \begin{bmatrix} (2t+1)e^t \\ 0 \end{bmatrix}$

4. If $\mathbf{U}$ is a fundamental matrix of the homogeneous system of differential equations
$$\mathbf{X}' = \mathbf{M}\mathbf{X}, \qquad (13.33)$$
show that $\mathbf{U}$ satisfies the differential equation $\mathbf{U}' = \mathbf{M}\mathbf{U}$. [Recall that a fundamental matrix of (13.33) has columns composed of vectors that are solutions of (13.33).]

5. Given
$$\mathbf{A}(t) = \begin{bmatrix} t & e^t \\ 3 & 3t^2 \end{bmatrix}, \qquad \mathbf{B}(t) = \begin{bmatrix} e^{3t} & e^{2t} \\ e^{-t} & e^{4t} \end{bmatrix}.$$

(a) Compute $t^2 \mathbf{A}(t)$.
(b) Compute $\mathbf{A}'(t)$.
(c) Verify that $[t^2 \mathbf{A}(t)]' = 2t\mathbf{A}(t) + t^2 \mathbf{A}'(t)$.
(d) Compute $e^{-2t}\mathbf{B}(t)$.
(e) Compute $\mathbf{B}'(t)$.

(f) Verify that $[e^{-2t}\mathbf{B}(t)]' = -2te^{-2t}\mathbf{B}(t) + e^{-2t}\mathbf{B}'(t)$.

6. Prove that every fundamental matrix has an inverse.

7. Consider the situation in which the concentration of salt in the solution entering container A in Example 13.1 is given by $4 - 2e^{-t/50}$. If all the other specifications of that example remain the same, solve the following initial value problem, $\mathbf{X}' = \mathbf{MX} + \mathbf{V}$, where
$$\mathbf{M} = \begin{bmatrix} -1/25 & 1/100 \\ 1/25 & -1/25 \end{bmatrix}, \qquad \mathbf{V} = \begin{bmatrix} 12 \\ 0 \end{bmatrix} - \begin{bmatrix} 6e^{-t/50} \\ 0 \end{bmatrix},$$
subject to the initial condition
$$\mathbf{X}(0) = \begin{bmatrix} 0 \\ 0 \end{bmatrix}.$$

13.3 Higher Order Systems

Thus far in this chapter we have considered solution methods for systems of only two linear differential equations in two unknown functions. There are many applications that require more differential equations and more unknown functions, so we consider such situations in this section. We start with an example that leads to a system of four differential equations in four unknown functions and discover that our previous techniques for solving systems still work. The only additional requirement is to find the determinant of a matrix that has four rows and four columns.

We start by reworking Example 8.23 — Two Spring-Masses — but now we will use matrix notation and solve a system of four first order differential equations.

Example 13.6 : Two Spring-Masses

The physical example we consider now is that of two masses suspended on two springs, as shown on the left-hand side of Figure 13.8.

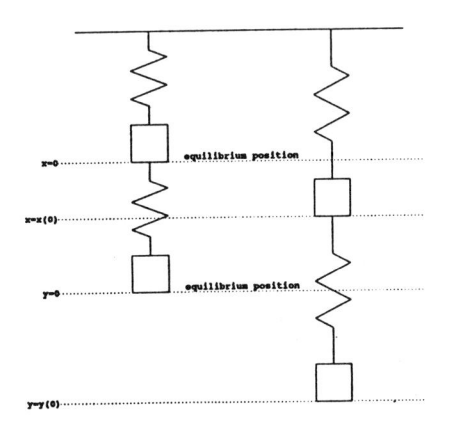

Figure 13.8 Two springs and two masses—in equilibrium and displaced

If the two masses are m_1 and m_2, and the two springs obey Hooke's law with respective spring constants k_1 and k_2, then this system is modeled by
$$\begin{cases} m_1 x'' &= -k_1 x + k_2(y - x), \\ m_2 y'' &= -k_2(y - x). \end{cases}$$

where x and y are the vertical displacements of the upper and lower masses, respectively, from equilibrium. Positive values of x and y are in the downward direction. To simplify the resulting algebra, we take the specific values $m_1 = m_2 = 1$, $k_1 = 5$, and $k_2 = 6$. We want to solve this system of equations subject to the initial conditions $x(0) = x'(0) = y'(0) = 0$, $y(0) = y_0$.

One way of solving this system is to convert it to a system of four first order differential equations by defining two new variables u and v, by $x' = u$ and $y' = v$. This gives us the fourth order system

$$\begin{cases} x' &= u, \\ y' &= v, \\ u' &= -5x + 6(y-x), \\ v' &= -5(y-x), \end{cases}$$

which may be put in the matrix form

$$\mathbf{X'} = \mathbf{MX}, \tag{13.34}$$

where

$$\mathbf{M} = \begin{bmatrix} 0 & 0 & 1 & 0 \\ 0 & 0 & 0 & 1 \\ -11 & 6 & 0 & 0 \\ 6 & -6 & 0 & 0 \end{bmatrix}.$$

Because (13.34) has the form of the system of differential equations we solved in the previous sections — the only difference is the size of the vectors and matrix in (13.34) — we seek a solution in the form of a vector of constants times an exponential function of t.

Now we substitute the vector

$$\mathbf{X} = \begin{bmatrix} A \\ B \\ C \\ D \end{bmatrix} e^{rt}$$

into (13.34) and obtain

$$\begin{bmatrix} rA \\ rB \\ rC \\ rD \end{bmatrix} = \begin{bmatrix} 0 & 0 & 1 & 0 \\ 0 & 0 & 0 & 1 \\ -11 & 6 & 0 & 0 \\ 6 & -6 & 0 & 0 \end{bmatrix} \begin{bmatrix} A \\ B \\ C \\ D \end{bmatrix},$$

or

$$\begin{bmatrix} -r & 0 & 1 & 0 \\ 0 & -r & 0 & 1 \\ -11 & 6 & -r & 0 \\ 6 & -6 & 0 & -r \end{bmatrix} \begin{bmatrix} A \\ B \\ C \\ D \end{bmatrix} = \begin{bmatrix} 0 \\ 0 \\ 0 \\ 0 \end{bmatrix}. \tag{13.35}$$

This system of algebraic equations will have a nontrivial solution only when the determinant of the coefficient matrix is zero. This gives $r^4 + 17r^2 + 30 = (r^2 + 15)(r^2 + 2) = 0$, which is called the characteristic equation associated with the matrix $\mathbf{M}$. The roots of this equation are called eigenvalues and are given by $r = \pm i\sqrt{2}$ and $\pm i\sqrt{15}$.

If we substitute the eigenvalue $r = i\sqrt{2}$ into (13.35), we obtain the dependent system of algebraic equations

$$\begin{bmatrix} -i\sqrt{2} & 0 & 1 & 0 \\ 0 & -i\sqrt{2} & 0 & 1 \\ -11 & 6 & -i\sqrt{2} & 0 \\ 6 & -6 & 0 & -i\sqrt{2} \end{bmatrix} \begin{bmatrix} A \\ B \\ C \\ D \end{bmatrix} = \begin{bmatrix} 0 \\ 0 \\ 0 \\ 0 \end{bmatrix}.$$

This means that an eigenvector may be chosen with $A = 1$, $B = 3/2$, $C = i\sqrt{2}$, and $D = i3\sqrt{2}/2$. Thus, our solution is

$$\mathbf{X} = \begin{bmatrix} 1 \\ 3/2 \\ i\sqrt{2} \\ i3\sqrt{2}/2 \end{bmatrix} e^{i\sqrt{2}t} = \left(\begin{bmatrix} 1 \\ 3/2 \\ 0 \\ 0 \end{bmatrix} + i \begin{bmatrix} 0 \\ 0 \\ \sqrt{2} \\ 3\sqrt{2}/2 \end{bmatrix} \right) \left(\cos\sqrt{2}t + i\sin\sqrt{2}t \right),$$

or

$$\mathbf{X} = \begin{bmatrix} \cos\sqrt{2}t \\ (3/2)\cos\sqrt{2}t \\ -\sqrt{2}\sin\sqrt{2}t \\ -(3\sqrt{2}/2)\sin\sqrt{2}t \end{bmatrix} + i \begin{bmatrix} \sin\sqrt{2}t \\ (3/2)\sin\sqrt{2}t \\ \sqrt{2}\cos\sqrt{2}t \\ (3\sqrt{2}/2)\cos\sqrt{2}t \end{bmatrix}. \tag{13.36}$$

Because the original differential equation $\mathbf{X}' = \mathbf{MX}$ contained all real numbers, both the real and imaginary parts of our solution in (13.36) must be solutions of (13.34).

Next we consider the eigenvalue $r = i\sqrt{15}$, which may be substituted into (13.35) to yield

$$\begin{bmatrix} -i\sqrt{15} & 0 & 1 & 0 \\ 0 & -i\sqrt{15} & 0 & 1 \\ -11 & 6 & -i\sqrt{15} & 0 \\ 6 & -6 & 0 & -i\sqrt{15} \end{bmatrix} \begin{bmatrix} A \\ B \\ C \\ D \end{bmatrix} = \begin{bmatrix} 0 \\ 0 \\ 0 \\ 0 \end{bmatrix}.$$

This dependent system of algebraic equations has a solution consisting of $A = 1$, $B = -2/3$, $C = i\sqrt{15}$, and $D = -i2\sqrt{15}/3$. Thus, our solution is

$$\mathbf{X} = \begin{bmatrix} 1 \\ -2/3 \\ i\sqrt{15} \\ -i2\sqrt{15}/3 \end{bmatrix} e^{i\sqrt{15}t} = \left(\begin{bmatrix} 1 \\ -2/3 \\ 0 \\ 0 \end{bmatrix} + i \begin{bmatrix} 0 \\ 0 \\ \sqrt{15} \\ -2\sqrt{15}/3 \end{bmatrix} \right) \left(\cos\sqrt{15}t + i\sin\sqrt{15}t \right),$$

or

$$\mathbf{X} = \begin{bmatrix} \cos\sqrt{15}t \\ -(2/3)\cos\sqrt{15}t \\ -\sqrt{15}\sin\sqrt{15}t \\ (2\sqrt{15}/3)\sin\sqrt{15}t \end{bmatrix} + i \begin{bmatrix} \sin\sqrt{15}t \\ -(2/3)\sin\sqrt{15}t \\ \sqrt{15}\cos\sqrt{15}t \\ -(2\sqrt{15}/3)\cos\sqrt{15}t \end{bmatrix}.$$

Again, both the real and imaginary components of this solution are solutions of our original system of differential equations. This allows us to form a fundamental matrix and give the general solution of (13.34) as $\mathbf{X}(t) = \mathbf{UC}$, where

$$\mathbf{U} = \begin{bmatrix} \cos\sqrt{2}t & \sin\sqrt{2}t & \cos\sqrt{15}t & \sin\sqrt{15}t \\ (3/2)\cos\sqrt{2}t & (3/2)\sin\sqrt{2}t & -(2/3)\cos\sqrt{15}t & -(2/3)\sin\sqrt{15}t \\ -\sqrt{2}\sin\sqrt{2}t & \sqrt{2}\cos\sqrt{2}t & -\sqrt{15}\sin\sqrt{15}t & \sqrt{15}\cos\sqrt{15}t \\ -(3\sqrt{2}/2)\sin\sqrt{2}t & (3\sqrt{2}/2)\cos\sqrt{2}t & (2\sqrt{15}/3)\sin\sqrt{15}t & -(2\sqrt{15}/3)\cos\sqrt{15}t \end{bmatrix},$$

and $\mathbf{C}$ is a vector containing four arbitrary constants. Notice that the top two rows in $\mathbf{UC}$ contain the solution for x and y and the bottom two rows give their derivatives.

If we now choose the constant in $\mathbf{C}$ using our initial conditions $x(0) = x'(0) = y'(0) = 0$, $y(0) = y_0$, we obtain $C_1 = -C_3 = (6/13) y_0$, $C_2 = C_4 = 0$. Thus, the displacement of the two springs is given by

$$\begin{cases} x(t) = (6/13) \left(\cos\sqrt{2}t - \cos\sqrt{15}t \right) y_0, \\ y(t) = \left[(9/13)\cos\sqrt{2}t + (4/13)\cos\sqrt{15}t \right] y_0. \end{cases} \tag{13.37}$$

The functions $x(t)$ and $y(t)$ in (13.37) are graphed in Figure 13.9 for $y_0 = 8$. $\square$

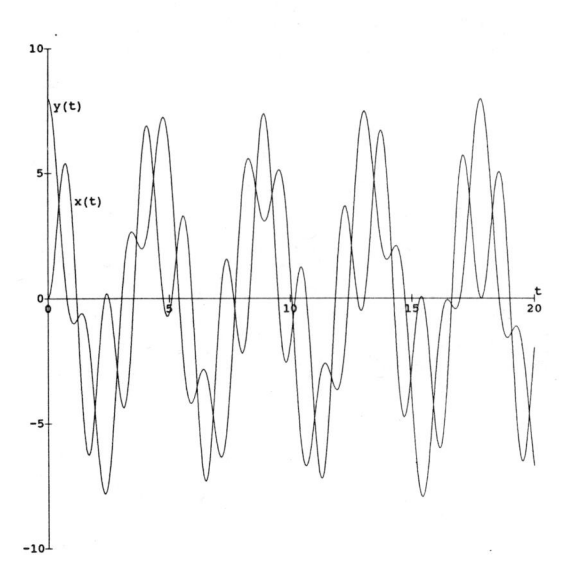

Figure 13.9 The functions $x(t) = (6/13)\left(\cos 2^{1/2}t - \cos 15^{1/2}t\right) y_0$, and $y(t) = \left[(9/13)\cos 2^{1/2}t + (4/13)\cos 15^{1/2}t\right] y_0$

We now give a general procedure for finding the solution of a system of linear differential equations with constant coefficients.

How to Solve $\mathbf{X}' = \mathbf{MX}$

Purpose: To find the general solution of the system of differential equations

$$\mathbf{X}' = \mathbf{MX}, \tag{13.38}$$

where $\mathbf{M}$ is a square matrix of constants and $\mathbf{X}$ is a vector of unknown functions. Both $\mathbf{M}$ and $\mathbf{X}$ have the same number of rows, denoted by n.

Process

1. Substitute the vector $\mathbf{X} = \mathbf{A}e^{rt}$, where $\mathbf{A}$ is a vector containing n unknown constants, into (13.38) and rearrange the result to obtain the system of algebraic equations

$$(\mathbf{M} - r\mathbf{I})\mathbf{A} = \mathbf{0}. \tag{13.39}$$

2. Solve for the eigenvalues. These are the solutions of the characteristic equation obtained by setting the determinant of the coefficient matrix in (13.39) to 0 — that is, $\det(\mathbf{M} - r\mathbf{I}) = 0$. There will be n eigenvalues if we include repeated roots of the characteristic equation.

3. Find a fundamental matrix for this system of differential equations. The process in this step depends on the nature of the eigenvalues: whether they are distinct or repeated, real or complex.

(a) For the case where we have n real, distinct eigenvalues $r_1, r_2, r_3, \cdots, r_n$, substitute each eigenvalue, in turn, into (13.39) and find the associated eigenvector. The columns in a fundamental matrix are formed by these eigenvectors times e^{rt}, for the appropriate value of r.

(b) For the case in which we have n complex, distinct eigenvalues, we need use only one half of these eigenvalues. (This is true because in this situation complex roots occur in pairs, one being the complex conjugate of the other.) Substitute one of each of the pairs of complex eigenvalues, in turn, into (13.39) and find the associated eigenvector. These eigenvectors will contain complex numbers and, when multiplied by the complex exponential function e^{rt}, will give a complex valued vector function whose real and imaginary parts each satisfy the original differential equation. Two columns of a fundamental matrix are formed by the components of these vector functions. Repeat this procedure for all pairs of complex eigenvalues.

(c) For the case in which all of the eigenvalues are distinct, either real or complex, follow the procedure in part (a) for the real eigenvalues and part (b) for the complex eigenvalues. A fundamental matrix may be formed, with its columns being the solutions found by each procedure.

(d) The case for repeated eigenvalues utilizes linear combinations of powers of t times e^{rt} times vectors of constants. (This case will not be extensively dealt with in this book. See Exercise 6 on page 310 for an example with repeated real eigenvalues.)

Comments about Solving $\mathbf{X}' = \mathbf{MX}$

- One way to check for mistakes in forming a fundamental matrix is to compute the determinant of the matrix. If this determinant is zero, you have made a mistake, because all fundamental matrices have nonzero determinants. Another way is to see if it satisfies $\mathbf{U}' = \mathbf{MU}$. It should!

Example 13.7 :

To illustrate the use of this procedure, we consider the problem of finding the general solution of $\mathbf{X}' = \mathbf{MX}$, where

$$\mathbf{M} = \begin{bmatrix} 2 & 0 & 9 \\ 0 & 3 & 0 \\ 1 & 0 & 2 \end{bmatrix} \text{ and } \mathbf{X} = \begin{bmatrix} u \\ v \\ w \end{bmatrix}.$$

We start by substituting the vector $\mathbf{X} = \mathbf{A}e^{rt}$, where

$$\mathbf{A} = \begin{bmatrix} A \\ B \\ C \end{bmatrix},$$

and A, B, C, and r are unknown constants, into the differential equation. Rearranging the result, we obtain the system of algebraic equations

$$(\mathbf{M} - r\mathbf{I})\mathbf{A} = \begin{bmatrix} 2-r & 0 & 9 \\ 0 & 3-r & 0 \\ 1 & 0 & 2-r \end{bmatrix} \begin{bmatrix} A \\ B \\ C \end{bmatrix} = \begin{bmatrix} 0 \\ 0 \\ 0 \end{bmatrix}. \quad (13.40)$$

From the determinant of this coefficient matrix, we obtain the characteristic equation

$$(2-r)(3-r)(2-r) - 9(3-r) = (3-r)[(2-r)(2-r) - 9] = 0.$$

Expanding the term in brackets on the right in this last equation gives the eigenvalues as $r = 3, -1$, and 5.

Substituting the eigenvalue $r = 3$ into (13.40) gives

$$\begin{bmatrix} -1 & 0 & 9 \\ 0 & 0 & 0 \\ 1 & 0 & -1 \end{bmatrix} \begin{bmatrix} A \\ B \\ C \end{bmatrix} = \begin{bmatrix} 0 \\ 0 \\ 0 \end{bmatrix},$$

so an eigenvector may be chosen as

$$\begin{bmatrix} 0 \\ 1 \\ 0 \end{bmatrix}.$$

Similar calculations for $r = -1$ and $r = 5$ yield the eigenvectors

$$\begin{bmatrix} 3 \\ 0 \\ -1 \end{bmatrix} \text{ and } \begin{bmatrix} 3 \\ 0 \\ 1 \end{bmatrix},$$

respectively.

Thus, we have a fundamental matrix

$$\mathbf{U} = \begin{bmatrix} 0 & 3e^{-t} & 3e^{5t} \\ e^{3t} & 0 & 0 \\ 0 & -e^{-t} & e^{5t} \end{bmatrix}, \tag{13.41}$$

and the general solution is given by $\mathbf{X}(t) = \mathbf{UC}$, where $\mathbf{U}$ is our fundamental matrix and $\mathbf{C}$ is a vector of arbitrary constants. □

We now consider the task of finding particular solutions of nonhomogeneous systems of linear differential equations. The methods used for higher order systems are identical in form to those used in Section 13.1. To illustrate these methods, we first consider the nonhomogeneous system

$$\mathbf{X}' = \mathbf{MX} + \mathbf{V}, \tag{13.42}$$

where $\mathbf{M}$ is an n by n matrix of constants and $\mathbf{V}$ has the form of a vector of constants times $e^{\omega t}$. If ω is not an eigenvalue of the matrix $\mathbf{M}$, then a proper form for a particular solution is that of a vector containing n unknown constants times $e^{\omega t}$ — namely, $\mathbf{K}e^{\omega t}$. Substitution of such a vector into (13.42) will result in a system of n algebraic equations in n unknowns. The solution of this algebraic system gives the components of the vector $\mathbf{K}$ in the particular solution.

Example 13.8 :

As an example of this method, we find a particular solution of $\mathbf{X}' = \mathbf{MX} + \mathbf{V}$, where

$$\mathbf{M} = \begin{bmatrix} 2 & 0 & 9 \\ 0 & 3 & 0 \\ 1 & 0 & 2 \end{bmatrix} \text{ and } \mathbf{V} = \begin{bmatrix} 4 \\ 12 \\ 4 \end{bmatrix} e^{-3t}. \tag{13.43}$$

If we substitute the trial particular solution

$$\mathbf{X_p}(t) = \begin{bmatrix} K_1 \\ K_2 \\ K_3 \end{bmatrix} e^{-3t}$$

into the differential equation and rearrange the resulting expression, we obtain

$$\begin{bmatrix} -3K_1 \\ -3K_2 \\ -3K_3 \end{bmatrix} = \begin{bmatrix} 2 & 0 & 9 \\ 0 & 3 & 0 \\ 1 & 0 & 2 \end{bmatrix} \begin{bmatrix} K_1 \\ K_2 \\ K_3 \end{bmatrix} + \begin{bmatrix} 4 \\ 12 \\ 4 \end{bmatrix}.$$

The solution of this system of algebraic equations is $K_1 = 1$, $K_2 = -2$, and $K_3 = -1$, so our particular solution is

$$\mathbf{X_p}(t) = \begin{bmatrix} 1 \\ -2 \\ -1 \end{bmatrix} e^{-3t}. \tag{13.44}$$

However, if $\mathbf{V}$ is not of the form of a vector of constants times $e^{\omega t}$, or if it is and ω is an eigenvalue of $\mathbf{M}$, this method of finding a particular solution will not work. In this case we need to use the method of variation of parameters — it always works.

To use the method of variation of parameters in the current case, we would write our particular solution as

$$\mathbf{X_p}(t) = \mathbf{U}(t) \int \mathbf{U}^{-1}(t) \mathbf{V}(t) \, dt,$$

where $\mathbf{U}$ is given by (13.41) and $\mathbf{V}$ is given by (13.43). We compute the inverse of $\mathbf{U}$ as

$$\mathbf{U}^{-1} = \frac{1}{6} \begin{bmatrix} 0 & 2e^{-3t} & 0 \\ e^{t} & 0 & -3e^{t} \\ e^{-5t} & 0 & 3e^{-5t} \end{bmatrix},$$

so the integral in the expression for $\mathbf{X_p}(t)$ is given by

$$\int \frac{1}{6} \begin{bmatrix} 0 & 2e^{-3t} & 0 \\ e^{t} & 0 & -3e^{t} \\ e^{-5t} & 0 & 3e^{-5t} \end{bmatrix} \begin{bmatrix} 4 \\ 12 \\ 4 \end{bmatrix} e^{-3t} \, dt = \int \frac{1}{6} \begin{bmatrix} 24e^{-6t} \\ -8e^{-2t} \\ 16e^{-8t} \end{bmatrix} dt = \frac{1}{6} \begin{bmatrix} -4e^{-6t} \\ 4e^{-2t} \\ -2e^{-8t} \end{bmatrix}.$$

To obtain $\mathbf{X_p}(t)$, we multiply this vector by $\mathbf{U}$ to obtain

$$\mathbf{X_p}(t) = \begin{bmatrix} 0 & 3e^{-t} & 3e^{5t} \\ 3e^{3t} & 0 & 0 \\ 0 & -e^{-t} & e^{5t} \end{bmatrix} \begin{bmatrix} -(2/3)e^{-6t} \\ (2/3)e^{-2t} \\ -(1/3)e^{-8t} \end{bmatrix}.$$

Performing the indicated multiplications yields the same particular solution we found in (13.44). □

Exercises

1. Find the general solution $\mathbf{X}' = \mathbf{MX}$, in terms of a fundamental matrix.

(a) $\mathbf{M} = \begin{bmatrix} 12 & -3 & -3 \\ -3 & 9 & 0 \\ -3 & 0 & 9 \end{bmatrix}$

(b) $\mathbf{M} = \begin{bmatrix} 1 & 2 & 1 \\ 0 & -1 & 0 \\ 1 & 2 & 1 \end{bmatrix}$

(c) $\mathbf{M} = \begin{bmatrix} 2 & 0 & 9 \\ 0 & 3 & 0 \\ 1 & 0 & 2 \end{bmatrix}$

(d) $\mathbf{M} = \begin{bmatrix} 3 & -1 & -1 \\ -1 & 3 & -1 \\ -1 & -1 & 3 \end{bmatrix}$

(e) $\mathbf{M} = \begin{bmatrix} 2 & -2 & 0 \\ 1 & -2 & -1 \\ -2 & 1 & -2 \end{bmatrix}$

(f) $\mathbf{M} = \begin{bmatrix} 2 & 0 & 5 \\ 0 & 1 & 2 \\ -4 & 5 & 0 \end{bmatrix}$

(g) $\mathbf{M} = \begin{bmatrix} -1 & 0 & 0 \\ 2 & -1 & 0 \\ 3 & 5 & -1 \end{bmatrix}$

(h) $\mathbf{M} = \begin{bmatrix} 0 & 3 & 3 \\ 2 & -1 & -3 \\ 0 & -1 & -1 \end{bmatrix}$

2. For the RLC parallel circuit of Figure 13.8, consider the situation in which the voltage source is shifted from below the inductor to below the resistor.

   (a) Show that the resulting circuit is described by
   $$\begin{cases} LI'_L + RI_R &= V(t), \\ RI'_R + (1/C)I_C &= V'(t), \\ I_R &= I_C + I_L. \end{cases}$$

   (b) Solve the system of equations in part (a) if $R = 50$, $C = 0.02$, $L = 0.004$, $V(t) = 100$, and $I_R(0) = I_L(0) = 0$. Show that the system has an equilibrium point, and determine its stability.

   (c) Solve the system of equations if R, L, C, and the initial conditions are as in part (b), and $V(t) = 100\sin 100t$.

3. Solve the following initial value problems.

   (a) Exercise 1(c) with $\mathbf{X}(0) = \begin{bmatrix} 1 \\ 0 \\ 1 \end{bmatrix}$

   (b) Exercise 1(d) with $\mathbf{X}(0) = \begin{bmatrix} 1 \\ 0 \\ 0 \end{bmatrix}$

   (c) Exercise 1(e) with $\mathbf{X}(0) = \begin{bmatrix} 0 \\ 1 \\ 1 \end{bmatrix}$

   (d) Exercise 1(g) with $\mathbf{X}(0) = \begin{bmatrix} 3 \\ 0 \\ 7 \end{bmatrix}$

4. If the springs and masses in Figure 13.8 are given by $m_1 = m_2 = 2$, $k_1 = 6$, $k_2 = 4$, find solutions of (13.34) that satisfy the initial conditions $x_1(0) = 0$, $x'_1(0) = 4$, $x_2(0) = 0$, $x'_2(0) = -7$.

5. Show that for positive values of m_1, m_2, k_1, and k_2, the solutions of (13.34) will always be oscillatory.

6. Consider the system of differential equations $\mathbf{X}' = \mathbf{MX}$, where
   $$\mathbf{M} = \begin{bmatrix} 1 & 1 & 0 & 0 \\ 0 & 1 & 2 & 0 \\ 0 & 0 & 1 & 0 \\ 0 & -2 & 0 & 1 \end{bmatrix} \text{ and } \mathbf{X} = \begin{bmatrix} x \\ y \\ u \\ v \end{bmatrix}.$$

(a) By substituting an appropriate expression for $\mathbf{X}$ as a vector of constants times e^{rt}, find the eigenvalues of the resulting characteristic equation. (They will be repeated.)

(b) Discover the two linearly independent eigenvectors associated with this repeated eigenvalue.

(c) Find the two remaining columns of a fundamental matrix by trying a solution of the form $\mathbf{X}(t) = \left(t^2 \mathbf{A}_2 + t \mathbf{A}_1 + \mathbf{A}_0\right) e^{rt}$, where r is the repeated eigenvalue from part (a). $\mathbf{A}_2$, $\mathbf{A}_1$, and $\mathbf{A}_0$ are vectors, each containing four constants that are to be determined so that $\mathbf{X}(t)$ is a solution.

(d) Combine the answers from parts (b) and (c) to find a fundamental matrix, and give the general solution of the original system of differential equations.

7. Find the general solution of $\mathbf{X}' = \mathbf{M}\mathbf{X} + \mathbf{V}$, where

(a) $\mathbf{M} = \begin{bmatrix} 12 & -3 & -3 \\ -3 & 9 & 0 \\ -3 & 0 & 9 \end{bmatrix}$, $\mathbf{V} = \begin{bmatrix} e^{-t} \\ e^{-t} \\ e^{-t} \end{bmatrix}$

(b) $\mathbf{M} = \begin{bmatrix} 1 & 2 & 1 \\ 0 & -1 & 0 \\ 1 & 2 & 1 \end{bmatrix}$, $\mathbf{V} = \begin{bmatrix} 0 \\ 1 \\ e^t \end{bmatrix}$

(c) $\mathbf{M} = \begin{bmatrix} 2 & 0 & 9 \\ 0 & 3 & 0 \\ 1 & 0 & 2 \end{bmatrix}$, $\mathbf{V} = \begin{bmatrix} \sin 2t \\ 0 \\ 0 \end{bmatrix}$

(d) $\mathbf{M} = \begin{bmatrix} 3 & -1 & -1 \\ -1 & 3 & -1 \\ -1 & -1 & 3 \end{bmatrix}$, $\mathbf{V} = \begin{bmatrix} e^t \\ 0 \\ 0 \end{bmatrix}$

(e) $\mathbf{M} = \begin{bmatrix} 2 & 0 & 9 \\ 0 & 3 & 0 \\ 1 & 0 & 2 \end{bmatrix}$, $\mathbf{V} = \begin{bmatrix} \cos 2t \\ 0 \\ 0 \end{bmatrix}$

(f) $\mathbf{M} = \begin{bmatrix} 2 & 0 & 9 \\ 0 & 3 & 0 \\ 1 & 0 & 2 \end{bmatrix}$, $\mathbf{V} = \begin{bmatrix} e^{-t} \\ e^{3t} \\ e^{5t} \end{bmatrix}$

(g) $\mathbf{M} = \begin{bmatrix} -1 & 0 & 0 \\ 2 & -1 & 0 \\ 3 & 5 & -1 \end{bmatrix}$, $\mathbf{V} = \begin{bmatrix} \cos 2t \\ 0 \\ \sin 2t \end{bmatrix}$

8. Solve the following initial value problems.

(a) Exercise 7(c) with $\mathbf{X}(0) = \begin{bmatrix} 1 \\ 0 \\ 0 \end{bmatrix}$

(b) Exercise 7(d) with $\mathbf{X}(0) = \begin{bmatrix} 1 \\ 0 \\ 1 \end{bmatrix}$

(c) Exercise 7(e) with $\mathbf{X}(0) = \begin{bmatrix} 0 \\ 1 \\ 1 \end{bmatrix}$

(d) Exercise 7(f) with $\mathbf{X}(0) = \begin{bmatrix} 3 \\ 0 \\ 7 \end{bmatrix}$

9. **A Two-Body Problem.**[1] A linear spring with spring constant k has two masses, m and M, attached at either end. The system lies in a line on a horizontal surface in equilibrium. All motion takes place along this line and is assumed frictionless. The two masses are now moved to initial positions along the line and given initial velocities along the line. If $x(t)$ is the distance traveled by the mass m from its equilibrium position, and $y(t)$ is the corresponding position for M, then the motion of this system is governed by $mx'' = k(y-x)$, $My'' = -k(y-x)$. Solve these equations for $x(t)$ and $y(t)$.

10. **Coupled Pendulums — A Simple Experiment.**[2] Construct the following piece of apparatus, which is a coupled pendulum. Suspend two identical pendulums (same length wire, same mass) and join the vertical wires with a horizontal light rod. Make sure that the points of suspension of the pendulums are the same distance apart as the length of the rod so that the wires hang vertically.[3] We are first going to perform two experiments, and then we are going to explain what we have found. All motion is assumed to be in the plane of the wires.

    (a) Perform the following experiment. Displace the two pendulums through the same small initial angle and simultaneously release both from rest. Describe what happens.

    (b) Perform the following experiment. Displace the two pendulums through the same small initial angle, but in opposite directions, and simultaneously release both from rest. Describe what happens.

    (c) Perform the following experiment. Displace one of the pendulums through a small initial angle while holding the other pendulum vertical. Now simultaneously release both from rest. Describe what happens.

    (d) It can be shown that if $x(t)$ and $y(t)$ are the horizontal displacements of the two pendulums from vertical, then the linearized differential equations governing the subsequent motion, neglecting air resistance, are $x'' + \lambda^2 x = \lambda^2 \Delta$, $y'' + \lambda^2 y = \lambda^2 \Delta$, where $\lambda^2 = g/H$, and $\Delta = \frac{1}{2}(x+y)\left(1 - \frac{H}{L}\right)$. Here L is the length of the pendulums from the pivot and H is the length of the pendulums from the rod. Solve these differential equations subject to $x(0) = x_0$, $x'(0) = u_0$, $y(0) = y_0$, $y'(0) = v_0$.

    (e) What are the solutions of the differential equations in part (d) subject to the initial conditions $x(0) = y(0) = x_0$ and $x'(0) = y'(0) = 0$? How does this agree with part (a)?

    (f) What are the solutions of the differential equations in part (d) subject to the initial conditions $x(0) = x_0$, $y(0) = -x_0$, and $x'(0) = y'(0) = 0$? How does this agree with part (b)?

    (g) What are the solutions of the differential equations in part (d) subject to the initial conditions $x(0) = x_0$, $y(0) = 0$, and $x'(0) = y'(0) = 0$? How does this agree with part (c)? Show that according to this model, the period, T, of the beats — that is, the time it takes for the mass at rest to return to rest — is $T = 2\pi/(\lambda - \mu)$, which can be written as $1/T = \sqrt{g}\left(1/\sqrt{H} - 1/\sqrt{L}\right)/(2\pi)$. Thus, a plot of $1/T$ (the frequency) versus $1/\sqrt{H}$ should yield a straight line of slope $\sqrt{g}/(2\pi)$ and intercept $-\sqrt{g}/\left(\sqrt{L}2\pi\right)$. Conduct an experiment where the

---

[1] "A simple example for the two body problem" by E. Maor, *The Physics Teacher*, February 1973, pages 104–105.

[2] "Teaching physics with coupled pendulums" by J. Priest and J. Poth, *The Physics Teacher*, February 1982, pages 80–85.

[3] An interesting in-class version of this pendulum can be constructed from two bowling balls suspended with two steel wires about 7 feet long. For the horizontal rod, a 4-foot dowel rod attached to the wires by rubber bands can be used.

beat period, T, is measured for various values of H. Then plot $1/T$ versus $1/\sqrt{H}$. Do you get a straight line? Table 13.1 and Figure 13.10 shows the results of such an experiment.[4]

Table 13.1 Frequency $1/T$ of coupled pendulums as a function of $H^{-1/2}$

$1/\sqrt{H}$ (cm$^{-1/2}$)	$1/T$ (sec^{-1})
0.247	0.044
0.256	0.097
0.266	0.141
0.280	0.209
0.295	0.276
0.315	0.375
0.340	0.502
0.364	0.633

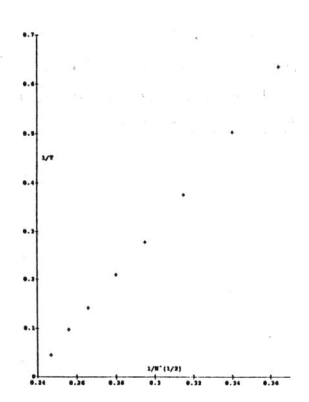

Figure 13.10 Frequency $1/T$ of coupled pendulums as a function of $H^{-1/2}$

11. **A Falling Spring.**[5] A linear spring with spring constant k has two masses, m and M, attached at either end. The end with mass m is held so the spring hangs vertically. When it has reached its equilibrium position, mass M is a distance d below m. The spring is now allowed to fall under gravity. If $x(t)$ is the distance traveled by m and $y(t)$ is the distance traveled by M, then the motion of this system is governed by $mx'' = mg + \frac{k}{\ell}(y - x - \ell)$, $My'' = Mg - \frac{k}{\ell}(y - x - \ell)$, where $\ell = d/(1 + Mg/k)$. Solve these equations for $x(t)$ and $y(t)$.

12. **Lift.** When a projectile flies through the air, it experiences three forces — gravitation, drag, and lift — the latter due to backspin in the case of a golf ball, and the position of the skis in the case of a ski jumper.[6] If this force is linear in the velocity, then the equations of motion — including the linear drag terms — are $mx'' = -kx' - cy'$ and $my'' = -mg - ky' + cx'$, where k and c are positive constants, k is associated with the drag, and c is a measure of the lift. Solve these equations subject to the initial conditions $x(0) = y(0) = 0$, and $x'(0) = w_0 \cos \alpha$, $y'(0) = w_0 \sin \alpha$, to find

$$x(t) = e^{-\lambda t}(A \sin \mu t - D \cos \mu t) + Ct + D,$$

---

[4] "Teaching physics with coupled pendulums" by J. Priest and J. Poth, *The Physics Teacher*, February 1982, pages 80–85, Figure 6.

[5] "Oscillations of a falling spring" by P. Glaster, *Physics Education* 20, 1993, pages 329–331.

[6] "Maximum projectile range with drag and lift, with particular application to golf" by H. Erichson, *American Journal of Physics* 51, 1983, pages 357–362; and "The Flight of a Ski Jumper" by E. True, *C·ODE·E*, Spring 1993, pages 5–8.

$$y(t) = e^{-\lambda t}(-D\sin\mu t - A\cos\mu t) + \frac{1}{\mu}[-C - \lambda(Ct+D) + w_0\cos\alpha],$$

where $\lambda = k/m$, $\mu = c/g$, $A = (w_0\cos\alpha - D\lambda - C)/\mu$, $C = \mu g/(\mu^2 + \lambda^2)$, and $D = [w_0(\lambda\cos\alpha - \mu\sin\alpha)(\mu^2 + \lambda^2) - 2\lambda\mu g]/(\mu^2 + \lambda^2)^2$.

(a) Does the projectile experience a terminal velocity in the x- and y-directions? Does the projectile experience a terminal velocity?

(b) **Computer Experiment.** Does the projectile experience a velocity lower than the terminal velocity?

(c) **The Flight of a Golf Ball with Drag and Lift.**[7] For a golf ball, the values of λ and μ are $\lambda = 0.25$ sec^{-1} and $\mu = 0.247$ sec^{-1}. These are the same to two decimal places, so we will consider the case where $\lambda = \mu = 0.25$, $w_0 = 200$ ft/sec, and $g = 32$ ft/sec^2. In this case,

$$x(t) = e^{-t/4}(A\sin t/4 - D\cos t/4) + Ct + D,$$

$$y(t) = e^{-t/4}(-D\sin t/4 - A\cos t/4) - 4C - D - Ct + 800\cos\alpha,$$

where $A = 400(\cos\alpha + \sin\alpha)$, $C = 64$, and $D = 400(\cos\alpha - \sin\alpha) - 256$. In Figure 13.11 we show the (X, Y) trajectory for $\alpha = 5°$ to $\alpha = 30°$ in intervals of $5°$. Notice that for these values of α, the greatest range occurs when $\alpha = 10°$ and is a little over 600 ft.

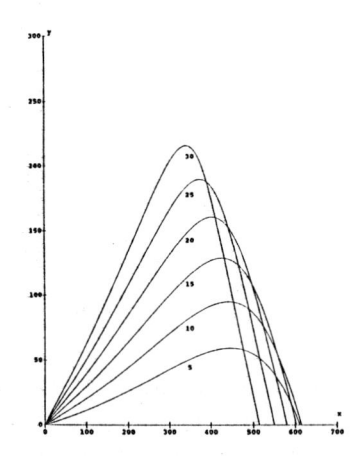

Figure 13.11 Golf ball trajectories with drag and lift for $\alpha = 5°$ to $\alpha = 30°$

    i. **Computer Experiment.** Plot the trajectory of a golf ball for $\alpha = 5°$ to $\alpha = 15°$ in intervals of $1°$. Which value of α gives the greatest range?

    ii. Consider what happens to the flight of a golf ball with lift but no drag. Experiment with different values of μ to see whether it is possible to obtain unrealistic trajectories.

What Have We Learned?

Main Ideas

- In this chapter we discovered ways of solving linear differential equations $\mathbf{X}' = \mathbf{MX} + \mathbf{V}$, where $\mathbf{X}$ is a vector of n independent variables, $\mathbf{M}$ is an n by n matrix of constants, and $\mathbf{V}$ is a vector of n given functions.

---

[7] "Maximum projectile range with drag and lift, with particular application to golf" by H. Erichson, *American Journal of Physics* 51, 1983, pages 357–362.

- A fundamental matrix, $\mathbf{U}$, is a square matrix whose columns are solution vectors of the homogeneous differential equation $\mathbf{X}' = \mathbf{MX}$. The general solution of such equations is $\mathbf{X}(t) = \mathbf{UC}$, where $\mathbf{C}$ is a vector of arbitrary constants.

- Eigenvalues of the matrix $\mathbf{M}$ are the zeros of the polynomial in r found by expanding the determinant of the matrix obtained by subtracting the parameter r from each of the elements on the diagonal of $\mathbf{M}$.

- We used three methods of finding particular solutions of $\mathbf{X}' = \mathbf{MX} + \mathbf{V}$.

  - If $\mathbf{V}$ has the form of a linear function of t, then

  $$\mathbf{X_p}(t) = \begin{bmatrix} A + Bt \\ C + Dt \end{bmatrix}.$$

  - If $\mathbf{V}$ has the form of a constant vector times $e^{\omega t}$ and ω is not an eigenvalue of $\mathbf{M}$, then $\mathbf{X_p}(t) = \mathbf{K}e^{\omega t}$, where $\mathbf{K}$ is a constant vector (ω may be real or complex).

  - If $\mathbf{V}$ does not have either of the previous forms or if ω is an eigenvalue of $\mathbf{M}$, then variation of parameters will give a particular solution.

- The proper trial solution if $\mathbf{V}$ contains a polynomial of degree n is

$$\mathbf{X_p}(t) = \begin{bmatrix} \sum_{j=0}^{n} A_j t^j \\ \sum_{j=0}^{n} B_j t^j \end{bmatrix}.$$

While this will work, it will obviously lead to a considerable amount of algebra.

The entire chapter is summarized by the following.

How to Solve Nonhomogeneous Systems of Linear Differential Equations with Constant Coefficients

Purpose: To find the general solution of the nonhomogeneous system of linear differential equations of the form

$$\mathbf{X}' = \mathbf{MX} + \mathbf{V}, \tag{13.45}$$

where $\mathbf{M}$ is an n by n matrix of real valued constants, $\mathbf{X}$ is a vector of dependent variables, and $\mathbf{V}$ is a forcing function.

Process

1. Find the general solution of the associated homogeneous differential equation. See *How to Solve* $\mathbf{X}' = \mathbf{MX}$ in Section 9.5 for $n = 2$, and page 306 for higher order equations.

2. Find a particular solution of (13.45). Here we have a choice.

   (a) If the forcing function has the form of a vector of constants times $e^{\omega t}$, where ω is not an eigenvalue of $\mathbf{M}$, then the proper form for a particular solution is $\mathbf{X_p}(t) = \mathbf{K}e^{\omega t}$, where $\mathbf{K}$ is a vector of unknown constants. Substitute this expression into (13.45) and choose the constants so the system of differential equations is satisfied.

   (b) If the forcing function is a polynomial of degree n, $\mathbf{X_p}(t)$ should be a polynomial of degree n for each dependent variable.

(c) If the forcing function is not of the form mentioned in part (a), or if in that form ω is an eigenvalue of the matrix $\mathbf{M}$, we use variation of parameters. Here a particular solution is given by

$$\mathbf{X_p}(t) = \mathbf{U}(t) \int \mathbf{U}^{-1}(t)\mathbf{V}(t)\,dt,$$

where $\mathbf{U}$ is the fundamental matrix found in the general solution of the homogeneous problem in step 1.

3. The general solution is given by

$$\mathbf{X}(t) = \mathbf{UC} + \mathbf{X_p}(t),$$

where $\mathbf{C}$ is a vector of arbitrary constants.